Olena O. Bulatsyk,
Boris Z. Katsenelenbaum,
Yury P. Topolyuk,
and Nikolai N. Voitovich

Phase Optimization Problems

Related Titles

Hartmann, A. K., Weigt, M.

Phase Transitions in Combinatorial Optimization Problems
Basics, Algorithms and Statistical Mechanics

2005
ISBN 978-3-527-40473-5

Hartmann, A. K., Rieger, H.

Optimization Algorithms in Physics

2002
ISBN 978-3-527-40307-3

Chandru, V., Hooker, J.

Optimization Methods for Logical Inference

1999
ISBN 978-0-471-57035-6

Olena O. Bulatsyk,
Boris Z. Katsenelenbaum,
Yury P. Topolyuk,
and Nikolai N. Voitovich

Phase Optimization Problems

Applications in Wave Field Theory

WILEY-VCH

WILEY-VCH Verlag GmbH & Co. KGaA

The Authors

Dr. Olena O. Bulatsyk
Dept. of Numerical Methods
Inst. for Appl. Probl. of Mech. and Math., NASU
Lviv, Ukraine
bul@iapmm.lviv.ua

Prof. Dr. Boris Z. Katsenelenbaum
Nahariya, Israel

Dr. Yury P. Topolyuk
Dept. of Numerical Methods
Inst. for Appl. Probl. of Mech. and Math., NASU
Lviv, Ukraine
top@iapmm.lviv.ua

Prof. Dr. Nikolai N. Voitovich
Dept. of Numerical Methods
Inst. for Appl. Probl. of Mech. and Math., NASU
Lviv, Ukraine
voi@iapmm.lviv.ua

Library of Congress Card No.: applied for

British Library Cataloguing-in-Publication Data:
A catalogue record for this book is available from the British Library.

Bibliographic information published by the Deutsche Nationalbibliothek
The Deutsche Nationalbibliothek lists this publication in the Deutsche Nationalbibliografie; detailed bibliographic data are available on the Internet at http://dnb.d-nb.de.

Composition le-tex publishing services GmbH, Leipzig
Printing and Binding Strauss GmbH, Mörlenbach
Cover Design Schulz Grafik-Design, Fußgönheim

Printed in the Federal Republic of Germany
Printed on acid-free paper

ISBN: 978-3-527-40799-6

Contents

Phase Optimization Problems. O. O. Bulatsyk, B. Z. Katsenelenbaum, Y. P. Topolyuk, and N. N. Voitovich

ISBN: 978-3-527-40799-6

Preface

But beyond this, my son, be warned:
the writing of many books is endless,
and excessive devotion to books
is wearying to the body.

Ecclesiastes 12:12, New American Standard Bible

The book is devoted to the optimization problems arising in various applications of the wave field theory which are based on the use of phase distributions as the optimization functions. Freedom of a choice of phase can be caused by two factors. First, this is a phase field distribution which can be used in cases when the field intensity distribution is only the subject of interest; this is the case in the antenna synthesis problems and in energy transmission problems. Second, the phase functions can describe the correction which should be provided by appropriate devices (phase correctors) in order to create fields of the desired structure. The multi-element phase field converters are samples of such devices.

This line of investigation is not sufficiently covered in the literature mainly, because of the nonlinearity of the mathematical problems arising there. They have, as a rule, nonunique solutions which change quantity as the physical parameters vary. This fact is often a feature for practice as it provides an additional freedom of choice of a solution from an existing set with the purpose of satisfying certain additional requirements.

On the other hand, the nonlinearity of the problem complicates its solution and demands additional investigation of the process of solution branching with changing physical parameters, which requires the development of special numerical methods and their justification.

In this book, a special class of analytically solvable nonlinear integral equations, arising in such problems, is described and analyzed in detail. Such cases seldom occur in practice and can, in particular, be used for illustrative purposes while studying the theory of nonlinear equations.

The book involves also nonstandard inverse problems that extend the scope given by the book title. In particular, the problems related to minimization of the back

Phase Optimization Problems. O. O. Bulatsyk, B. Z. Katsenelenbaum, Y. P. Topolyuk, and N. N. Voitovich

ISBN: 978-3-527-40799-6

scattering are also considered. A nonstandard epilogue regarding ethical aspects of the scientific work concludes the book.

The book is written on the basis of the extended and renewed course which has long been taught by one of the authors at Ivan Franko National University in Lviv to the students in applied mathematics. This affected the form in which the material is explained. For methodical reasons, the physical and mathematical formulations of the problems, as well as the numerical methods and results of their solution are located in different chapters which enables, in particular, their association and a more compact statement. This allows each chapter to be read almost independently.

The book is intended for experts working in the field of research, design and optimization of radiating and transmitting systems, and also for mathematicians interested in the theory of nonlinear integral equations. It will be also useful for students and graduate students in appropriate fields.

The authors are profoundly grateful to their colleagues O. M. His', P. O. Savenko, V. P. Tkachuk, and O. F. Zamorska whose results are partially used in the book. A special gratitude is expressed to M. I. Andriychuk and S. A. Yaroshko who wrote separate subsections in the book, as well as to V. V. Shevchenko for his contribution into the compilation of the reference list. The authors recall with sorrow V. V. Semenov who stood at beginning of this direction and sadly passed away before his time.

Authors also thank Halyna Issayeva and Oleg Kusyi for their contributions to the translation and technical preparation of the book. Special thanks are due to Ulrike Werner and Anja Tschörtner for permanent attention to the project, and to le-tex publishing services for thorough typesetting of the book.

Lviv, Ukraine
Nahariya, Israel
January 2010

Olena Bulatsyk
Boris Katsenelenbaum
Yury Topolyuk
Nikolai Voitovich

1 Introduction

1.1 Topics covered

The problems involving an undefined phase of a wave field arise in various applications and are well known in the literature. The so-called *phase problem* is the most common of these (see, e.g. [1–4]). It consists in the reconstruction of the phase distribution of the Fourier transform of a finite (compactly supported) function if its amplitude distribution is given (measured) on the entire real axis. This problem belongs to the classical *reconstruction* (identification) problems and requires the existence of a unique solution.

This book considers another class of inverse problems, which can be called the *optimization* (design) problems. In terms of the Fourier transform they could be formulated as problems in determining a finite complex function (or its amplitude and phase distribution only if the other is given) such that the modulus (amplitude) of its Fourier transform satisfies certain demands (e.g. it is close to a given (desired) positive (non-negative) function). As a rule such demands are formulated in the variational form, as the minimization of certain functionals. It can be seen that this formulation does not require the uniqueness of the solution. On the contrary, nonuniqueness is often desired because it provides additional degrees of freedom for making a decision.

The main applications of the considered phase optimization problems are the theory of the transmitting lines, field transformers, antennas and resonators. Of course, the theory developed in the book is not confined only to these applications.

The first publications on nonlinear inverse problems of the type considered have most likely appeared in the nineteen fifties and sixties [5–10].

In terms of mathematics, the problems are reduced to the nonlinear integral equations of the Hammerstein type [11, 12]. They have a linear kernel and nonlinear factor depending on the unknown function. As a rule, the phase of this function appears in the integrand separately. Similar equations can be found in the literature in the context of the mentioned phase problem [13, 14]. They have nonunique solutions, and the investigation of their structure and the processes of their branching are important mathematical problems [15]. A class of analytically solvable equations of this type is found and investigated in this book.

Phase Optimization Problems. O. O. Bulatsyk, B. Z. Katsenelenbaum, Y. P. Topolyuk, and N. N. Voitovich

ISBN: 978-3-527-40799-6

Due to nonlinearity, the problems considered require development and the application of specific numerical methods. Besides simple iterative methods having a descriptive physical interpretation, different modifications of the Newton method (see e.g. [16]) seem to be the most useful. One of these modifications uses the singular value decomposition of the Jacobi matrix. Information about the singular value distribution allows for the detection of the branching points while solving the problem.

The theory described in the book was started from the works [48, 49] and [52, 75] concerning the two-element and multi-element phase field transformers, respectively. V. V. Semenov's ideas on transferring this approach to the antenna problems was described in his last work [53]. They were carried out in [27], where the main nonlinear integral equation for the problem of linear antenna synthesis by the desired amplitude pattern was obtained and an algorithm for its solution was proposed. The mathematical theory describing the structure of solutions to this type of equation and their branching began from [47]. During the following years the approach was intensively developed and utilized for many concrete problems. The theory and results were summarized and described in [25] and [26].

A new impulse for developing the theory of nonlinear integral equations of the type considered was provided by the works [28] and [29] where analytical solutions to one of these equations were obtained. The results were then generalized for some classes of nonlinear equations of the Hammerstein type [54] and [55].

The book is mostly based on the results obtained by the authors and their colleagues and published in journals or conference proceedings over some time. Some of these publications are referred to in the book. Several results were obtained during preparation of the book. The references to the original works, where the methods or results are described, are given where they are explained.

Chapter 2 was written by B. Z. Katsenelenbaum. Sections 3.1.3, 3.1.5.2, and 3.1.8.3 were written by Yu. P. Topolyuk. Section 3.3 was written by N. N. Voitovich. The rest of Chapter 3 was written by Yu. P. Topolyuk and N. N. Voitovich. Section 4.1 was written by O. O. Bulatsyk and N. N. Voitovich, Sections 4.2 and 4.5 were written by O. O. Bulatsyk, Yu. P. Topolyuk and N. N. Voitovich. Sections 4.3 and 4.4 were written by O. O. Bulatsyk. Section 5.1.2.1 was written by Yu. P. Topolyuk, and N. N. Voitovich. The rest of Section 5.1 was written by Yu. P. Topolyuk. M. I. Andriychuk wrote Sections 5.2.2, 5.2.3.3 and contributed to 5.2.3.2. S. A. Yaroshko wrote Section 5.4. The rest of Chapter 5 was written by N. N. Voitovich. Section 6.2.3.2 was written by B. Z. Katsenelenbaum and N. N. Voitovich, Sections 6.2.3.3 and 6.2.3.4 were written by N. N. Voitovich. The rest of Chapter 6 as well as Epilogue where written by B. Z. Katsenelenbaum. The general editing of the book was made by B. Z. Katsenelenbaum and N. N. Voitovich.

1.2 Outline of the book

The physical meaning of the problems considered in the book is described in Chapter 2. The problems are classified into three groups according to the physical objects they describe: transmission lines and field transformers, antennas, waveguides and resonators. Nontypical physical systems such as a cavity resonant antenna and a complex multielement beam transformer are also considered.

The problems are formulated as variational ones using different integral criteria, such as the mean-square difference between the desired and obtained field distributions (as a rule, amplitude only), the transmission, excitation or coupling coefficients, etc. In all the problems the phase functions (the phase distribution of the fields or the phase correction provided by the lenses or mirrors) are the optimization parameters.

For different reasons, several interesting problems are not in fact considered in this book. They include minimizing the field in certain areas of the middle zone [17] or creating zeros in certain directions of the radiation pattern [18], using the geometrical parameters of the system (including its shape) as additional optimization functions [19, 20], etc. Some of these problems are considered in the books [24–26].

The physical problems described in Chapter 2 are classified and generalized in Chapter 3 in the form of several variational problems formulated in terms of the functional operators. They are divided into three groups depending on the type of information given and restrictions imposed on the solution: (a) the problems with no restrictions on the solutions; (b) the problems with given either moduli (amplitudes) or arguments (phases) of the complex functions to be determined; (c) the 'homogeneous problems' in which no desired functions are given, but certain integral characteristics described by the eigenvalues of the problems are optimized.

The Hilbert spaces of L_2-type are used, in which the isometric and compact operator are both considered. Some questions concerning the stability of the problems are investigated. All the variational problems are reduced to the Lagrange–Euler equation describing the necessary conditions for the functional to be extremal. They are nonlinear integral or matrix equations and have nonunique solutions. The homogeneous equations for the branching points of their solutions are obtained.

Simple iterative methods for solving the problems are proposed. Most of them are relaxational, that is, the values of the functional make up a monotonic sequence on the successive approximations.

Typical particular cases of the operators are considered. The specific peculiarities of the problems are analyzed on these examples.

Chapter 4 contains results concerning the analytically solvable nonlinear equations of Hammerstein type, obtained in the previous chapter, and their generalizations. The existence of such equations is an important mathematical phenomenon, more unexpected than predicted. Note that the first equation of this type was obtained about forty years ago in [27]. However, its analytical solutions were observed only at the end of the last century [28] and [29].

The mentioned solutions are described by a polynomial of finite degree with complex nonconjugate zeros. These zeros are determined from a set of transcendental equations. The situation is similar to that for the linear equations with degenerated kernel, the solutions of which depend on a limited number of numeral parameters determined from the linear equation system. Indeed, the kernel of the equations considered here has a degenerated multiplier consisting of two addends. However, the polynomial degree defining the number of unknown parameters is bounded from above by a value connected with the number of oscillations of the functions contained in this multiplier. In the case when a solution of the considered equation is analytically continued to an entire function, then this function has only a limited number of the complex nonconjugate zeros, coinciding with zeros of the mentioned polynomial.

A general equation, for which such solutions exist, is given and its properties are investigated. Several theorems are proved. They describe the necessary and sufficient conditions for the function of the considered form to be a solution to this equation, the boundedness of the polynomial degree and the existence of the equivalent groups of solutions. Since the solutions with a polynomial of different degree exist simultaneously, the general structure of the set of solutions as well as the process of their branching can be completely investigated. The particular cases, connected with the Fourier transform and related to the above formulated physical problems, are analytically and numerically analyzed.

The theory is extended to a more complicated equation, with solutions which are determined by a polynomial of the described type and a real positive function. In this case the zeros of the polynomial together with the mentioned function are determined from a set of transcendental equations plus one integral equation. Both the transcendental and integral equations depend linearly on this unknown function. Numerical results for a concrete problem are presented.

The numerical methods and algorithms intended for solving the nonlinear problems of the considered types are described in Chapter 5. In the theoretical part of the chapter, two methods, namely, the steepest descent method for the direct minimization of the functionals and the simple iteration method for the solution of nonlinear equations, are investigated. In both cases the problems are complicated by the nonuniqueness of the solutions. The theorems about the convergence of the method and the accuracy of its estimations are proved. It is shown that the convergence rate decreases infinitely when approaching the branching points.

The rest of the chapter concerns the numerical solutions of the concrete problems formulated in Chapter 2. Of course, not all these problems can be solved in the context of one book. However, the authors try to overview all the key problems and show the methods which can be applied to their solution.

The simplest methods which can be applied to almost all of the problems considered in the book are the iterative methods described in the previous chapters. They are simply realizable, have clear physical interpretation and, owing to their monotonicity, almost always converge to an extremum (at least, to the local one). However, their disadvantages are the mentioned slow convergence near the branching points and nonapplicability when finding other solutions of the nonlinear equations, except those describing the extrema of the functional generating these equations.

The most powerful technique is the modified Newton method based on the singular value decomposition, permitting its application also near the branching points, by passing through them along a solution branch with almost no loss of the convergence rate. As the numerical experiments show, these points are indicated with a satisfactory accuracy. The disadvantages of this method are its sensitivity to the initial approximations and more costly result.

Combining both the above methods allows us to solve the nonlinear integral equations obtained in almost all problems considered. In this book, the combination was applied to problems of the amplitude-phase synthesis of the circular antenna, the two-elements phase field transformers and phased antenna arrays.

The most complicated problems concerning multi-element field (beam) transformers are solved by the modified iterative method in combination with the method of local variations, called together as the method of opposite directions. When optimizing an element of the system, this method compactly (in volume and time) uses and transforms the information about the other two parts, located on both sides of the optimized element.

An original method of the generalized separation of variables, intended for solving, multi-dimensional linear and nonlinear problems is described and applied to the amplitude-phase optimization of the two-dimensional antenna arrays. The method expresses the solution as a sum of functions of separated variables, which are determined successively from a one-dimensional system of nonlinear equations. The number of equations equals the dimensionality of the initial problem.

Two 'homogeneous' problems concerning the optimal shapes of the lenses in the waveguide and open resonator are solved and analyzed.

The main peculiarities of the methods and properties of the solution are mostly illustrated by two-dimensional problems. For some problems, the reader is referred to the literature where they are solved.

Finally, Chapter 6 somewhat extends the topic given by the title of the book. It considers nonstandard inverse problems formulated with respect to other (different from the phase) physical characteristics of the objects considered. They can be constructive parameters, such as impedance, orientation or the transparency of the gratings, etc.

Greatest attention is paid to the problem about the minimization of the back scattering from bodies with irregular complex impedance. A new nonlinear integral equation system is given for the impedance distribution providing the zero back scattering for all orientations of the body with respect to the source. The method is illustrated by examples of circular and elliptic cylinders.

Problems in matching the variable impedance with the incident field and applying the short-periodical grids and arrays for transformation of the linear polarization into a circular one are also considered. The chapter is unconnected with previous ones and can be read independently.

In Epilogue an attempt is made to systemize different types of the interpersonal relations in the scientific collectives. Some painful questions related to coauthorship, supervision, scientific discussions and consultations etc., are considered. These questions have the same important meaning for the scientists, as the concrete mathematical and physical problems considered in the main part of the book.

2 Formulation of Physical Problems

In this chapter we formulate optimization problems inherent to different applications of wave field theory. Usually, they are aimed at forming physical fields of the desired structures. All the problems are variational and consist in minimizing (or maximizing) certain functionals that have a quite clear physical meaning. As a rule, the functionals describe the mean square difference between the desired and obtained fields (to be minimized), or their coupling coefficient (to be maximized). In all cases the phase of the desired field is not given; it is a free function that can be chosen arbitrarily when solving the problem. In certain problems, phase functions appear that describe the optimizing system elements (phase correctors) and they are the arguments in functionals which are to be minimized (maximized).

The problems are divided into three groups which are presented in corresponding sections. The division is for the sake of convenience and certain similar problems will be found in different sections.

The problems in the first two groups describe the energy transmission from one area into another. The groups differ in the type of areas and in the structure of fields realizing this transmission. In Section 2.1 the systems transmitting the fields between two parallel planes are considered. The transmission is realized by a wide beam (with the transverse size much larger than the wavelength) in free space. In systems of the first type the transmission is described by the integral Fourier transform. The systems differ in the existence or absence of restrictions on the field distribution in the radiating domain. Section 2.1.1 describes the systems with an arbitrary input field which is varied in optimization problems. In the systems considered in Section 2.1.2, certain restrictions are imposed on the amplitude distribution of the input field: it is assumed that this distribution is given. The optimization of these systems consists in finding an appropriate phase distribution of the fields that can be realized by phase correctors located on the apertures. The multi-element field transformers are also considered.

The problems of the second group are related to different types of radiation system. Besides antennas of 'classical' type: aperture antennas (Section 2.2.1) and antenna arrays (Section 2.2.2), nonstandard radiation systems, such as multi-element beam transformers (Section 2.2.3) and cavity resonant antennas (Section 2.2.4) are also considered. The problems formulated for the aperture antennas and antenna arrays are similar to those for energy transmission lines. They differ only in the fact

Phase Optimization Problems. O. O. Bulatsyk, B. Z. Katsenelenbaum, Y. P. Topolyuk, and N. N. Voitovich

ISBN: 978-3-527-40799-6

that a field created on a planar domain is transformed into the radiation pattern, being the angular distribution of the field in the far zone. This process is also described by the Fourier transform of a finite function (the field on the aperture) into a function of angles. The definition domain of the Fourier image contains not only real angles but also complex ones, describing a nonradiated field near the antenna. In the case of the antenna array, the discrete Fourier transform is used, which acts from the finite-dimensional space of fields (currents) on the radiators into the space of the angular-dependent functions. If the equidistant array is considered, when the images are periodical functions, then only their values over a period are considered as the image domain. This domain can also contain complex angles besides the real ones.

A multi-element beam transformer is a hybrid of the multi-element field transformer considered in Section 2.1.2 and the antenna array. Each element of such a system is a semi-transparent curved lens-mirror that partially reflects the incoming field and partially transmits it; both the reflection and transmission are performed with appropriate phase corrections. The superposition of all reflected fields forms a wide well-directed beam. The optimizing functions are complex variable reflection and transmission coefficients of the elements or, more specifically, the constructive parameters of the electrodynamic structure (e.g. the short-periodical arrays) realizing these elements. This type of system can stand alone, or function as elements of an array. They can be used either as antennas or focusing systems. As a possible application, the radiating or receiving systems for transportation of the solar energy from a satellite to the Earth can be considered.

The cavity resonant antennas have the form of open resonators with slightly transparent walls. Such a resonator works at one of its resonant frequencies; the radiation pattern created by the current of the eigenoscillation should be approximated in modulus (amplitude). As the optimization parameters, the shape of the resonator walls and their transparency distribution are used. An impedance inclusion can be located inside the resonator and the distribution of its impedance can be used as an additional parameter of the optimization. The generalized method of eigenoscillations is used in the formulation of the problem and its solution.

The third group includes optimization problems for closed and open resonators and beam waveguides. The problems formulated in Section 2.3.1 are similar to those from Section 2.1.2. In both cases the field given on the input aperture is transformed into the desired field on the output one by means of the phase correctors located in the domain between the apertures. The problems differ only in the structure of the fields and in the domain where they propagate. In the first case, the field represents a directed beam in free space; in the second one, it is a superposition of the eigenwaves in the closed waveguide. Since both apertures are located in two cross-sections of the waveguide, the size of the phase correctors should not obligatory be large in comparison with the wavelength. The calculations are performed exactly, only omitting the transformations of traveling waves near the first aperture. The next problem (Section 2.3.2) is also related to the phase correctors (lenses) in the waveguides. The lenses are identical and they are located periodically inside the waveguide. The problem is to find the shape of the lenses providing minimal heat losses in the walls. The system, considered in the last subsection is an open quasi-optical resonator formed by two convex mirrors in free space. The shape of the mirrors (more precisely, the phase correction provided by them) must be found such that the radiation losses of the main eigenwave are minimal. Two different situations are considered: the mirrors are located exactly on the axis of the system, or one of them is shifted in the transverse direction.

Depending on restrictions imposed on the functions that are optimized, almost all the problems have three variants:

1) The amplitudes and phases of the sought functions may not be subject to any restrictions (amplitude-phase optimization – APO);
2) The amplitudes are given and the phases must be found (phase optimization – PO);
3) The phases are given and the amplitudes must be found (amplitude optimization – AO).

Different physical problems (corresponding to different functionals) are denoted by bold capital letters from **A** to **Z**. Single letters denote the APO problems and, as a rule, only these are formulated here. Unless otherwise stated, it is implied that each of these problems is accompanied by corresponding PO and AO problems and we will refer to these problems again. References to these problems will have the additional small letters **p** and **a**, respectively. For instance, if Problem **A** is formulated as: "maximize the transmission coefficient (2.10) with respect to the generating field $u_0(\vec{r}_0)$," then the problem addressed as Problem **Ap** is: 'maximize (2.10) with respect to phase distribution $\arg(u_0(\vec{r}_0))$ of the generating field given its amplitude distribution $|u_0(\vec{r}_0)| = v_0(\vec{r}_0)$'; similarly, Problem **Aa** is: 'maximize (2.10) with respect to amplitude distribution $|u_0(\vec{r}_0)|$ of the generating field given its phase distribution $\arg(u_0(\vec{r}_0)) = \psi_0(\vec{r}_0)$.' Sometimes, PO or AO problems are formulated as stand-alone ones.

2.1 Forming fields of a given structure

2.1.1 Optimization of beam power transmission lines

We start with the investigation of a two-element system for transmitting the energy by means of a long beam of high-frequency electromagnetic waves. The system consists of the transmitting and receiving antennas (named below as antenna and rectenna, respectively, see Figure 2.1) with apertures large in comparison with the wavelength. The distance between the antenna and rectenna is large in comparison with their sizes (the rectenna lies in the middle zone of the antenna). In this case, the wave beam does not widen and the radiation pattern is not yet formed. The vector nature of the field appears weak and the geometrical polarization is kept, so

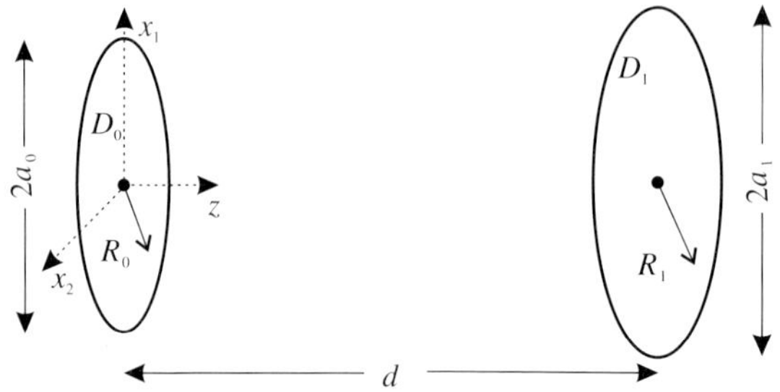

Figure 2.1 Antenna–rectena transmitting line.

that the field can be described by a scalar function (usually one of its transverse electrical components).

In optimization problems for the transmission lines, the field created on the antenna should be chosen in such a way that the field on the rectenna and in its entire plane satisfies different (sometimes contradictory) demands. The first such demand is that the energy received by the rectenna should be maximal. More precisely, the energy transmission coefficient (the ratio of the energy received by the rectenna to the total energy radiated by the antenna) is maximally close to unity. The second demand follows from the fact that the area of the rectenna should be utilized with maximal efficiency. For this, the incoming energy should be distributed on the rectenna as uniformly as possible.

According to the first demand, the energy bypassing the rectenna should be minimal. Besides these demands, the additional, ecological restrictions are imposed on the energy distribution. If the rectenna is located near a city or great manufacturing center then it is required that the energy should be concentrated close to the rectenna. If the density of population near the rectenna is approximately the same over a large area, then the minimal harm to the people occurs in the case when the energy is uniformly distributed over this area.

All these demands would be satisfied if such a field could be created on the antenna that the field on the rectenna completely coincides with a certain desired 'ideal' field. Unfortunately, such a field cannot be created on an antenna of limited size, if it does not belong to some very restricted class. However, it is possible to create a field that is maximally close to this ideal one, the so-called 'quasi-solution' of the problem. This quasi-solution is calculated by explicit formulas (see [36], Section 5.3.5). It depends not only on the amplitude of the field on the rectenna, but also on its phase, which influences neither the energy nor the ecological parameters of the transmitting line. Therefore, it would be necessary to find the quasi-solutions for all possible phase distributions of the 'ideal' field.

The freedom in choosing the phase of the field on the rectenna leads to more complicated formulations of the variational problems than for the 'ideal' field. For different examples these formulations will be specified below in Problems **A**–**E**. In all cases the square-integrated field U_0 in the antenna domain D_0, optimizing an appropriate functional, is found. Below, we briefly explain the physical meaning of these functionals. Note that, near the antenna, the normal (to its plane) derivative of the field is approximately proportional to the field itself. Therefore, the energy flux is proportional to $|U_0)|^2$. The same property is inherent in the field near the rectenna.

The functional L in Problem **A** describes the energy transmission coefficient, that is, the ratio of the energy received by the rectenna to that radiated by the antenna. In Problem **B**, the functional σ is used, which characterizes the deviation of the amplitude of the field in the rectenna plane from the amplitude of the desired field. The value of σ should be minimal for all phases of the field in the rectenna plane. The functional σ_0, minimized in Problem **C**, describes the deviation of the amplitude distribution of the field from the given one in the rectenna domain only. It is required that the desired field should be created only on the rectenna itself. The ecological demands are nonessential and they are given only by the integral restriction on the energy bypassing the rectenna. This energy should not be larger than a certain number N. It is implied that $N \ll 1$. Note that $N = 1 - L$, that is, all the energy radiated by the antenna falls on the rectenna plane. In contrast to Problem **C**, the integral 'ecological harm' (total energy bypassing the rectenna) is minimized in Problem **D**. The 'incorrectness' of the field on the rectenna, that is, the deviation in amplitude of the obtained field from the desired one must not exceed a given value only. The weight factor t, introduced in Problem **E**, regulates the proportion between the energy and ecological criteria. If t is small, then the energy distribution on the rectenna (smallness of σ_0) dominates (as in Problem **C**). On the contrary, if t is large, then the ecological demands have a larger priority.

We denote by $U_0(\vec{R}_0)$ and $U_1(\vec{R}_1)$ the transverse fields on the antenna aperture D_0 and in the rectenna plane $z = d$, respectively. Here $\vec{R}_0 = (x, y)|_{z=0}$ and $\vec{R}_1 = (x, y)|_{z=d}$ are the radius-vectors of points in corresponding planes, d is the distance between these planes, the symbol '·' denotes the scalar product of the geometrical vectors and k is the wave number. The rectenna domain is denoted by D_1 (see Figure 2.1).

In a usual approximation, the field U_1 in the rectenna plane is calculated (with accuracy to a nonessential constant phase factor) by the field U_0 on the antenna, as follows (see, e.g., [36], Section 5.3.1):

$$U_1\left(\vec{R}_1\right) = \frac{k}{2\pi d}\iint_{D_0} U_0(\vec{R}_0)\exp\left(-i\frac{k}{2d}\left|\vec{R}_0 - \vec{R}_1\right|^2\right)d\vec{R}_0 . \tag{2.1}$$

In order to simplify formulas and calculations, it is often worthwhile introducing new functions u_0, u_1 instead of U_0, and U_1 by the expressions

$$u_0\left(\vec{R}_0\right) = U_0\left(\vec{R}_0\right)\exp\left(-ik\frac{\left|\vec{R}_0\right|^2}{2d}\right), \tag{2.2a}$$

$$u_1\left(\vec{R}_1\right) = U_1\left(\vec{R}_1\right)\exp\left(ik\frac{\left|\vec{R}_1\right|^2}{2d}\right); \tag{2.2b}$$

for simplicity, we will call the functions u_0, u_1 'fields', keeping in mind that they differ from the actual fields U_0, U_1 by quadratic phase factors. Then formula (2.1) obtains the form

$$u_1\left(\vec{R}_1\right) = \frac{k}{2\pi d} \iint_{D_0} u_0\left(\vec{R}_0\right) \exp\left(i\frac{k}{d}\left(\vec{R}_0 \cdot \vec{R}_1\right)\right) \mathrm{d}\vec{R}_0 \,. \tag{2.3}$$

For the points in the antenna and rectenna plane we introduce the dimensionless coordinates

$$\vec{r}_0 = \vec{R}_0 / a_0 \,, \tag{2.4a}$$

$$\vec{r}_1 = \vec{R}_1 / a_1 \,, \tag{2.4b}$$

where a_0 and a_1 are the characteristic sizes of the domains D_0 and D_1, respectively. In fact, these transformations change the domains D_0 and D_1 proportionally. In order not to complicate the derivations, we keep the same notation for new domains, if this does not lead to any misunderstanding. After the transformations, (2.3) connecting the fields u_0 and u_1 is rewritten as

$$u_1\left(\vec{r}_1\right) = \frac{c}{2\pi}\frac{a_0}{a_1} \iint_{D_0} u_0\left(\vec{r}_0\right) \exp\left(ic\vec{r}_0 \cdot \vec{r}_1\right) \mathrm{d}\vec{r}_0 \,, \tag{2.5}$$

where

$$c = \frac{k a_0 a_1}{d} \tag{2.6}$$

is the main physical parameter of the transmitting line. In these notations the above-mentioned relations between the electrical and geometrical parameters of the system are written as

$$k a_0 \gg 1, \qquad k a_1 \gg 1, \qquad d \gg a_0, \qquad d \gg a_1 \,. \tag{2.7}$$

We introduce the scalar products

$$(u_0, v_0)_0 = \iint_{D_0} u_0\left(\vec{r}_0\right) \bar{v}_0\left(\vec{r}_0\right) \mathrm{d}\vec{r}_0 \,, \tag{2.8a}$$

$$(u_1, v_1)_1 = \iint_{D_1} u_1\left(\vec{r}_1\right) \bar{v}_1\left(\vec{r}_1\right) \mathrm{d}\vec{r}_1 \,, \tag{2.8b}$$

$$(u_1, v_1)_\infty = \iint_{-\infty}^{\infty} u_1\left(\vec{r}_1\right) \bar{v}_1\left(\vec{r}_1\right) \mathrm{d}\vec{r}_1 \,, \tag{2.8c}$$

defined on the antenna, rectenna, and in the rectenna plane, respectively, and denote the norms generated by them as

$$\|u_0\|_0^2 = (u_0, u_0)_0 \,, \qquad \|u_1\|_1^2 = (u_1, u_1)_1 \,, \qquad \|u_1\|_\infty^2 = (u_1, u_1)_\infty \,. \tag{2.9}$$

Here u_0, v_0 and u_1, v_1 are arbitrary fields (with finite energy) defined on the antenna and in the rectenna plane, respectively.

Different optimization problems can be formulated for the described transmitting lines. Here we formulate problems in which some phase functions are unknown. The domains of antenna D_0 and rectenna D_1 and the distance between them are given in these problems.

Problem A. Find the generating field u_0 on the antenna which maximizes the transmission coefficient

$$L(u_0) = \frac{\|u_1\|_1^2}{\|u_0\|_0^2} . \tag{2.10}$$

For the mathematical formulation and numerical methods see Section 3.3.1.

Problem B. Given the desired amplitude distribution v_1 of the field in the rectenna plane, find the generating field u_0 on the antenna D_0 which minimizes the normalized mean square difference

$$\sigma(u_0) = \frac{\||u_1| - v_1\|_\infty^2}{\|u_0\|_0^2 \, ||v_1||_\infty^2} \tag{2.11}$$

between the amplitude distributions of the obtained and desired fields in the whole rectenna plane.

For the mathematical formulation and numerical methods for Problems **B** and **Bp** see Sections 3.1.5 and 3.2.1, respectively. The two-dimensional cases for Problems **B** are considered analytically and numerically in Section 4.2.

Problem C. Given the desired amplitude distribution v_1 of the field in the rectenna domain D_1 and a positive number N, find the generating field u_0 on the antenna D_0 which minimizes the normalized mean square difference

$$\sigma_0(u_0) = \frac{\||u_1| - v_1\|_1^2}{\|u_0\|_0^2 \, ||v_1||_1^2} , \qquad u_0 \in L_2(D_0) , \tag{2.12}$$

between the amplitude distributions of the obtained and desired fields in the rectenna domain D_1, with the restriction

$$\sigma_1(u_0) \le N , \tag{2.13}$$

where

$$\sigma_1(u_0) = \frac{\|u_1\|_\infty^2 - \|u_1\|_1^2}{\|u_0\|_0^2} \tag{2.14}$$

is the energy bypassing the rectenna.

Problem D. Given a positive number M, find the generating field u_0 on the antenna D_0 which minimizes the energy $\sigma_1(u_0)$ bypassing the rectenna, with the restriction

$$\sigma_0(u_0) \le M . \tag{2.15}$$

Problem E. Given the desired amplitude distribution v_1 of the field in the rectenna domain D_1 and a positive number t, find the generating field u_0 on the antenna D_0 which minimizes the weighted functional

$$\sigma_t(u_0) = \sigma_0(u_0) + t\sigma_1(u_0) \tag{2.16}$$

which describes a compromise between two contradictory demands: decreasing the difference between the amplitude distributions of the obtained and desired fields in the rectenna domain D_1, and the energy bypassing the rectenna (two-criterion optimization).

Problems **C**, **D** and **E** can be united by (2.16). For a mathematical formulation see Section 3.1.6; the two-dimensional case is considered analytically and numerically in Section 4.3.

In all the Problems **A–E**, after obtaining the optimal function u_0 we must calculate the actual field distribution U_0 on the antenna using (2.2a).

2.1.2
Multi-element phase field transformers

As in the previous subsection, the problems considered here deal with the transformation of the field structure from one aperture to another by means of the wave beams and wave correctors. In contrast to the previous case, the field on the second aperture must be matched with the desired one, not only in amplitude, but also in phase. This demand is due to the fact that the second aperture can be used, not as the energy receiver (rectenna), but as an input of another transmitting line or certain antenna; then the closeness of the obtained field to the desired one is defined by their coupling coefficient described by the normed scalar product of this field. It is assumed that certain phase correctors (e.g., lenses) can be placed on the output of the first aperture and on the input of the second one; the functions describing their phase corrections are sought in the problem. If needed, additional correctors can be placed in the domain between the apertures. The sizes of the correctors and the distances between them satisfy the quasi-optical conditions (2.7).

Let us consider a system that consists of the transmitting aperture D_0 on which an incident (in general, complex) field U_0 is given, and the receiving one D_1 on which the obtained field U_1 should be maximally close to the desired one V_1 (Figure 2.2) [30]. The parameters of the system are subject to conditions (2.7). The phase correctors (e.g. lenses) can be placed on both apertures, and will add the variable phase factors $\exp(i\Psi_0(\vec{r}_0))$ and $\exp(i\Psi_1(\vec{r}_1))$ to the fields U_0 and U_1, respectively.

Since we are now essentially interested in the phase distribution of the fields, the notation (2.2) is not needed; it is more convenient to include the quadratic phase factor in the sought phase corrections. Introducing the dimensionless coordinates (2.4) into (2.1) and notations

$$\psi_0\left(\vec{r}_0\right) = \Psi_0\left(\vec{r}_0\right) - \frac{c}{2}\left|\vec{r}_0\right|^2 , \tag{2.17a}$$

$$\psi_1\left(\vec{r}_1\right) = \Psi_1\left(\vec{r}_1\right) - \frac{c}{2}\left|\vec{r}_1\right|^2 , \tag{2.17b}$$

we write the field U_1 on the aperture D_1 (after phase correction) in the form

$$U_1\left(\vec{r}_1\right) = \exp\left(i\psi_1\left(\vec{r}_1\right)\right) u_1\left(\vec{r}_1\right) , \tag{2.18}$$

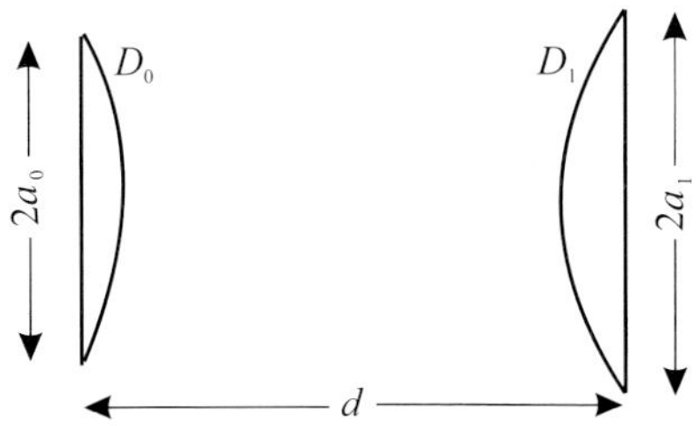

Figure 2.2 Two-element field transformer.

where, according to (2.5),

$$u_1(\vec{r}_1) = \frac{c}{2\pi}\frac{a_0}{a_1}\iint_{D_0} U_0(\vec{r}_0)\exp(i\psi_0(\vec{r}_0))\exp(ic\vec{r}_0\cdot\vec{r}_1)\,d\vec{r}_0\,, \tag{2.19}$$

the parameter c is given by (2.6).

The measure of similarity of the fields U_1 and V_1 can be described by the functional

$$\chi(\psi_0,\psi_1) = \frac{|(U_1,V_1)|}{\|U_0\|\,\|V_1\|}\,, \tag{2.20}$$

where the dependence of the field U_1 on the optimization functions is given by (2.18) and (2.19). This quantity can be interpreted as the 'coupling coefficient' between the fields U_1 and V_1 (or the excitation coefficient in the case when V_1 is a field of a mode of some transmitting system, e.g. an open or closed waveguide). It is known that the maximization of $\chi(\psi_0,\psi_1)$ minimizes the summand in U_1 orthogonal to V_1.

The optimization problem is formulated as

Problem Fp. Given the apertures D_0, D_1, distance d between them, the generating field U_0 on D_0 and desired field V_1 on D_1, find the real functions ψ_0, ψ_1 which provide maximum to the functional (2.20).

For mathematical formulation and numerical methods and results see Sections 3.2.2 and 5.2.3, respectively.

The system may have several phase correctors all, in general, of different sizes, which are placed on the apertures D_0, D_1, ..., D_M with distances d_m, $m=1,\dots,M$ between them (Figure 2.3). In this case the relations (2.18) and (2.19) are simply generalized for each pair of the aperture, as follows

$$u_0(\vec{r}_0) = U_0(\vec{r}_0)\,, \tag{2.21a}$$

$$u_m(\vec{r}_m) = \frac{c_m}{2\pi}\frac{a_{m-1}}{a_m}\iint_{D_{m-1}} u_{m-1}(\vec{r}_{m-1})\exp(i\psi_{m-1}(\vec{r}_{m-1})) \times \exp(ic_m\vec{r}_{m-1}\cdot\vec{r}_m)\,d\vec{r}m-1\,,\qquad m=1,2,\dots,M\,, \tag{2.21b}$$

$$U_M(\vec{r}_M) = u_M(\vec{r}_M)\exp(i\psi_M(\vec{r}_M))\,, \tag{2.21c}$$

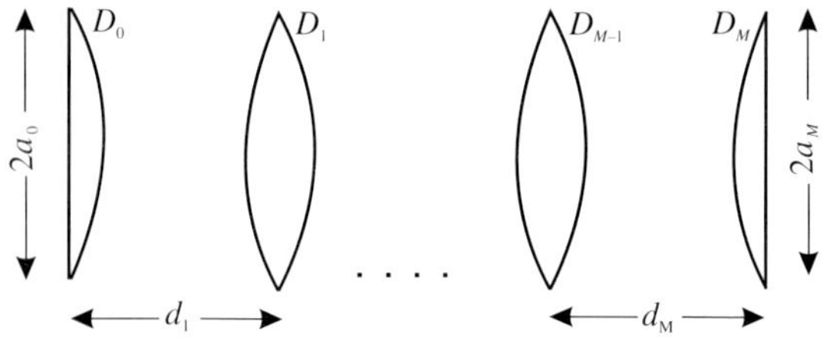

Figure 2.3 Multi-element field transformer.

where

$$c_m = \frac{k a_{m-1} a_m}{d_m} . \tag{2.22}$$

However, the relations between the phase corrections $\Psi_m(\vec{r}_m)$ and auxiliary functions $\psi_m(\vec{r}_m)$ are different for the middle apertures and the first and last ones:

$$\psi_0\left(\vec{r}_0\right) = \Psi_0\left(\vec{r}_0\right) - \frac{c_1}{2}\left|\vec{r}_0\right|^2 , \tag{2.23a}$$

$$\psi_m\left(\vec{r}_m\right) = \Psi_m\left(\vec{r}_m\right) - \frac{c_m + c_{m+1}}{2}\left|\vec{r}_m\right|^2 , \quad m = 1, 2, \ldots, M-1 , \tag{2.23b}$$

$$\psi_M\left(\vec{r}_M\right) = \Psi_M\left(\vec{r}_M\right) - \frac{c_M}{2}\left|\vec{r}_M\right|^2 . \tag{2.23c}$$

The desired field distribution on the aperture D_M is denoted by V_M. The optimization problem for this system is formulated as:

Problem Gp. Given the apertures $D_0, D_1, \ldots, D_M$ with distances d_m, $m = 1, \ldots, M$, between them, and the generating field U_0 on D_0 and desired field V_M on D_M, find the real functions $\psi_0, \psi_1, \ldots, \psi_M$ which provide a maximum for the functional

$$\chi(\psi_0, \psi_1, \ldots, \psi_M) = \frac{|(U_M, V_M)|}{\| U_0 \| \, \| V_M \|} \tag{2.24}$$

being the modulus of the coupling coefficient between the obtained and desired fields on D_M. The dependence of the field U_M on the optimization functions ψ_m is given by (2.21).

The system shown in Figure 2.3 can also be considered as a generalization of the two-element transmission line (see Figure 2.1). In particular, the multi-element line can be used for maximization of the transmission coefficient (2.10) rewritten for the field u_M on the aperture D_M in the form

$$L(u_0, \psi_1, \ldots, \psi_{M-1}) = \frac{\| u_M \|_M^2}{\| u_0 \|_0^2} . \tag{2.25}$$

Then a problem, analogous to Problem **A**, can be formulated for such a system as:

Problem H. Given the apertures $D_0, D_1, \ldots, D_M$, with distances d_m, $m = 1, \ldots, M$, between them, find the real functions $\psi_0, \psi_1, \ldots, \psi_M$ which maximize the transmission coefficient (2.25).

In the variant **Hp** of this problem, the amplitude $|u_0|$ is given and the function ψ_0 replaces u_0 on the left-hand side of (2.25).

For a mathematical formulation of Problems **Gp**, **Hp** see Section 3.2.2. For numerical methods and results see Section 5.3.1.

2.2 Antenna synthesis problems

2.2.1 Synthesis of aperture antennas by amplitude radiation pattern

Consider the so-called aperture antenna for which the radiation pattern (i.e. the angular dependence of the radiated field in the far zone) is assumed to be uniquely determined by the transverse electrical field distribution on the aperture D in the plane $z = 0$ (Figure 2.4). The influence of this field outside D on the pattern is neglected. Using the notations introduced in Section 2.1.1, we denote by $U_0(\vec{R}_0)$ a transverse component of the electric field on D. According to the radiation condition, this component in the far-field zone is written in the form

$$U(r, \vartheta, \varphi)|_{r\to\infty} \simeq \frac{\exp(-ikr)}{kr} f(\vartheta, \varphi), \tag{2.26}$$

where $\{r, \vartheta, \varphi\}$ is the spherical coordinate system with the axis directed perpendicular to the aperture, f is the same component as in the radiation pattern (below we call the field U and the pattern f).

In the chosen approximation, the pattern f is expressed by the field U_0 on the aperture D as follows

$$f\left(\vec{d}\right) = k^2 \cos\vartheta \iint_D U_0\left(\vec{R}_0\right) \exp\left[ik\left(\vec{R}_0 \cdot \vec{d}\right)\right] d\vec{R}_0, \tag{2.27}$$

where $\vec{d}$ is a vector on the unit sphere with the angular coordinates ϑ, φ. In Cartesian coordinates this formula takes the form

$$f(\xi, \eta) = k^2 \cos\vartheta \iint_D U_0(X_0, Y_0) \exp[ik(X_0\xi + Y_0\eta)]\, dX_0 dY_0 \tag{2.28}$$

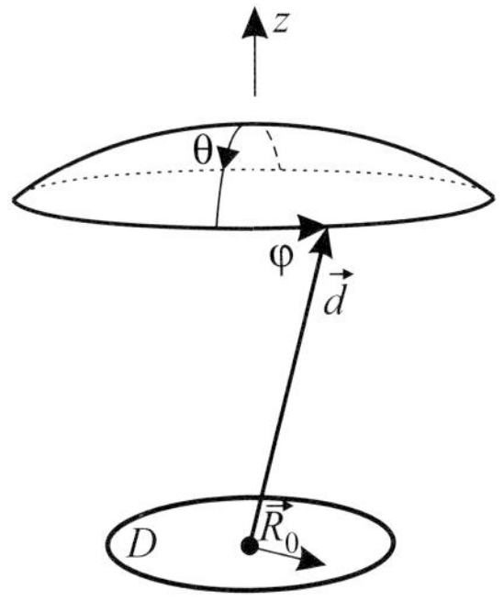

Figure 2.4 Aperture antenna.

where

$$\xi = \sin\vartheta\cos\varphi, \qquad \eta = \sin\vartheta\sin\varphi \tag{2.29}$$

are the horizontal Cartesian components of the vector $\vec{d}$. In antenna theory, ξ and η are often called the generalized angle coordinates in the far-field zone. The values $|\xi| \le 1$, $|\eta| \le 1$ correspond to the real angles ϑ, φ, whereas the larger $|\xi|$, $|\eta|$ correspond to the so-called complex angles. The values of $f(\xi,\eta)$ at complex angles describe the part of the power not radiated into the pattern ('reactive power').

The function $f(\xi,\eta)$ has the main lobe that is concentrated around the radiation direction (as a rule, the direction $\vartheta = 0$ corresponding to $\xi = \eta = 0$) and surrounded by the line of the local minimum of the radiation intensity, and the side lobes located outside this line. The objectives of optimization are the antenna directivity (minimization of the main lobe size), expansion of the main lobe and of the intensity distribution inside it, the level of the side lobes and their location, the level of the reactive power, etc. In this book main attention is paid to the problems in which the amplitude radiation pattern is close to a given positive function F that is assumed to be zero outside an angular domain Ω.

Before formulating the optimization problems for aperture antennas, we introduce scalar products that will be used in the functionals

$$(U,V)_D = \iint_D U(\vec{R}_0)\bar{V}(\vec{R})\mathrm{d}\vec{R}\,, \tag{2.30a}$$

$$(f_1, f_2)_\Omega = \iint_\Omega f_1(\xi,\eta)\,\bar{f}_2(\xi,\eta)\,\mathrm{d}\xi\mathrm{d}\eta\,, \tag{2.30b}$$

$$(f_1, f_2)_\infty = \int_{-\infty}^{\infty}\int_{-\infty}^{\infty} f_1(\xi,\eta)\,\bar{f}_2(\xi,\eta)\,\mathrm{d}\xi\mathrm{d}\eta\,, \tag{2.30c}$$

and the norms generated by them

$$\|U_0\|_0^2 = (U_0, U_0)_0\,, \quad \|f\|_\Omega^2 = (f,f)_1\,, \quad \|f\|_\infty^2 = (f,f)_\infty\,. \tag{2.31}$$

The integration domain Ω in (2.30b) is the part of the real angles for which the amplitude pattern $|f|$ has to be optimized (maximized or made close to a given function). The last product involves the integration over both the visible and complex angles.

The optimization problems with a free phase for aperture antennas are similar to those for power transmitting lines (see Section 2.1.1). In all problems the antenna domain D and angular domain Ω, where the pattern is prescribed, are given.

Problem I. Find the field distribution U_0 on the aperture D which maximizes the energy radiated into the angular domain Ω:

$$L(U_0) = \frac{\|f\|_\Omega^2}{\|U_0\|_0^2}\,, \tag{2.32}$$

For a mathematical formulation and numerical methods see Section 3.3.1.

Problem J. Given the desired amplitude pattern F in the whole (real and complex) angular space, $F \equiv 0$ outside Ω, find the field distribution U_0 on the aperture D which minimizes the normalized mean square difference

$$\sigma(U_0) = \frac{\| |f| - F \|_\infty^2}{\| U_0 \|_0^2 \, \| F \|_\Omega^2} \tag{2.33}$$

between the obtained and desired amplitude patterns in the whole angular space.

For a mathematical formulation and numerical methods for Problems **J** and **Jp** see Sections 3.1.5 and 3.2.1, respectively. Two-dimensional cases of Problem **J** are considered analytically and numerically in Section 4.2.

Problem K. Given the desired amplitude pattern F in the angular domain Ω and a positive number N, find the field U_0 on the aperture D which minimizes the normalized mean-square difference

$$\sigma_0(U_0) = \frac{\| |f| - F \|_\Omega^2}{\| U_0 \|_0^2 \, \| F \|_\Omega^2}, \qquad U_0 \in L_2(D_0)\,, \tag{2.34}$$

between the given and obtained amplitude patterns in this domain, in the class of functions satisfying the restriction

$$\sigma_1(U_0) \le N\,, \tag{2.35}$$

where

$$\sigma_1(U_0) = \frac{\| f \|_\infty^2 - \| f \|_\Omega^2}{\| U_0 \|_0^2} \tag{2.36}$$

is the energy bypassing the domain Ω (including the complex angles).

Problem L. Given a positive number M. Find the field U_0 on the aperture that minimizes the energy $\sigma_1(U_0)$ bypassing the domain Ω, with the restriction

$$\sigma_0(U_0) \le M\,. \tag{2.37}$$

Problem M. Given the desired amplitude pattern F in the angular domain Ω and a positive number t, find the field U_0 on the aperture D that minimizes the weighted functional

$$\sigma_t(U_0) = \sigma_0(U_0) + t\sigma_1(U_0) \tag{2.38}$$

which describes a compromise between two contradictory demands: decreasing the difference between the obtained and desired amplitude patterns in the area Ω, and the energy of the pattern bypassing this area (two-criterion optimization).

Problems **K**, **L**, **M** can be united by (2.38). For a mathematical formulation see Section 3.1.6; the two-dimensional case is considered analytically and numerically in Section 4.3. See also Section 5.2.2.

2.2.2
Synthesis of antenna arrays by amplitude radiation pattern

The antenna array is a system of elements (radiators) located in a bounded area of the space. The radiation pattern of the array is defined by the patterns of the radiators and their positions. As a rule, radiators are identical and they are located in a plane domain (Figure 2.5). The radiators are fed in such a way that their patterns differ only in a complex coefficients (we call them the currents on the radiators). In general, these coefficients are not proportional to the values of power supplied to the elements. This fact is explained by the mutual influence of the radiators and it is often neglected.

The radiation pattern of such an array with P radiators is

$$f\left(\vec{d}\right) = f_0\left(\vec{d}\right)\sum_{p=1}^{P} u_p \exp\left[ik\left(\vec{R}_p\cdot\vec{d}\right)\right], \tag{2.39}$$

where $f_0\left(\vec{d}\right)$ (not necessarily a scalar) is the pattern of a single radiator with the unite current, located at the coordinate origin, u_p is the complex current on the pth radiator, $\vec{R}_p$ is the vector describing its position and the sum in (2.39) is called the array factor. For simplicity, we put $f_0 \equiv 1$ and identify the array factor with the radiation pattern; this fact should be kept in mind when interpreting the problem formulations and results.

In the Cartesian coordinates in the array plane and the generalized angular coordinates in the far-field zone this formula (with the above assumption) is rewritten as

$$f(\xi,\eta) = \sum_{p=1}^{P} u_p \exp\left[ik\left(X_p\xi + Y_p\eta\right)\right], \tag{2.40}$$

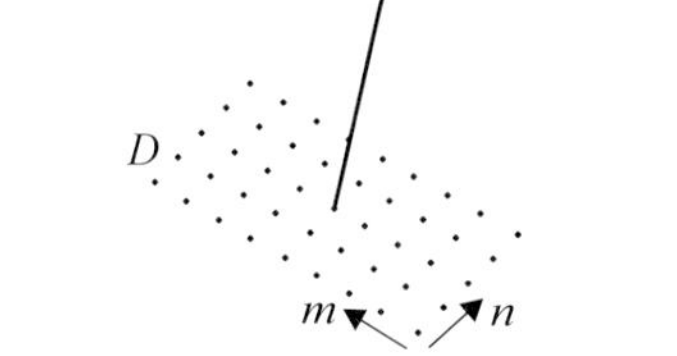

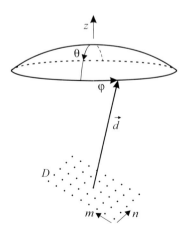

Figure 2.5 Antenna array.

where X_p, Y_p are the Cartesian coordinates of the pth radiator and ξ and η are given in (2.29).

In practice, the rectangular equidistant arrays are commonly used, for which $P = NM$, $p = \{n, m\}$ and the radiator coordinates are $X_n = nh_x$, $n = 1, 2, \ldots, N$; $Y_m = mh_y$, $m = 1, 2, \ldots, M$, so that

$$f(\xi, \eta) = \sum_{n=1}^{N} \sum_{m=1}^{M} u_{nm} \exp\left[ik\left(nh_x\xi + mh_y\eta\right)\right] . \tag{2.41}$$

In the particular case $u_{nm} = v_n w_m$, we have $f(\xi, \eta) = f_\xi(\xi) f_\eta(\eta)$, where

$$f_\xi(\xi) = \sum_{n=1}^{N} v_n \exp\left(iknh_x\xi\right) , \qquad f_\eta(\eta) = \sum_{m=1}^{M} w_m \exp\left(ikmh_y\eta\right) \tag{2.42}$$

are the patterns of linear arrays directed along the x- and y-axes, respectively. The linear arrays are widely used themselves and their optimization can be considered as an independent problem.

An important property of the antenna array is its ability to scan the radiation pattern by changing the phase distribution of the currents on its elements (the arrays with this property are called phased antenna arrays). For this reason the phase optimization is very useful for such antennas.

The main distinction between the antenna array and aperture antennas is that the radiation pattern (2.40) as a function of the generalized coordinates ξ, η is not square integrated over the entire plane. Therefore, in general, the optimization problems should be formulated in a different way. In particular, the radiation patterns of equidistant arrays are double-periodical functions with the periods $T_x = 2\pi/(kh_x)$, $T_y = 2\pi/(kh_y)$ in the ξ- and η-directions, respectively. Usually, the distances between the radiators are subject to the physical restriction $h_{x,y} > \lambda/2$, so that one period fully contains all visible angles. The Parseval equality

$$\sum_{n=1}^{N} \sum_{m=1}^{M} |u_{nm}|^2 = \frac{k^2 h_x h_y}{4\pi^2} \int_{-T_x}^{T_x} \int_{-T_y}^{T_y} |f(\xi, \eta)|^2 \mathrm{d}\eta \mathrm{d}\xi \tag{2.43}$$

is valid for the rectangular equidistant arrays. For the linear array, it is

$$\sum_{n=1}^{N_x} |v_n|^2 = \frac{kh_x}{2\pi} \int_{-T_x}^{T_x} |f(\xi)|^2 \mathrm{d}\xi . \tag{2.44}$$

These equalities are used in the formulation of optimization problems for antenna arrays.

We denote the vectors of different current distributions on the radiators as

$$U^{(j)} = \left\{u_{nm}^{(j)}\right\} , \tag{2.45}$$

and the patterns created by $U^{(j)}$ as f_j. We introduce the scalar products

$$\left(U^{(1)}, U^{(2)}\right)_0 = \sum_{n=1}^{N} \sum_{m=1}^{M} u_{nm}^{(1)} \bar{u}_{nm}^{(2)} , \tag{2.46a}$$

$$(f_1, f_2)_1 = \int_{-1}^{1}\int_{-1}^{1} f_1(\xi, \eta)\, \bar{f}_2(\xi, \eta)\, d\xi\, d\eta\,, \tag{2.46b}$$

$$(f_1, f_2)_\Omega = \iint_\Omega f_1(\xi, \eta)\, \bar{f}_2(\xi, \eta)\, d\xi\, d\eta\,, \tag{2.46c}$$

$$(f_1, f_2)_T = \int_{-T_x/2}^{T_y/2}\int_{-T_x/2}^{T_x/2} f_1(\xi, \eta)\, \bar{f}_2(\xi, \eta)\, d\xi\, d\eta\,. \tag{2.46d}$$

and the norms generated by them as

$$\|U\|_0^2 = (U, U)_0,\ \|f\|_1^2 = (f, f)_1,\ \|f\|_\Omega^2 = (f, f)_\Omega,\ \|f\|_T^2 = (f, f)_T. \tag{2.47}$$

The integration is made over the all visible angles in (2.46b), over the part Ω of the real angles where the amplitude pattern $|f|$ should be optimized (maximized or made close to a given function) in (2.46c), over one period of the radiation pattern (for equidistant arrays only) in (2.46c).

Problem N. Maximize the radiation into a given angle area

$$L(U) = \frac{\|f\|_\Omega^2}{\|U\|_0^2} \tag{2.48}$$

with respect to complex current distribution U on the array.

For a mathematical formulation and numerical methods see Section 3.3.1.

Problem O. Minimize the normalized mean-square approximation

$$\sigma(U) = \frac{\||f| - F\|_1^2}{\|U\|_0^2\,\|F\|_\Omega^2} \tag{2.49}$$

of the given (desired) amplitude pattern F in the entire real angles with respect to the complex current distribution U on the array. The function F is assumed to be zero in the complement of the angle domain Ω to the entire domain of the real angles.

For a mathematical formulation and numerical methods for Problems **O** and **Op** see Sections 3.1.5 and 3.2.1, respectively. Two-dimensional cases of Problem **O** are considered analytically and numerically in Section 4.2. A numerical method for the three-dimensional case is described and applied in Section 5.4.

Problem P. Minimize the normalized mean-square approximation

$$\sigma_0(U) = \frac{\||f| - F\|_\Omega^2}{\|U\|_0^2\,\|F\|_\Omega^2} \tag{2.50}$$

of the given (desired) amplitude pattern F with the restriction

$$\sigma_1(U) \le N\,, \tag{2.51}$$

where

$$\sigma_1(U) = \frac{\|f\|_1^2 - \|f\|_\Omega^2}{\|U\|_0^2} \tag{2.52}$$

is the energy part outgoing into the side lobes of the pattern (more exactly, into the real angle area outside the controlled domain Ω), with respect to the complex current distribution U on the array.

Problem Q. Minimize $\sigma_1(U)$ with the restriction

$$\sigma_0(U) \leq M\,, \tag{2.53}$$

with respect to the complex current distribution U on the array.

Problem R. Minimize the weighted functional

$$\sigma_t(U) = \sigma_0(U) + t\sigma_1(U) \tag{2.54}$$

at a given $t > 0$ which describes a compromise demand of simultaneously decreasing both the approximation of a given amplitude pattern in the angular area Ω and the outgoing energy outside it (two-criterion optimization), with respect to the complex current distribution U on the array.

Problems **P**, **Q** and **R** can be united by (2.54). For a mathematical formulation see Section 3.1.6; the two-dimensional case is considered analytically and numerically in Section 4.3.

Problems **O–R** can be substituted by others having similar physical meaning but using the squared norm of currents $\| U\|_0^2$ instead of $\sigma_1(U)$ in formulas (2.51) and (2.54). Such a formulation cannot be applied to the case when the amplitude current distribution on the radiators is given, because this norm is fixed in this case.

For equidistant arrays only the formulations that use the values of the radiation pattern in one period can be applied. Further, we denote the corresponding problems by the primed letters.

Problem O.′ Minimize the normalized mean-square approximation

$$\sigma(U) = \frac{\||f| - F\|_T^2}{\| U\|_0^2 \, \|F\|_\Omega^2} \tag{2.55}$$

of the given (desired) amplitude pattern F in the angle period with respect to the complex current distribution U on the array. The function F is assumed to be zero in the complement of the angular domain Ω to the period.

For a mathematical formulation and numerical methods for Problems **O**′ and **O**′**p** see Sections 3.1.5 and 3.2.1, respectively. The two-dimensional variant of Problem **O** is considered analytically and numerically in Section 4.2.

Problem P.′ Minimize the normalized mean-square approximation

$$\sigma_0(U) = \frac{\||f| - F\|_\Omega^2}{\| U\|_0^2 \, \|F\|_\Omega^2} \tag{2.56}$$

of the given (desired) amplitude pattern F with the restriction

$$\sigma_1(U) \leq N\,, \tag{2.57}$$

where

$$\sigma_1(U) = \frac{\|f\|_T^2 - \|f\|_\Omega^2}{\|U\|_0^2} \tag{2.58}$$

is the outgoing energy part outside the angle domain Ω (including the complex angles up to the period), with respect to the complex current distribution U on the array.

Problem Q.$'$ Minimize $\sigma_1(U)$ with the restriction

$$\sigma_0(U) \leq M\,, \tag{2.59}$$

with respect to the complex current distribution U on the array.

Problem R.$'$ Minimize the weighted functional

$$\sigma_t(U) = \sigma_0(U) + t\sigma_1(U) \tag{2.60}$$

at a given $t > 0$ which describes a compromise demand of simultaneously decreasing both the approximation of a given amplitude pattern in the angular area Ω and the outgoing energy outside it (two-criterion optimization), with respect to the complex current distribution U on the array.

Problems $\mathbf{P}'$, $\mathbf{Q}'$ and $\mathbf{R}'$ can be united by (2.60). For a mathematical formulation see Section 3.1.6; the two-dimensional case is considered analytically and numerically in Section 4.3.

All problems considered are easily reformulated for the linear antennas. For this, expression (2.41) should by replaced by one of the formulas (2.42) and the scalar products (2.46) by their one-dimensional analogs.

2.2.3
Multi-element beam transformers

In this subsection we consider a multi-element quasioptical system which is a long antenna array fed by a narrow energy beam coming from the side wall of the antenna ([31, 32]) (Figure 2.6). The system can be used for two purposes: either as an antenna in the antenna–rectenna transmitting line of the type considered in Section 2.2.1, or as a linear antenna array creating the field with radiation pattern of desired structure. In its second role it can also be used as an element of a bigger array focusing the energy into a bounded domain in the middle zone. Here we formulate the optimization problems for such a system considered as an independent antenna in both mentioned roles. For using it as an array element, problems similar to those formulated in the previous subsection should be solved.

Since the system is nontypical and has not been considered in the wide-accessible literature, we give short derivations of the basic relations connecting the fields considered in the problems. The system consists of the vertical aperture D_0 located at its input, a series of parallel inclined semi-transparent mirrors D_m, $m = 1, \ldots, M-1$, and the inclined opaque mirror D_M at its end. A phase corrector

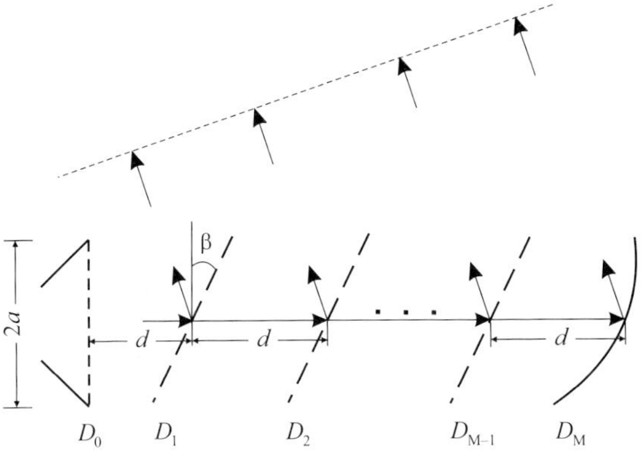

Figure 2.6 Multi-element beam transformer.

can be located on the input aperture. Each semi-transparent mirror partially transmits the incoming field, and partially reflects it in the direction determined by the system geometry. The transmitted and reflected fields have appropriate phase corrections. The last mirror only reflects the field with a phase correction.

For simplicity, we assume that all mirrors are the same vertical size $2a$ and the distance between their centers is identical and equal to d. In order that all reflected fields compose a wide beam forming a well-directed radiation pattern, the inclination angle β should be consistent with the mirror sizes as

$$\sin(2\beta) = \frac{2a}{d} \,. \tag{2.61}$$

The input phase correction as well as the reflection and transmission coefficients of the inclined mirrors are variable along their planes. They are the optimized functions within the problems. The incoming field may also be not given and should be optimized.

The quasioptical conditions of type (2.7) hold, which means, that the sizes of all elements are much larger than the wavelength and all distances are much larger than these sizes. Both large parameters ka and d/a are of the same order. According to condition (2.61), the angle β is small, of inverse order to the above large parameters.

We denote by U_m, $m = 0, \ldots, M$, the fields on the left of apertures D_m (before phase corrections). The phase corrector located on the input aperture D_0 adds the phase factor $\exp(i\psi_0)$ to the field U_0. Actions of the semitransparent phase-correcting mirrors are described by the variable complex reflection coefficients R_m, $m = 1, \ldots, M$, and the transmission ones T_m, $m = 1, \ldots, M-1$, supplying the fields $U_m(\vec{R}_m)$, $m = 1, \ldots, M$, when calculating the reflected and transmitted fields, respectively (do not confuse the vector $\vec{R}_m$ with the scalar function R_m).

In order to calculate the field transformations along the line with inclined apertures, we must modify (2.1) for a transformation of the fields on the vertical apertures. The general relation between the fields on two arbitrary oriented apertures is

$$U_{m+1}\left(\vec{R}^{(3)}_{m+1}\right) = \frac{k}{2\pi}\iint_{D_m} \frac{\exp\left(-ik\left|\vec{R}^{(3)}_m - \vec{R}^{(3)}_{m+1}\right|\right)}{\left|\vec{R}^{(3)}_m - \vec{R}^{(3)}_{m+1}\right|} \times U_m\left(\vec{R}^{(3)}_m\right) T_m\left(\vec{R}^{(3)}_m\right) \mathrm{d}\vec{R}^{(3)}_m , \quad (2.62)$$

where $\vec{R}^{(3)}_m$, $\vec{R}^{(3)}_{m+1}$ are the three-dimensional radius-vectors of the points on the apertures D_m and D_{m+1}, respectively, in the global coordinate system $\{x, y, z\}$. We simplify this relation, using the above quasioptical conditions. First, we write the distance between points on different mirrors in the form

$$\left|\vec{R}^{(3)}_m - \vec{R}^{(3)}_{m+1}\right| = \sqrt{\left|\vec{R}_{mx} - \vec{R}_{m+1,x}\right|^2 + (Z_m - Z_{m+1})^2} , \quad (2.63)$$

where $\vec{R}_{mx}$, $\vec{R}_{m+1,x}$ are the two-dimensional radius-vectors (in local coordinates) of projections of the points onto the vertical planes passing through the mirror centers, so that

$$\vec{R}^{(3)}_m = \vec{R}_{mx} + \vec{Z}_m , \qquad \vec{R}^{(3)}_{m+1} = \vec{R}_{m+1,x} + \vec{Z}_{m+1} . \quad (2.64)$$

As usual, under the quasioptical conditions, the denominator in (2.62) can be replaced by d owing to the smallness of a/d. However, in the numerator, under the exp symbol, this value should be calculated more exactly. Since in our problem both apertures are inclined in the (X, Z)-plane, we have

$$Z_{m+1} - Z_m = d + (X_{m+1}\tan\beta_{m+1} - X_m \tan\beta_m) . \quad (2.65)$$

Here we assume that the inclination angles β_m, β_{m+1} of the apertures may be not identical (as for D_0 and D_1). Taking out the factor d under the radical and remaining two first terms in the Taylor expansion of the second factor, we obtain (see Figure 2.7)

$$\begin{aligned}\left|\vec{R}^{(3)}_m - \vec{R}^{(3)}_{m+1}\right| &= d\left(1 + \frac{\left|\vec{R}_m - \vec{R}_{m+1}\right|^2}{2d^2} + \frac{X_{m+1}\tan\beta_{m+1} - X_m\tan\beta_m}{d}\right) \\ &= d + \frac{\left|\vec{R}_{mx} - \vec{R}_{m+1,x}\right|^2}{2d} + X_{m+1}\tan\beta_{m+1} - X_m\tan\beta_m . \end{aligned} \quad (2.66)$$

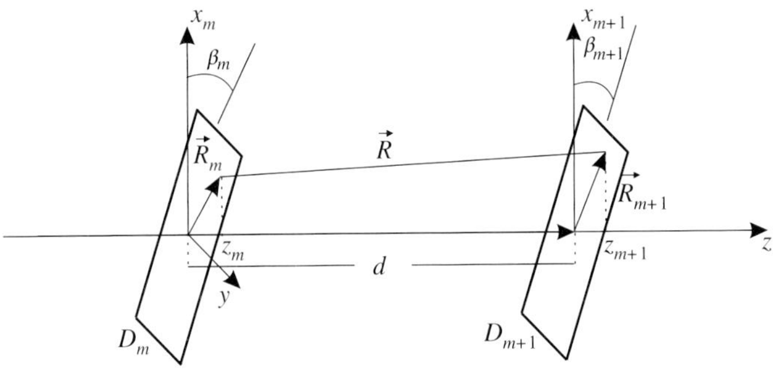

Figure 2.7 Illustration for expression (2.66).

Substituting obtained approximations into (2.62) and introducing a new integration variable $\vec{R}_m$ instead of $d\vec{R}_m^{(3)}$, so that $d\vec{R}_m^{(3)} = (\cos\beta_m)^{-1} d\vec{R}_m$, yield

$$U_{m+1}\left(\vec{R}_{m+1,x}\right) = \frac{k\exp(-ikd)}{2\pi\cos\beta_m} \iint_{\tilde{D}_m} \exp\left(-ik\left(\frac{\left|\vec{R}_{mx} - \vec{R}_{m+1,x}\right|^2}{2d}\right)\right)$$
$$\times \exp\left(-ik\left(X_{m+1}\tan\beta_{m+1} - X_m\tan\beta_m\right)\right) U_m\left(\vec{R}_{mx}\right) T_m\left(\vec{R}_{mx}\right) \mathrm{d}\vec{R}_{mx}\,,$$
$$m = 0,\ldots,M-1\,. \qquad (2.67a)$$

Here the integration domains $\tilde{D}_m$ are the projections of the apertures D_m onto the vertical planes, $\beta_0 = 0$; $\beta_m = \beta$, $m = 1,2,\ldots,M$.

Let the system be the antenna in the transmitting line. The rectenna occupies the domain Ω located perpendicular to the direction of the reflected rays in the geometro-optical approximation. This direction lies in the plane (x,z) and makes up the angle 2β with the axis $-z$. The distance ρ_0 from the center of the first mirror to the rectenna is large in comparison with the antenna width, $\rho_0 \gg Md$.

The total field on the antenna is a sum of the fields created by its elements

$$V(\omega) = \sum_{m=1}^{M} V_m(\omega)\,, \qquad m = 1,\ldots,M\,, \qquad (2.68)$$

At any point ω of the rectenna, the field V_m radiated by the m th mirror, is calculated by a formula, similar to (2.62),

$$V_m(\vec{\omega}) = \frac{k}{2\pi}\iint_{D_m} \frac{\exp\left(-ik\left|\vec{R}\right|\right)}{\left|\vec{R}\right|} U_m\left(\vec{R}_m^{(3)}\right) R_m\left(\vec{R}_m^{(3)}\right) \mathrm{d}\vec{R}_m^{(3)}\,, \qquad (2.69)$$

where $\vec{R}$ is the vector connecting the points $\vec{R}_m$ on the mth aperture and $\vec{r}$ on the rectenna,

$$\vec{R} = \vec{\rho}_0 + \left(\vec{r} - \vec{R}_m^{(3)}\right)\,, \qquad (2.70)$$

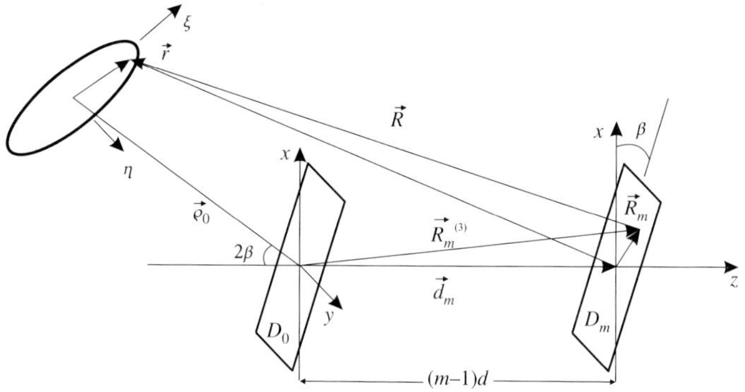

Figure 2.8 Illustration for expression (2.72).

$\vec{r}$ is the two-dimensional radius-vector of the point ω on the rectenna, $\vec{R}_m^{(3)} = \vec{d}_m + \vec{R}_m$ (see Figure 2.8). As before, we put $\left|\vec{R}\right| = \rho_0$ in the denominator of (2.70), and keep the first two terms in its expansion with respect to a small value ρ_0^{-1} in the exponent of the numerator.

According to the generalized cosine theorem,

$$\left|\vec{R}\right| = \sqrt{\left|\vec{\rho}_0\right|^2 - 2\vec{\rho}_0 \cdot \vec{R}_m^{(3)} + \left|\vec{r} - \vec{R}_m^{(3)}\right|^2} \,. \tag{2.71}$$

Here it is taken into account that the vectors $\vec{\rho}_0$ and $\vec{r}$ are orthogonal. Taking out the factor ρ_0 from the radical and performing obvious derivations, we obtain

$$\begin{aligned} \left|\vec{R}\right| &= \rho_0 \sqrt{1 - \frac{2\vec{\rho}_0}{\rho_0^2} \cdot \vec{R}_m^{(3)} + \frac{\left|\vec{r} - \vec{R}_m^{(3)}\right|^2}{\rho_0^2}} \\ &\simeq \rho_0 - \left(\vec{R}_m^{(3)}\right)_{\rho_0} + \frac{\left|\vec{r} - \vec{R}_m^{(3)}\right|^2}{2\rho_0} \,, \end{aligned} \tag{2.72}$$

where the index ρ_0 means the projection onto the vector $\vec{\rho}_0$. Finally,

$$\begin{aligned} V_m(\vec{r}) = \frac{k \exp(-ik\rho_0)}{2\pi\rho_0} \iint_{D_m} \exp\left(ik\left(\vec{R}_m^{(3)}\right)_{\rho_0} - \frac{\left|\vec{r} - \vec{R}_m^{(3)}\right|^2}{2\rho_0}\right) \\ \times\, U_m\left(\vec{R}_m^{(3)}\right) R_m\left(\vec{R}_m^{(3)}\right) \mathrm{d}\vec{R}_m^{(3)} \,. \end{aligned} \tag{2.73}$$

In the global Cartesian system, the components of the vector $\vec{r}$ are

$$r_x = \xi \cos(2\beta) \,, \tag{2.74a}$$

$$r_y = \eta \,, \tag{2.74b}$$

$$r_z = \xi \sin(2\beta) \,, \tag{2.74c}$$

where ξ, η are the components of the vector $\vec{r}$ in the local Cartesian system in Ω with the ξ-axis lying in the plane (x, z) and directed into its first quadrant, and η coinciding with y. If the components of the vector $\vec{R}_m^{(3)}$ in the local Cartesian system which is originated at the center of the mth mirror are $\{x_m, y_m, z_m\}$, then its point projection onto $\vec{\rho}_0$ is

$$\left(\vec{R}_m^{(3)}\right)_{\rho_0} = x_m \sin(2\beta) - \left[(m-1)d + z_m\right] \cos(2\beta) . \tag{2.75}$$

From the antenna point of view, the system is a linear antenna; its elements are the inclined mirrors radiating the field reflected by them. The direction, in which the antenna forms the radiation pattern, is indicated by the vector $\vec{\rho}_0$. For this reason, the radiation pattern is calculated in the spherical coordinate system (ρ, θ, φ) originating at the center of the first mirror, with the axis coinciding with direction $\vec{\rho}_0$.

The pattern of the entire system is the sum of partial patterns of its elements

$$f(\omega) = \sum_{m=1}^{M} f_m(\omega) , \qquad m = 1, \ldots, M , \tag{2.76}$$

where $\omega = (\theta, \varphi)$ is the aggregated angular coordinate of a point on the removed sphere; we denote the radius-vector of this point by $\vec{\omega}$. The technique for obtaining the expression for $f_m(\omega)$ is similar to that applied for obtaining (2.73). As a result, we obtain

$$f_m\left(\vec{\omega}\right) = \frac{k}{2\pi} \exp\left(ik\vec{\omega}_0 \cdot \vec{d}_m\right) \tag{2.77}$$

$$\times \iint_{D_m} U_m\left(\vec{R}_m\right) R_m\left(\vec{R}_m\right) \exp\left(ik\vec{\omega}_0 \cdot \vec{R}_m\right) \mathrm{d}\vec{R}_m . \tag{2.78}$$

If the vector $\vec{\omega}_0$ is given by its components θ, φ in the local spherical system with the axis $\vec{\rho}_0$, then its Cartesian components are

$$\omega_x = -\xi \cos(2\beta) - \cos\theta \sin(2\beta) , \tag{2.79a}$$

$$\omega_y = \eta , \tag{2.79b}$$

$$\omega_z = \xi \sin(2\beta) - \cos\theta \cos(2\beta) , \tag{2.79c}$$

where $\xi = \sin\theta \cos\varphi$, $\eta = \sin\theta \sin\varphi$.

The optimization problems formulated for the system considered, are different depending on the purpose for which the system is used. If it is used as the antenna in the transmitting line, then problems analogous to those described in Section 2.1.2 can be formulated. If it is intended for use as the antenna array creating the field with a desired radiation pattern, then the problems similar to those described in Section 2.2.2 can be formulated. Below we formulate only three of these problems. Some other formulations will be considered in the next chapters where the mathematical properties of the problems will be investigated and algorithms for their solution will be described.

In all problems, the optimization functions are the reflection coefficients $R_m(\vec{R}_m)$, $m = 1, 2, \ldots, M$, and the transmission ones $T_m(\vec{R}_m)$, $m = 1, 2, \ldots, M-1$. These coefficients are variable, depending on the point on the mirrors. If there are no losses in the mirrors, then these functions fulfill the conditions

$$\left|R_m\left(\vec{R}_m\right)\right|^2 + \left|T_m\left(\vec{R}_m\right)\right|^2 = 1\,, \qquad m = 1, \ldots, M-1\,, \tag{2.80a}$$

$$\left|R_M\left(\vec{R}_M\right)\right| = 1\,. \tag{2.80b}$$

We introduce the scalar products

$$\left(U_0^{(1)}, U_0^{(2)}\right)_0 = \iint_{D_0} U_0^{(1)}\left(\vec{R}_0\right) \bar{U}_0^{(1)}\left(\vec{R}_0\right) \mathrm{d}\vec{R}_0\,, \tag{2.81a}$$

$$\left(f^{(1)}, f^{(2)}\right)_\Omega = \iint_\Omega f_1(\omega)\, \bar{f}_2(\omega)\, \mathrm{d}\omega\,, \tag{2.81b}$$

on the input aperture and on the rectenna or in the far-field zone, respectively, and the norms generated by them

$$\| U_0 \|_0^2 = (U_0, U_0)_0\,, \qquad \| f \|_\Omega^2 = (f, f)_\Omega\,. \tag{2.82}$$

For simplicity, we use the same notation Ω for the domain in the middle zone and on the removed sphere for different problems. Below, we formulate optimization problems for the system considered only as the antenna array.

Problem S. Maximize the functional

$$L(U_0, R_1, \ldots, R_M, T_1, \ldots, T_{M-1}) = \frac{\| f \|_a^2}{\| U_0 \|_0^2} \tag{2.83}$$

with respect to the generating field u_0, and the reflection R_m, $m = 1, \ldots, M$, and transmission coefficients T_m, $m = 1, \ldots, M-1$ being under conditions (2.80).

We do not consider here the input phase correction $\psi_0(x_0)$ as one of the optimization functions in this problem, because it can be added to the phase of the sought input field. This function should be included to the optimization functions in the **p**-variant of this problem (Problem **Sp**), in which $|u_0|$ is given.

Problem T. Given desired amplitude radiation pattern F, find the generating field u_0, both the reflection R_m, $m = 1, \ldots, M$, and transmission T_m, $m = 1, \ldots, M-1$, coefficients, under conditions (2.80), which provide the maximum of the functional

$$\chi(u_0, R_1, \ldots, R_M, T_1, \ldots, T_{M-1}) = \frac{(|f|, F)_\Omega}{\| u_0 \|_0}\,. \tag{2.84}$$

Problem U. Given desired amplitude radiation pattern F, find the generating field u_0, both the reflection R_m, $m = 1, \ldots, M$, and transmission T_m, $m = 1, \ldots, M-1$ coefficients, under conditions (2.80), which provide the minimum of the functional

$$\sigma_0(u_0, R_1, \ldots, R_M, T_1, \ldots, T_{M-1}) = \| |f| - F \|_\Omega^2 \tag{2.85}$$

with the restriction

$$\|u_0\|_0 \leq N \, . \tag{2.86}$$

In the **p**-variant of this problem (Problem **Up**), in which $|u_0|$ is given, restriction (2.86) is absent.

For a mathematical formulation and numerical methods see Section 3.2.2.4. For numerical methods and results see Section 5.3.2.

2.2.4
Cavity resonant antennas

The cavity resonant antenna is an open resonator with semi-transparent walls. It may contain an inclusion with metallic or impedance boundary (Figure 2.9). If the walls have a small transparency, then there are some discrete frequencies (close to eigenfrequencies of the interior domain) at which the interior and exterior fields (including the radiation pattern) weakly depend on the excitation. They are mainly determined by the geometry of the resonator walls and their transparency distribution; if the inclusion exists, then the frequencies also depend on the shape and impedance distribution of its boundary. These constructive parameters are the optimization functions in problems of the constructive synthesis of resonant antennas. Efficient methods for solving such synthesis problems are described in detail in the book [24]. Here we focus on the problems in which only the amplitude pattern is given and its phase distribution can be chosen arbitrarily.

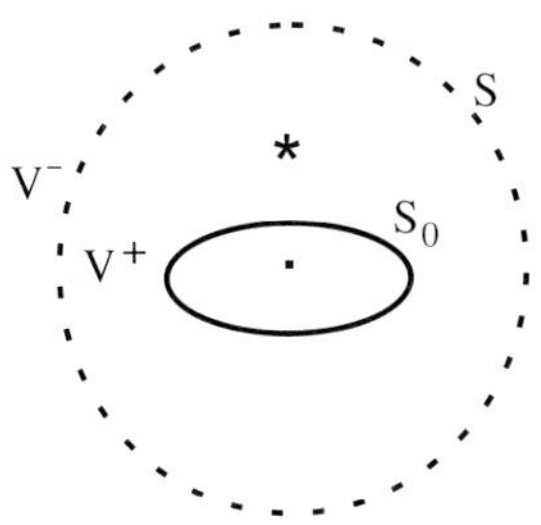

Figure 2.9 Cavity resonant antenna.

The main technique for the formulation of the resonant antenna problems and their solution is the *generalized method of eigenoscillations* [33]. The principal distinction of the method is that various parameters (not only frequency) are taken as the eigenvalues in auxiliary homogeneous problems. In the problem considered here, we use the variant of the method in which the wall transparency plays the role of eigenvalues. Its scalar two-dimensional formulation is briefly described below. For simplicity, we assume that the inclusion boundary is metallic (not impedance).

Let the sources be located inside the resonator, so that the field u satisfies the Helmholtz equations

$$\Delta u + k^2 u = \Phi \qquad \text{in } V^+ \, , \tag{2.87a}$$

$$\Delta u + k^2 u = 0 \qquad \text{in } V^- \, , \tag{2.87b}$$

inside and outside the resonator, respectively and the boundary condition

$$u + w(s)\frac{\partial u}{\partial N}\bigg|_{S_0} = 0 \tag{2.88}$$

on the impedance (in particular, metallic at $w(s) \equiv 0$) boundary S_0 of the inclusion, satisfies the matching conditions

$$u^+ - u^- = 0\,, \tag{2.89a}$$

$$\frac{\partial u^+}{\partial N} - \frac{\partial u^-}{\partial N} = \frac{1}{\rho(s)} u \tag{2.89b}$$

on the semi-transparent exterior boundary S, and the radiation condition

$$u|_{r\to\infty} \simeq \frac{\exp(-ikr)}{\sqrt{kr}} f(\varphi) \tag{2.90}$$

at infinity. Here the indices '+', '−' mark the values of the functions inside and outside the resonator, respectively, N is the interior normal to S, $\rho(s)$ is a real function describing the transparency distribution along the boundary S and $f(\varphi)$ is the radiation pattern. The solution to problem (2.87)–(2.90) is expressed in the form

$$u = U^0 + \sum_{n=1}^{\infty} A_n u_n\,, \tag{2.91}$$

where U^0 is the field of the same sources in the absence of the boundary S (i.e. U^0 satisfies (2.87a) for both V^+ and V^-, condition (2.88) on S_0, and the radiation condition (2.90) with a certain pattern f_0), u_n are the fields of generalized eigenoscillations which are solutions to the following homogeneous problem

$$\Delta u_n + k^2 u_n = 0 \qquad \text{in } V^+ + V^-\,, \tag{2.92}$$

$$u_n + w(s)\frac{\partial u_n}{\partial N}\bigg|_{S_0} = 0\,, \tag{2.93}$$

$$u_n^+ - u_n^-\Big|_S = 0\,, \tag{2.94a}$$

$$\frac{\partial u_n^+}{\partial N} - \frac{\partial u_n^-}{\partial N}\bigg|_S = \frac{1}{\rho_n(s)} u_n\,, \tag{2.94b}$$

$$u_n|_{r\to\infty} \simeq \frac{\exp(-ikr)}{\sqrt{kr}} f_n(\varphi)\,. \tag{2.95}$$

Here $\rho_n(s)$ are auxiliary functions connected with the transparency distribution $\rho(s)$ by the formula

$$\frac{1}{\rho_n(s)} = \eta(s) + \nu_n\left(\frac{1}{\rho(s)} - \eta(s)\right)\,, \tag{2.96}$$

$\eta(s)$ is an arbitrary (generally) real function, $\nu_n = \nu_n' + i\nu_n''$ are the complex eigenvalues of the problem and $f_n(\varphi)$ are the patterns of eigenoscillations.

Each coefficient A_n is determined by the incident field U^0, the eigenoscillation field u_n, the eigenvalue ν_n and the arbitrary function $\eta(s)$ (which can be chosen in the concrete algorithm). The dependence of A_n on ν_n in the form of the resonant factor $(\nu_n - 1)^{-1}$ is absolutely essential. If ν_n is close to unity at a certain frequency, then the appropriate term in the sum (2.91) becomes dominant in the total field so that the other terms (including U^0) can be neglected (with an accuracy which is inversely proportional to the Q-factor of the resonator). This fact allows us to formulate the optimization problem not for the total pattern $f(\varphi)$, but only for the pattern $f_n(\varphi)$ of the resonant eigenoscillation. Assume that the resonant oscillation corresponds to $n = 1$ and denote its field and pattern by u_1 and $f_1(\varphi)$, respectively. Of course, the fields of eigenoscillations are different for different functions $\eta(s)$ in (2.96). However, for the resonant oscillation, this difference is as small as the neglected terms in (2.91).

Various optimization problems can be formulated for the antennas considered here. They differ in optimization functions, such as the shapes of boundaries S and S_0, their transparency and impedance distributions, etc. Here we consider the conformal antennas, that is, the shape of the exterior boundary S is given. The problems in which S is determined are considered in [24].

We denote the desired amplitude pattern of the resonant eigenoscillation by $F(\varphi)$, and introduce the scalar products as

$$\left(u^{(1)}, u^{(2)}\right)_S = \int_S u^{(1)}(s)\bar{u}^{(2)}(s)\mathrm{d}s\,, \tag{2.97a}$$

$$\left(f^{(1)}, f^{(2)}\right)_\infty = \int_0^{2\pi} f^{(1)}(\varphi)\bar{f}^{(2)}(\varphi)\mathrm{d}\varphi\,, \tag{2.97b}$$

and the norms generated by them as

$$\|u\|_S^2 = (u, u)_S, \qquad \|f\|_\infty^2 = (f, f)_\infty\,. \tag{2.98}$$

The functional to be minimized is taken in the form

$$\sigma_t(S_0, w, \rho) = \|F - |f_1|\|_\infty^2 + t\,\|u_1\|_S^2\,, \tag{2.99}$$

where t is the given real positive weight factor. This functional implicitly depends on the arguments S_0, w, ρ. To calculate its value, first we should solve the homogeneous problem (2.92)–(2.95), determine the field $u_1(s)$ on the semi-transparent boundary S and its pattern $f_1(\varphi)$. It is assumed that the frequency k is chosen such that this oscillation is resonant, that is, the eigenvalue ν_1 is close to unity.

Problem Vs. Given desired amplitude radiation pattern F, the shapes of the exterior boundary S and interior one S_0, find the distribution $\rho(s)$ of the transparency on S and the distribution $w(s)$ of the impedance on S_0 which provides a minimum for (2.99).

Problem Vw. Given desired amplitude radiation pattern F, the shapes of the exterior boundary S and constant impedance w of the interior boundary S_0, find the distribution $\rho(s)$ of the transparency on S and the shape of the boundary S_0 which provides a minimum for (2.99).

Since at resonance the field u_1 inside V^+ is close to the appropriate eigenoscillation of the closed resonator, that is, in-phase, and condition (2.94a) provides the continuity of this field on S, it is natural to require this field also to be in-phase (i.e. without loss of generality, real) on S. This fact allows us to solve the above problems in two steps. In the first step the following auxiliary problem should be solved to determine the optimal field distribution on S.

Problem Va. Given desired amplitude radiation pattern F, and the shape of the exterior boundary S, find the real field distribution u_1 on S, which provides a minimum for

$$\sigma_t(u_1) = \| F - |f_1| \|_\infty^2 + t \, \| u_1 \|_S^2 \,. \tag{2.100}$$

This problem is an amplitude optimization problem; in its statement, it is similar to those for other types of antennas. For the mathematical formulation and numerical method see Section 3.2.3. For the numerical algorithms and results see [21, 22].

In the second step of the algorithm, the constructive parameters of the antenna must be found. They provide (exactly or approximately) the field distribution u_1 on the antenna boundary determined in the first step. The procedures which can be applied in this step are different for Problems **Vs** and **Vw** and are beyond the scope of this book. We refer readers to the books [24] and [25] for more details of them.

2.3
Phase optimization problems in waveguides and open resonators

2.3.1
Waveguide mode convertors

There is a problem in waveguide theory which consists of the conversion of one mode (or combination) into another mode (or combination). This problem is usually solved by nonhomogeneous filling of a segment of a regular waveguide or by creating an irregularly shaped homogeneous segment (see, e.g., [34, 35]). The first work concerning the application of quasi-optical elements for this purpose was as expected [38]. Here we consider one possibility of solving this problem by placing two (or more) lenses of complicated shapes in a regular waveguide segment.

As before, such a system transforms the field on one aperture into a field on another. The fields on the apertures are given. The sought functions are the phase corrections of the lenses. The problem differs from previous ones in the manner of transmission of the field between the apertures, and, consequently, in the formulas for their recalculation. In the systems considered above, the fields were transmitted by means of the wave beams in free space and the transmission was described by the Fourier transform. Here the transmission is realized by a system of traveling waves (modes) of the waveguide, which are characterized by waves (not only trav-

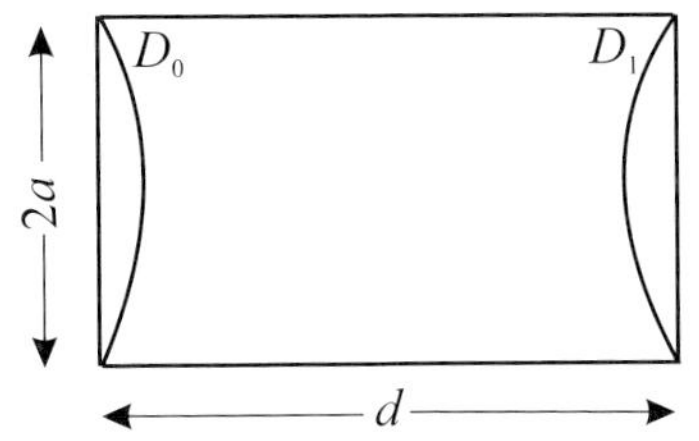

Figure 2.10 Two-element waveguide mode convertor.

eling) with different structures and wave numbers. The fields of these waves are orthogonal.

The field between the apertures is transformed in the following way. The field before the first lens is given in the form of an expansion by the eigenwaves of the waveguide. The first lens transforms the incident field into a superposition of modes that propagate up to the end of the segment. The second lens transforms the incoming field up to the end into a field of the desired structure. As well as the two functions describing the phase corrections of the lenses, the segment length can also be considered as an additional minimization parameter.

Consider a segment of a regular closed waveguide with arbitrary cross-section of characteristic size a (Figure 2.10). At the input and output apertures D_0 and D_1 the thin lenses are located which provide phase corrections $\exp(i\psi_0(\vec{r}_0))$ and $\exp(i\psi_1(\vec{r}_1))$, respectively. It is assumed that the waveguide is wide, $ka \gg 1$, so that a large number N of traveling modes can propagate along it. In this case the reflection from the lenses is small and can be neglected. The length d of the convertor is sufficiently large so that the nonpropagating modes which arise after the aperture D_0 do not reach the aperture D_1. Then the field in any cross-section is uniquely defined by the transverse components $E_{n,x}$, $E_{n,y}$, $H_{n,x}$, $H_{n,y}$ of the traveling modes. The index n is aggregated, it contains the mode type (*TM* or *TE*) and its two-digits index. The propagating constant of the mode is denoted by h_n.

Denote by $u_n(\vec{r})$ and $U(\vec{r}, z)$ the 'vectors' consisting of the above four components of the nth mode and the field in a cross-section, respectively; here $\vec{r}$ is the transverse radius-vector of the point in the cross-section. At any cross-section,

$$U(\vec{r}, z) = \sum_{n=1}^{N} C_n(z) u_n(\vec{r}) . \tag{2.101}$$

Introduce the scalar products of two 'vectors' as

$$(U_1, U_2) = \int_S \left[\vec{E}_1 \times \vec{H}_2\right]_z \mathrm{d}S , \tag{2.102}$$

where $\vec{E}_1$, $\vec{H}_2$ are the transversal electric and magnetic fields of the first and second 'vectors' respectively, and the integral is taken over the cross-section of the waveguide. According to the relation (3.18) from [36], the 'vectors' u_n are orthogonal,

$$(u_n, u_m) = \delta_{nm} . \tag{2.103}$$

Here the normalization different from [36] is used.

Denote by $U_0^-(\vec{r})$ and $U_0^+(\vec{r})$ the 'vectors' at the left- and right-hand sides of the lens D_0, respectively. In similar way the 'vectors' $U_1^-(\vec{r})$ and $U_1^+(\vec{r})$ at the both sides on the lens D_1 are introduced. Determine the field U_1^+ in the output of the system by the field U_0^- in its input. Denote the coefficients in representations (2.101) of these fields by $C_n^{0,-} = C_n(-0)$, $C_n^{0,+} = C_n(+0)$, $C_n^{1,-} = C_n(d-0)$, $C_n^{1,+} = C_n(d+0)$, respectively; here d is the distance between the correctors. It can be seen that

$$C_n^{1,-} = C_n^{0,+}\exp(-ih_n d)\,, \tag{2.104}$$

where h_n is the propagation constant of the nth mode. By the definition of the function $\psi_0(\vec{r})$,

$$\sum_{n=1}^{N} C_n^{0,+}u_n(\vec{r}) = \exp(i\psi_0(\vec{r}))\sum_{n=1}^{N} C_n^{0,-}u_n(\vec{r})\,, \tag{2.105}$$

from which, according to (2.103),

$$C_m^{0,+} = \sum_{n=1}^{N} C_n^{0,-}\left(\exp(i\psi_0)u_n, u_m\right)\,. \tag{2.106}$$

The relations (2.104), (2.106), and the obvious formula $U_1^+ = \exp(i\psi_1)U_1^-$, allow us to calculate the coefficients $\{C_n^{1,+}\}$ by $\{C_n^{0,-}\}$, and therefore, U_1^+ by U_0^-.

Let the fields described by the 'vectors' U_0^- and V_1be given. The phase corrections ψ_0 and ψ_1 should be found which provide maximal similarity of the field outgoing from the aperture D_1 to $\vec{v}_1(\vec{r}_1)$. As the optimization criterion we use the functional

$$\chi(\psi_0, \psi_1) = \frac{\left|\left(U_1^+, V_1\right)\right|}{\left[\left(U_0^-, U_0^-\right)(V_1, V_1)\right]^{1/2}}\,. \tag{2.107}$$

Problem Wp. Given incident and desired field distributions u_0^-, v_1, find the functions ψ_0, ψ_1 which provide a maximum for the functional (2.105).

A multi-element transformer similar to that in Section 2.1.2 can be considered and the problem can be formulated for a functional similar to (2.24). Some numerical results are presented in [37].

2.3.2
Minimization of losses in waveguide walls

The losses of the modes in a closed waveguide can be caused by irregularities of the cross-section, the absorption in the filling, or the imperfections in the wall material. The last type of loss (usually called heat loss) is considered in this subsection. Heat loss is proportional to the density of the current in the wall. In order to reduce it, the current must be decreased. For instance, the field of the mode can be adjusted to

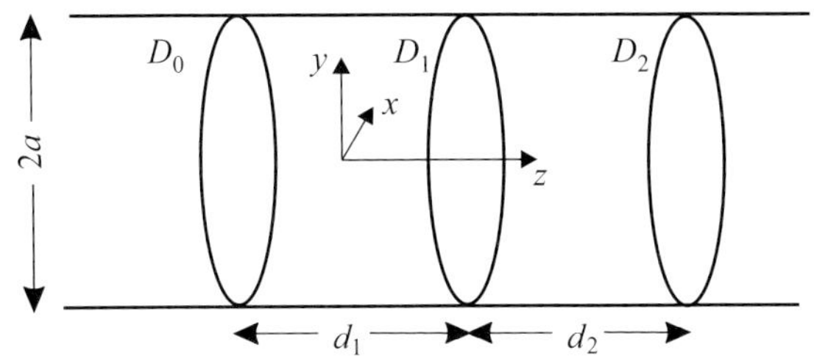

Figure 2.11 Lens line in the waveguide.

the waveguide axis. This can be done by periodically placing the phase correctors (lenses) inside the waveguide (see Figure 2.11, $d_n = L$), so that the field of the mode is held away from the walls. Of course, the lens material itself must not cause any loss comparable with the heat loss. Such a system is no longer a classical waveguide, it is a periodical hybrid of the waveguide and an open quasi-optical line. In this system, the eigenwaves are formed in a different way from the modes in the hollow waveguide. Each of these waves is a wave beam, that is, a superposition of the waveguide modes. Similar to the open line, the lenses focus the field of the beam in the middle of the cross-section as if transmitting the modes to each other. The phase front of the beam incoming to a lens is divergent; the lens 'corrects' it by making it convergent. In terms of the waveguide modes, the lenses 'mix' the mode superposition changing the coefficients of their components. The field distribution over the cross-section must be repeated before each lens with accuracy to a constant wave factor. In contrast to the open waveguide, there is no diffraction or radiation loss in this system.

The fields on both sides of the lens placed at $z = L/2$ are related by

$$\left\{\vec{E}^{+}, \vec{H}^{+}\right\} = \left\{\vec{E}^{-}, \vec{H}^{-}\right\} \exp\left(i\psi\left(\vec{R}\right)\right), \tag{2.108}$$

where $\psi(\vec{R})$ is the phase correction of the lens; here the indices '+' and '−' denote the values of the fields at $z = L/2 \pm 0$, respectively. The coordinate system is introduced with the origin in the middle of the considered cell of the periodical system. Besides, the field distributions at the ends of the period (at $z = \pm L/2$) must satisfy the periodicity conditions

$$\left\{\vec{E}^{+}, \vec{H}^{+}\right\}\Big|_{z=-L/2} = \left\{\vec{E}^{+}, \vec{H}^{+}\right\}\Big|_{z=L/2} \exp(-ihL), \tag{2.109}$$

where h is the propagation factor in the considered transmission system. Similar to the hollow waveguide, different modes satisfying condition (2.109) can exist. If there are losses in the waveguide, then h is complex, $h = h' + ih''$, $h'' < 0$. If, for instance, the complex impedance of the wall is given, then the optimization problem may consist in minimization of $|h_1''|$ by choosing the phase correction provided by the lenses. The index '1' means that we consider the first (main) eigenwave of the hybrid system, that is, the losses of the main eigenwave should be minimized.

If the impedance is unknown, then a simplified problem can be formulated, based on the fact that the heat loss is approximately proportional to the squared modulus of the current in the walls integrated over their surface. If the impedance is small, then this current is almost the same as that in a waveguide with perfectly conducting walls. Then the optimization criterion can be chosen in the form of the normed value of this quantity averaged over the length of a cell in the periodical system considered:

$$J(\psi) = \frac{\frac{1}{L}\int_{-L/2}^{L/2}\int_C |H|^2 \mathrm{d}C\mathrm{d}z}{\int_{S_0} \mathrm{Re}\left(\left(\vec{E}\times\vec{H}\right)_z\right)\mathrm{d}S}, \tag{2.110}$$

where C is the contour of the waveguide cross-section. For definiteness, this value is normed by the energy flux over the central cross-section. If losses are present, then this flux is different for different sections; however, this difference is negligible in our approximation. For the same reasons, the imaginary parts of the propagation factors h_n of the traveling waveguide modes can be neglected in all calculations. At a given geometry, the functional J depends only on the function $\varphi(\vec{R})$ considered as its argument.

Problem X. For a given geometry of the waveguide cross-section S_0 and length L of the periodicity cell, find the function $\varphi(\vec{R})$ providing a minimum for (2.110) under conditions (2.108) and (2.109).

For numerical results see Section 5.5.1.

2.3.3
Mirror shape optimization in a quasi-optical resonator

A problem, similar to that considered in the previous subsection, can also be formulated for the open beam waveguide or for the open quasi-optical resonator with correctors in the form of lenses or mirrors. The only difference is that in this problem the radiation losses must be minimized instead of the head losses in the previous one. The mathematical description of the eigenwaves in beam waveguides and the eigenoscillations in quasi-optical resonators is the same, and therefore we confine ourselves only to the resonator case.

In a geometrical sense, the open quasi-optical resonator is similar to the beam transmission line considered in Section 2.1.1. In the resonator, two mirrors are located in place of the antenna and rectenna (Figure 2.12). In this system, 'quasi-eigenwaves' may exist, damping in time but keeping their space structure. The damping is caused by radiation losses, because not all the incoming energy is con-

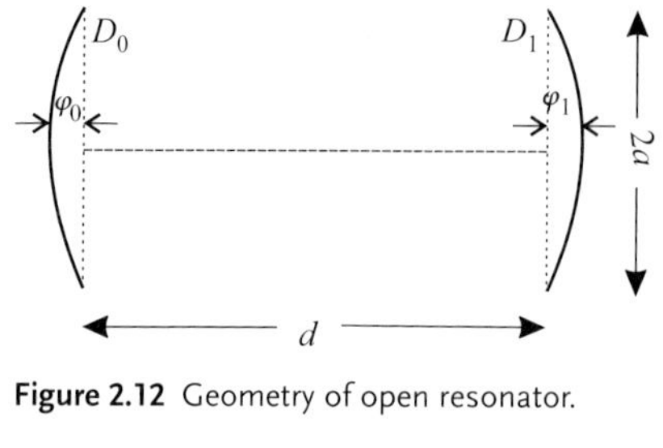

Figure 2.12 Geometry of open resonator.

centrated on the mirror; part of it bypasses the mirror area. In order that such a wave can exist without damping, it must be fed by a source. As before, we call these waves eigenwaves or modes.

The loss in the modes depends on the shape of the mirrors (more exactly, on the phase correction provided by them). The mode with the smallest loss is called the main mode; as a rule, its field has the simplest structure in cross section. The optimization problems formulated for the systems considered consist in finding the shape of the mirrors (their phase corrections) such that the main mode has certain optimal properties. Two types of problem can be formulated: to minimize the loss of the main mode, or to maximize the difference between this loss and the losses of all other modes. Here we consider only the first problem.

In contrast to the theory of transmission lines, it is convenient here to formulate the problem in terms of the functions $U_j(\vec{R}_j)$ without using substitution (2.3). For simplicity, we assume that both mirrors have the same shape ($D_0 = D_1 = D$), size ($a_0 = a_1 = a$), and that they provide the same phase correction $\exp(i\psi(\vec{R}_j))$. The geometrical parameters satisfy the quasi-optical conditions (2.7), in our case having the form

$$ka \ll 1, \quad d \ll a\,. \tag{2.111}$$

The field distributions $v_n(\vec{R}_j)$ of the modes over the mirrors satisfy the homogeneous integral equation (see, e.g. [39])

$$\lambda_n v_n(\vec{r}_1) = \frac{c}{2\pi} \iint_D v_n(\vec{r}_0) \exp\left(\frac{ic}{2}\left[\vec{r}_0 - \vec{r}_1\right]^2\right) \times \exp\left(i\left[\psi(\vec{r}_0) - \psi(\vec{r}_1)\right]\right) d\vec{r}_0\,, \tag{2.112}$$

where $c = ka^2/d$, $n = 1, 2, \ldots$. The radiation losses of the nth mode are determined by its eigenvalue λ_n as follows: $\Delta_n = 1 - |\lambda_n|^2$.

Problem Y. Given shape D of the mirrors and the parameter c, find the phase correction $\psi(\vec{r}_j)$, providing a minimum radiation loss Δ_1 of the main ($n = 1$) mode.

For a particular case, this problem was probably first solved in [23]. We will obtain this solution using the common approach described in this book. For a mathematical formulation see Section 3.3.2.1.

The mirrors in the resonator (or, essentially, the lenses in the mathematically equivalent system – a quasi-optical beam waveguide) can be randomly shifted (Figure 2.13). Such a shift modifies (2.112) as follows:

$$\lambda_n v_n(\vec{r}_1) = \frac{c}{2\pi} \iint_D v_n(\vec{r}_0) \exp\left(\frac{ic}{2}\left[\vec{r}_0 + \vec{\delta} - \vec{r}_1\right]^2\right) \times \exp\left(i\left[\psi(\vec{r}_0) - \psi(\vec{r}_1)\right]\right) d\vec{r}_0\,, \tag{2.113}$$

where $\vec{\delta}$ is the shift vector. The problem is to minimize the loss of the main mode at a given (average or maximal) value of $|\vec{\delta}|$. The solution should not depend on the shift direction.

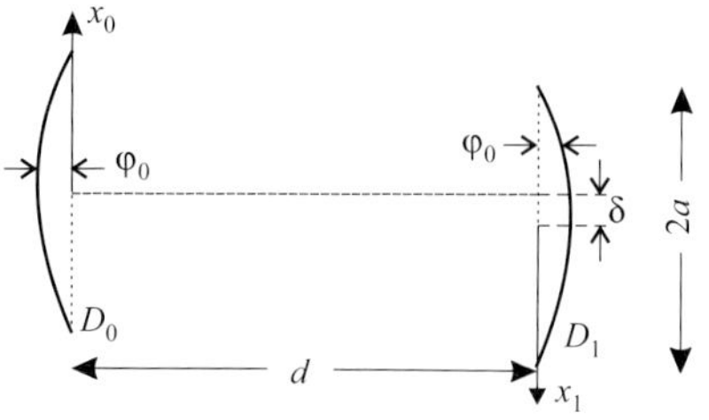

Figure 2.13 Resonator with shifted mirrors.

Problem Z. Given shape D of the mirrors, the parameter c and shift value $|\vec{\delta}|$, find the phase correction $\psi(\vec{r}_j)$, providing a minimum radiation loss Δ_1 of the main mode.

For a mathematical formulation and numerical results see Section 5.5.2.

3
Mathematical Formulation of the Problems

The physical problems formulated in the previous chapter can be combined in several groups, united by one common mathematical formulation. For instance, Problems **B** and **J** are similar not only in the objective functions, but also in the operators of appropriate direct problems (cf. (2.3) and (2.27)). It can be seen that, by applying the Lagrange multiplier technique, we can reduce the conventional minimization Problems **C** and **D** to a problem of nonconventional minimization similar to Problem **E**. Also, different modifications of the same problem (e.g. Problems **A**, **Aa** and **Ap**) cannot be described by one mathematical formulation, because they have different restrictions on the required functions.

On the other hand, physical problems similar to those considered in the previous chapter do not only arise in electromagnetic wave theory. They may differ only in the operator describing the devices or processes to be optimized. Here we give a general mathematical formulation of the problems.

The problems considered in this chapter are divided onto three groups located in three sections. The optimization criteria in the problems of the first two sections contain the mean-square difference between the moduli of the given (desired) functions and their approximations. The problems with no aplitude-phase restrictions on the sought functions and those having such restrictions are considered in Sections 3.1 and 3.2, respectively. Section 3.3 focuses on the problems in which the desired functions are not given. In these problems certain integral quantities such as the total energy, losses etc., are extremized (maximized or minimized). As a rule, such problems are reduced to homogeneous equations; the modulus of their maximal or minimal eigenvalue describes the extreme value of the objective function. For this reason, we call them *homogeneous problems*, although the optimization functions may participate in their formulation nonlinearly (e.g. as arguments of some complex functions).

Each section contains a general formulation of the variational problem and its reduction to the Lagrange–Euler equation which is a necessary condition for the functional to be extremal. The equations are nonlinear; their solutions are, as a rule, nonunique. The simplest iterative procedures are proposed for their solution and the relaxation properties of these procedures are established. Homogeneous equations are obtained, which describe the branching of the solutions with respect to

Phase Optimization Problems. O. O. Bulatsyk, B. Z. Katsenelenbaum, Y. P. Topolyuk, and N. N. Voitovich

ISBN: 978-3-527-40799-6

numerical parameters in the operator of the problem. The general theory is illustrated by concrete examples for particular cases of the operators.

All the functionals use the mean-square metric in the optimization criteria. This fact suggests the functional spaces in which the operators act in the problems considered. This is a class of complex functional Hilbert spaces in which, besides the usual properties of the inner product, certain additional conditions hold. We call such spaces *spaces of L_2-type.*

In such a space H, multiplication by bounded functions is introduced, and

$$(u_1 v, u_2) = (u_1, u_2 \bar{v}) \tag{3.1}$$

for any $u_1 \in H$, $u_2 \in H$, $|v| \leq C < \infty$. The inequality

$$|(u_1, u_2)| \leq (|u_1|, |u_2|) \tag{3.2}$$

is assumed to be valid for this space. Inequality (3.2) becomes an equality if and only if $\arg(u_1) - \arg(u_2) = const$ almost everywhere in support of the functions from H.

3.1
Variational problems with no amplitude-phase restrictions on the required functions

In this section, those problems are considered in which neither the modulus or argument of the required complex function are given. This means that the permissible functions for the minimized functional fill the entire space in which the operator of the direct problem is acting. The optimization criteria are the functionals in the form of the mean-square difference between the moduli of the obtained and desired functions.

In variational calculus, an important role is played by the *Lagrange–Euler equation* [40]. This equation follows from the condition that the first variation of the functional (the linear part of its increment) equals zero for any permissible perturbation of its argument-function. The condition is analogous to the condition of equality with zero of the partial derivatives in the theory of many variable functions. The functional is stationary on the solutions to this equation, that is, such a solution can be a minimum, maximum, or a saddle point of the functional. Although the Lagrange–Euler equation describes only the necessary condition for the functional to be extremal, the reduction of the variational problem to such an equation allows us to simplify its theoretical investigation and numerical solution. Such equations will be derived for all functionals considered.

The isometric and compact linear bounded operators are considered. Since the variational problems for these two types of operator are formulated in different ways, they are investigated separately. For both cases, the question regarding stability of the solution is investigated. Various cases of the operators which arise in applications are considered, such as the one-dimensional Fourier transform of finite (compactly supported) functions, the discrete Fourier transform, the Hankel transform, etc. The technique and main results are analogous to those from [25] and [26]. The stability of minimizing sequences for the appropriate functionals is investigated in [45].

3.1.1 Problem formulation

Consider the following equation

$$|Au| = F\,, \tag{3.3}$$

where A is a linear bounded operator acting from the Hilbert space H_1 of L_2-type into another space H_2 of the same type and F is a given non-negative function from H_2. We introduce the inner products

$$(u_1, u_2)_1\,, \qquad u_1, u_2 \in H_1\,, \tag{3.4a}$$

$$(f_1, f_2)_2\,, \qquad f_1, f_2 \in H_2 \tag{3.4b}$$

in the spaces H_1 and H_2 and denote the norms associated with them by

$$\|u\|_1^2 = (u, u)_1\,, \qquad \|f\|_2^2 = (f, f)_1\,. \tag{3.5}$$

The problem is to find the function $u \in H_1$ which minimizes the functional

$$\sigma(u) = \||Au| - F\|_2^2\,. \tag{3.6}$$

3.1.2 Reducing to the Lagrange–Euler equation

We reduce the problem of the functional (3.6) minimization to a nonlinear Lagrange–Euler equation. For this we must calculate the first variation of the functional, that is, the linear part $\delta\sigma$ of the perturbed functional

$$\sigma(u + \delta u) = \sigma(u) + \delta\sigma(u, \delta u) + o(\delta u)\,, \tag{3.7}$$

caused by an arbitrary small perturbation $\delta u \in H_1$ of the function u. It is convenient to write the functional $\sigma(u)$ in the form

$$\sigma(u) = (Au, Au)_2 - 2\,(|Au|, F)_2 + \|F\|_2^2 \tag{3.8}$$

and introduce the notation

$$f = Au\,. \tag{3.9}$$

Replacing u by $u + \delta u$ in (3.8) and keeping the linear part of the perturbation, we obtain

$$\sigma(u + \delta u) = \sigma(u) + 2\,\mathrm{Re}\,(\delta f, f)_2 - 2\,(\delta|f|, F)_2\,, \tag{3.10}$$

where

$$\delta f = A[\delta u] , \tag{3.11}$$

$$\delta |f| = |f + \delta f| - |f| . \tag{3.12}$$

Multiplying the right-hand side of (3.12) by the unit factor $(|f + \delta f| + |f|)/|f + \delta f| + |f|)$ and making simple derivations (dropping the values of higher order), we obtain:

$$\delta |f| = \frac{|f + \delta f|^2 - |f|^2}{|f + \delta f| + |f|} = \frac{(f + \delta f)(\bar{f} + \delta \bar{f}) - f \cdot \bar{f}}{|f + \delta f| + |f|}$$
$$= \frac{\delta f \cdot \bar{f} + \delta \bar{f} \cdot f}{2|f|} = \mathrm{Re}\,(\delta f \cdot \exp(-i \arg f)) . \tag{3.13}$$

Since we are interesting only in the linear terms with respect to the small perturbation, here (and further in similar cases) we drop the values of the higher order and use the symbol '=' for the approximate equality.

Substituting (3.11), (3.13) into (3.10) and subtracting $\delta(u)$ from the result with using (3.9), we obtain

$$\delta \sigma(u, \delta u) = 2\,\mathrm{Re}((A[\delta u], Au)_2 - (A[\delta u] \cdot \exp(-i \arg(Au)), F)_2) . \tag{3.14}$$

According to property (3.1) of the space H_2,

$$(A[\delta u] \cdot \exp(-i \arg Au), F)_2 = (A[\delta u], F \exp(i \arg Au))_2 . \tag{3.15}$$

As a result, we obtain from (3.14)

$$\delta \sigma(u, \delta u) = 2\,\mathrm{Re}(\delta u, A^*[Au - F \exp(i \arg Au)])_1 , \tag{3.16}$$

where A^* is the adjoint operator to A, acting from H_2 to H_1, that is, the operator satisfying the identity

$$(Au_1, f_2)_2 = (u_1, A^* f_2)_1 \tag{3.17}$$

for any $u_1 \in H_1$, $f_2 \in H_2$.

The Lagrange–Euler equation is obtained from the condition

$$\delta \sigma(u, \delta u) = 0 \tag{3.18}$$

for all $\delta u \in H_1$. In particular, if δu is chosen as an arbitrary real function from H_1, then (3.18) yields

$$\mathrm{Re}(A^* Au - A^*[F \exp(i \arg Au)]) = 0 . \tag{3.19a}$$

Similarly, choosing δu as an arbitrary imaginary function from H_1, we obtain

$$\mathrm{Im}(A^* Au - A^*[F \exp(i \arg Au)]) = 0 . \tag{3.19b}$$

These two equations together give the sought Lagrange–Euler equation

$$A^* Au = A^*[F \exp(i \arg Au)] \tag{3.20}$$

for functional (3.6). Acting with the operator A on the both sides of (3.20), we have

$$AA^* f = AA^*[F \exp(i \arg f)] . \tag{3.21}$$

This equation is equivalent to (3.20) if the kernel of A is empty, that is, if there are no functions $u \in H_1$ such that $Au = 0$. If (3.21) is solved, then the solution to (3.20) is calculated from the linear equation

$$A^* Au = A^*[F \exp(i \arg f)] , \tag{3.22}$$

which is a consequence of (3.20) and (3.9). Equation (3.21) is simpler than (3.20) both for analytical investigation and numerical solution.

Remember, that the Lagrange–Euler equation is obtained from the necessary condition of a minimum of the functional $\sigma(u)$, so that its solutions are not only the minima of the functional, but also all its stationary points.

3.1.3 Stability of the functional

In this subsection the stability of functional (3.6) is investigated. In general, the minimum of $\sigma(u)$ may be unattainable, that is, there may be no function u on which the minimum value of σ is reached. As will be shown later, this fact is important for theoretical investigations. However, from a practical point of view, we are interested only in the greatest lower bound of σ:

$$\mu = \inf_{u \in H_1} \sigma(u) , \tag{3.23}$$

which always exists due to the lower boundedness of σ by zero. The most important question is the stability of μ, that is, its continuous dependence on the function F and the operator A.

If the function F is given, then it can be assumed to be known exactly, whereas in practical calculations the operator A is always substituted by an approximate operator A_δ:

$$\|A - A_\delta\| < \delta . \tag{3.24}$$

The continuous dependence of μ on A is equivalent to the assertion that the greatest lower bound μ_δ of the functional

$$\sigma_\delta (u) = \| |A_\delta u| - F\|^2 \tag{3.25}$$

approximates μ and that any minimizing sequence for the functional σ_δ converges to μ. The validity of this assertion depends on the type of the operators A and A_δ.

Definition 3.1 The problem of the functional σ minimization is *stable* on a set of operators A_δ, if the numerical sequence $\{\mu_{\delta_n}\}$:

$$\mu_{\delta_n} = \inf_{u \in H_1} \| |A_{\delta_n} u| - F \|^2 \qquad (3.26)$$

converges to μ as $n \to \infty$ for any $\{\delta_n\}$ converging to zero as $n \to \infty$ and any sequence of the approximate functionals $\{\sigma_{\delta_n}(u)\}$.

This definition is especially important for the case when the operator equation (3.3) is inconsistent, that is, if $\mu > 0$. This can be seen from the following lemma.

Lemma 3.1

The greatest lower bound

$$\mu_\delta = \inf_{u \in H_1} \| |A_\delta u| - F \|^2 \qquad (3.27)$$

of the perturbed functional $\sigma_\delta(u)$ has the property

$$\overline{\lim_{\delta \to 0}} \mu_\delta \le \mu . \qquad (3.28)$$

Proof: Let us first estimate the value

$$|\sigma(u) - \sigma_\delta(u)| = \left| (Au, Au)_2 - 2(|Au|, F)_2 + 2(|A_\delta u|, F)_2 - (A_\delta u, A_\delta u)_2 \right| . \qquad (3.29)$$

Using the Cauchy–Bunyakovski inequality, the obvious inequality $||A_\delta u| - |Au|| \le |A_\delta u - Au|$, and (3.24), we obtain

$$|(|A_\delta u|, F) - (|Au|, F)| \le (|(A_\delta - A) u|, F) \le \delta \|F\| \|u\| . \qquad (3.30)$$

In a similar way the estimation

$$\|(Au, Au) - (A_\delta u, A_\delta u)\| \le (\|A\| + \|A_\delta\|) \delta \|u\|^2 \qquad (3.31)$$

is obtained. Combining these inequalities and using the triangle inequality, we have

$$|\sigma(u) - \sigma_\delta(u)| \le \delta \Sigma(u) , \qquad (3.32)$$

where

$$\Sigma(u) = (\|A\| + \|A_\delta\|) \|u\|^2 + 2\|F\| \|u\| . \qquad (3.33)$$

By the greatest lower bound definition, for any $\varepsilon > 0$ there exists an element $u_\varepsilon \in H_1$, such that

$$\mu \le \| |Au_\varepsilon| - F \|^2 \le \mu + \varepsilon . \qquad (3.34)$$

On the other hand, using (3.32) and (3.34), we have

$$\mu_\delta \leq \| |A_\delta u_\varepsilon| - F\|^2 \leq \| |A u_\varepsilon| - F\|^2 + \delta \Sigma\,(u_\varepsilon) \leq \mu + \varepsilon + \delta \Sigma\,(u_\varepsilon)\,. \quad (3.35)$$

Passing to the limit $\delta \to 0$ and using the arbitrariness of ε and the fixedness of u_ε, we obtain (3.28). □

This result means that, in the general case, the greatest lower bound is only upper semicontinuous (i.e. the stability property is absent), and certain stabilization procedures should be used for its determination (see, e.g. [41]). One such procedure will be proposed in Theorem 3.2.

3.1.4 Alternative formulation

There exists an alternative formulation of the above variational problem. Let the function minimizing functional (3.6) be known with accuracy to a real constant factor C:

$$u = Cv\,. \quad (3.36)$$

Substituting (3.36) into (3.6) and solving the equation $d\sigma/dC = 0$ with respect to C, we obtain

$$C = \frac{(|Av|, F)_2}{\|Av\|_2^2} \quad (3.37)$$

which leads to

$$\sigma(Cv) = \|F\|_2^2 - \chi^2(v)\,, \quad (3.38)$$

where

$$\chi(v) = \frac{(|Av|, F)_2}{\|Av\|_2}\,. \quad (3.39)$$

If the function v maximizing $\chi(v)$ is found, then the required function u minimizing $\sigma(u)$ is calculated by formulas (3.36) and (3.37). Thus, the minimization problem for the functional $\sigma(u)$ is equivalent to the maximization problem for the functional $\chi(v)$.

To derive the Lagrange–Euler equation for $\chi(v)$, we temporarily denote $f = Av$, perturb the required function v by δv, and perform auxiliary derivations keeping in them only linear terms with respect to δv :

$$\chi(v + \delta v) = \frac{(|f + \delta f|, F)_2}{\|f + \delta f\|_2} = \frac{(|f|, F)_2 + (\delta|f|, F)_2}{\sqrt{(f + \delta f, \bar{f} + \delta \bar{f})_2}}$$

$$= \frac{(|f|, F)_2}{\|f\|_2} \cdot \frac{1 + (\delta|f|, F)_2/(|f|, F)_2}{1 + \text{Re}(\delta f, f)_2/\|f\|_2^2}$$

$$= \frac{(|f|, F)_2}{\|f\|_2} \left[1 + \frac{\text{Re}(\delta f \exp(-i \arg f), F)_2}{(|f|, F)_2} - \frac{\text{Re}(\delta f, f)_2}{\|f\|_2^2} \right]$$

$$= \chi(v) + \frac{\text{Re}(\delta f, F \exp(i \arg f) - C f)_2}{\|f\|_2}. \tag{3.40}$$

With accuracy to a nonessential constant factor, the first variation of χ is obtained from (3.40) as follows:

$$\delta\chi(v, \delta v) = \text{Re}\left(\delta v, A^* \left[F \exp(i \arg f) \right] - C \cdot A^* f \right), \tag{3.41}$$

where C is defined in (3.37). Using the considerations before (3.20), we get

$$C \cdot A^* A v = A^* \left[F \exp(i \arg A v) \right]. \tag{3.42}$$

In fact, the factor C is arbitrary, its variation changes only the scale of v (which does not change the right-hand side of (3.42)), whereas the functional $\chi(v)$ is homogeneous with respect to v. Without loss of generality, we can put $C = 1$. Then (3.42) coincides with the Lagrange–Euler equation (3.20) for the functional $\sigma(u)$.

3.1.5
The case for an isometric operator

3.1.5.1 The Lagrange–Euler equation
If the operator A is isometric, that is,

$$(A u_1, A u_2)_2 = (u_1, u_2)_1 \tag{3.43}$$

for any $u_1, u_2 \in H_1$, then

$$(A[\delta u], A u)_2 = (\delta u, u)_1. \tag{3.44}$$

Substituting (3.44) into (3.14), we obtain the Lagrange–Euler equation

$$u = A^* \left[F \exp(i \arg A u) \right] \tag{3.45}$$

instead of (3.20), and the equivalent one

$$f = A A^* \left[F \exp(i \arg f) \right] \tag{3.46}$$

instead of (3.21). Since the kernel of any isometric operator is empty, the equations (3.45) and (3.46) are equivalent. After (3.46) has been solved, the solution to (3.45) is calculated explicitly as

$$u = A^* \left[F e^{i \arg f} \right], \tag{3.47}$$

which is a consequence of (3.45) and (3.9).

Equality (3.38) remains of the same form for the alternative functional

$$\chi(u) = \frac{(|Au|, F)_2}{\|u\|_1} . \tag{3.48}$$

Both forms of Lagrange–Euler equation for this functional differ from (3.45) and (3.46) only in an indeterminate constant factor on the right-hand side, which obviously can be arbitrary in this case. Similar to the general case, the Lagrange–Euler equation for this functional has the form (3.45) with an indeterminate factor on the right-hand side. This factor changes only the coefficient in u which is canceled when calculating χ by (3.48).

3.1.5.2 **Stability of minimizing sequences**

A special case is the *isometric* operator A. For such operators the Parseval equality

$$\|Au\|_2^2 = \|v\|_1^2 \tag{3.49}$$

is valid. It is the only property of the isometric operator needed in further derivations. This equality allows us to rewrite the functional $\sigma(u)$ in the form

$$\sigma(u) = (F, F) - 2(|Au|, F) + (u, u) . \tag{3.50}$$

It turns out that no stabilizing procedure is required to determine the greatest lower bound of the functional σ in this case. This fact is established by the theorem below.

Definition 3.2 Sequence $\{u_n\}$ is *minimizing* for the functional $\sigma(u)$ if $\sigma(u_n) \to \mu$ as $n \to \infty$.

Theorem 3.1

If A is an isometric operator acting from H_1 into H_2, then any algorithm generating a minimizing sequence $\{u_n^\delta\}$ for the functional $\sigma_\delta(u)$ possesses the property $\sigma_\delta(u_n^\delta) \to \mu$ as $\delta \to 0,\ n \to \infty$.

Proof: Let us consider the elementary inequality

$$\left|\sigma_\delta\left(u_n^\delta\right) - \mu\right| \le \left|\sigma_\delta\left(u_n^\delta\right) - \mu_\delta\right| + |\mu_\delta - \mu| . \tag{3.51}$$

By the theorem condition, the first addend on the right-hand side tends to zero by the theorem condition. In order to prove the tendency of the second addend to zero, we estimate the difference between the exact functional and the perturbed one

$$\sigma_\delta(u) = (F, F) - 2(|A_\delta u|, F) + (u, u) . \tag{3.52}$$

After sequential use of the inequality

$$||A_\delta u| - |Au|| \leq |A_\delta u - Au| \tag{3.53}$$

valid for any function $u \in H_1$ at each point of its definition domain, the linearity of the operators A and A_δ, the non-negativity of the given function F, and the Cauchy–Bunyakovski inequality, we obtain

$$|\sigma(u) - \sigma_\delta(u)| \leq 2\delta \|F\| \|u\| \ . \tag{3.54}$$

Since the sequence $\{u_n^\delta\}$ is minimizing for the functional $\sigma_\delta(u)$, then

$$\mu_\delta \leq \sigma_\delta\left(u_n^\delta\right) . \tag{3.55}$$

In turn,

$$\begin{aligned} \sigma_\delta\left(u_n^\delta\right) &= (F, F) - 2\left(\left|A_\delta u_n^\delta\right|, F\right) + \left(u_n^\delta, u_n^\delta\right) \\ &= (F, F) - 2\left(\left|A_\delta u_n^\delta\right|, F\right) + \left(u_n^\delta, u_n^\delta\right) \\ &\quad + 2\left(\left|Au_n^\delta\right|, F\right) - 2\left(\left|Au_n^\delta\right|, F\right) \\ &\leq \sigma\left(u_n^\delta\right) + 2\delta\|F\|\left\|u_n^\delta\right\| . \end{aligned} \tag{3.56}$$

As a result, we have

$$\mu_\delta \leq \sigma\left(u_n^\delta\right) + 2\delta\|F\|\left\|u_n^\delta\right\| . \tag{3.57}$$

On the other hand,

$$\mu \leq \sigma\left(u_n^\delta\right) . \tag{3.58}$$

The estimation

$$|\mu_\delta - \mu| \leq 2\delta\|F\|\left\|u_n^\delta\right\| \tag{3.59}$$

follows from the two last inequalities.

If the minimizing sequence $\{u_n^\delta\}$ is bounded, then the right-hand side of (3.59) tends to zero as $\delta \to 0,\ n \to \infty$. To prove this fact, we apply Lemma 10.1 from [43] which claims that any minimizing sequence is bounded if the functional is real-valued ($\sigma_\delta(u) \in R$, where R is the real number set), lower bounded ($\sigma_\delta(u) > 0$) and growing ($\sigma_\delta(u) \to \infty$ as $\|u\| \to \infty$). The first two properties are obvious, the third one follows from the inequality $(|A_\delta u|, F) \leq \|A_\delta\|\|u\|\|F\|$ which is a consequence of the operator norm definition and the Cauchy–Bunyakovski inequality for the inner product. It follows from the boundedness of $\left\|u_n^\delta\right\|$ that $\mu_\delta \to \mu$ as $\delta \to 0,\ n \to \infty$. Finally, using (3.51), we obtain $\sigma_\delta\left(u_n^\delta\right) \to \mu$ as $\delta \to 0,\ n \to \infty$.

□

3.1.5.3 The iterative method

The simplest iterative method can be proposed for solving (3.46). In each step of the method, a new approximation of the function f is calculated explicitly by the previous one as follows:

$$f_{p+1} = AA^* \left[F \exp\left(i \arg f_p \right) \right] , \qquad p = 0, 1, 2, \ldots \tag{3.60}$$

The justification of this method will be given in Chapter 5. Here we only show its main property, namely, its reflexivity. This property means that the value of functional (3.6) does not increase in each step of the method.

From the methodical reasons, we divide each step of the method into two substeps and write its scheme shifted up by one substep, as

$$f_p = Au_p , \tag{3.61a}$$

$$u_{p+1} = A^* \left[F \exp\left(i \arg f_p \right) \right] . \tag{3.61b}$$

We introduce the auxiliary functional related to the $(p+1)$th step of the method:

$$\sigma_p(u) = \left\| Au - F \exp\left(i \psi_p \right) \right\|_2^2 , \tag{3.62}$$

where $\psi_p = \arg f_p$. The function ψ_p is assumed to be given in (3.62). This functional can be transformed in the following way:

$$\begin{aligned} \sigma_p(u) &= \| Au \|_2^2 - 2 \operatorname{Re} \left(Au, F \exp\left\{ i\psi_p \right\} \right)_2 + \| F \|_2^2 \\ &= \| u \|_2^2 - 2 \operatorname{Re} \left(Au, F \exp\left\{ i\psi_p \right\} \right)_2 + \| F \|_2^2 . \end{aligned} \tag{3.63}$$

Similar to Section 3.1.2, the Lagrange–Euler equation can be obtained for this functional. It has the form

$$u = A^* \left[F \exp(i\psi_p) \right] . \tag{3.64}$$

This is an explicit form of the function minimizing the functional $\sigma_p(u)$. Comparison of (3.64) with (3.61b) shows that the function u coincides with u_{p+1}.

Calculate the difference

$$\begin{aligned} \sigma_p(u) - \sigma(u) &= 2 \left(|f|, F \right)_2 - 2 \operatorname{Re} \left(f, F \exp\left\{ i\psi_p \right\} \right)_2 \\ &= 2 \operatorname{Re} \left(f \left\{ 1 - \exp\left[i \left(\arg f - \psi_p \right) \right] \right\}, F \right)_2 \\ &= 2 \left(|f| \left\{ 1 - \cos\left(\arg f - \psi_p \right) \right\}, F \right)_2 \geq 0 . \end{aligned} \tag{3.65}$$

Here f is given by (3.9). Note that, in (3.65), the equality is reached if and only if $\arg f$ coincides with ψ_p or differs from it by π almost everywhere.

The following chain of relations is valid:

$$\sigma(u_{p+1}) \leq \sigma_p(u_{p+1}) \leq \sigma_p(u_p) = \sigma(u_p) . \tag{3.66}$$

Indeed, the first inequality follows from (3.65) applied to $u = u_{p+1}$, the second one follows from the fact that u_{p+1} minimizes σ_p, and the third relation (equality) follows from (3.65) applied to $u = u_p$. Consequently,

$$\sigma(u_{p+1}) \leq \sigma(u_p) , \tag{3.67}$$

that is, each step of the iteration procedure (3.61) (and its analogue (3.60)) does not increase the value of σ. Since σ is bounded from below ($\sigma(u) \geq 0$), the numeral sequence $\sigma(u_p)$ converges.

3.1.6
Particular cases of the isometric operator

We give three important examples of the isometric operator and write the Lagrange–Euler equations (3.45) and (3.46) for each of them. These examples are related to the Fourier transform, the discrete Fourier transform and the Neuman transform.

3.1.6.1 The Fourier transform

Let the operator $A\colon L_2(-c, c) \to L_2(-\infty, \infty)$ describe the Fourier transform of the finite (compactly supported) function

$$f(\xi) = [Au](\xi) = \frac{1}{\sqrt{2\pi}} \int_{-c}^{c} u(x) \exp(i\xi x)\, dx , \tag{3.68}$$

where $u(x) \in L_2(-c, c)$, $c > 0$ is a given real parameter. The functional to be minimized is

$$\sigma(u) = \int_{-\infty}^{\infty} \{|[Au](\xi)| - F(\xi)\}^2\, d\xi , \tag{3.69}$$

where $F \in L_2(-\infty, \infty)$:

$$F(\xi) \begin{cases} \geq 0 , & |\xi| \leq 1, \\ = 0 , & |\xi| > 1 , \end{cases} \tag{3.70}$$

is a given non-negative compactly supported function. It is known that A defined by (3.68) is an isometric operator. It is easy to show that $A^* : L_2(-\infty, \infty) \to L_2(-c, c)$ is a contraction of the inverse Fourier transform operator

$$[A^* v](x) = \frac{1}{\sqrt{2\pi}} \int_{-\infty}^{\infty} v(\xi) \exp(-i\xi x)\, d\xi , \tag{3.71}$$

and the operator $AA^* : L_2(-\infty, \infty) \to L_2(-\infty, \infty)$ is defined as

$$\left[AA^* v\right](\xi) = \frac{1}{2\pi} \int_{-\infty}^{\infty} v(\xi') \frac{\sin[c(\xi - \xi')]}{\xi - \xi'} d\xi' . \tag{3.72}$$

Then equations (3.45) and (3.46) become

$$u(x) = \frac{1}{\sqrt{2\pi}} \int_{-1}^{1} F(\xi) \exp\{i \arg([Au](\xi) - ix\xi)\} \, d\xi , \quad (3.73)$$

$$f(\xi) = \frac{1}{\pi} \int_{-1}^{1} \frac{\sin(c(\xi - \xi'))}{\xi - \xi'} F(\xi') \exp\left[i \arg f(\xi')\right] d\xi' , \quad (3.74)$$

respectively. This equation was first obtained in another form in [27].

3.1.6.2 The discrete Fourier transform

The next example is related to the discrete Fourier transform operator A: $l_2(-M, M) \to L_2(-\pi/c, \pi/c)$,

$$f(\xi) = [Au](\xi) = \sqrt{\frac{c}{2\pi}} \sum_{n=-M}^{M} u_n \exp(icn\xi) , \quad (3.75)$$

where $l_2(-M, M)$ is a complex Euclidean $2M + 1$ dimensional space; $u = \{u_n\}$, $n = -M, \ldots, M$, is the $(2M + 1)$-dimensional vector of complex components, $0 < c \leq \pi$ is a given real parameter; for simplicity, we chose the vector space dimension as an odd number. The function $f(\xi)$ defined by (3.75) is periodical, its period is $2\pi/c$, and it defines the function value domain in the space L_2. The functional to be minimized is

$$\sigma(u) = \int_{-\pi/c}^{\pi/c} \{|[Au](\xi)| - F(\xi)\}^2 \, d\xi , \quad (3.76)$$

where the given function $F \in L_2(-\pi/c, \pi/c)$ satisfies the same condition (3.70) as in the previous example.

It could be directly checked that the operator A defined by (3.75) is isometric. The adjoint operator $A^* : L_2(-\pi/c, \pi/c) \to l_2(-M, M)$ is defined as

$$[A^* v]_n = \sqrt{\frac{c}{2\pi}} \int_{-\pi/c}^{\pi/c} v(\xi) \exp(-icn\xi) \, d\xi, \qquad n = -M, \ldots, M . \quad (3.77)$$

The operator $AA^* : l_2(-M, M) \to l_2(-M, M)$ is

$$\left[AA^* v\right](\xi) = \frac{c}{2\pi} \int_{-\pi/c}^{\pi/c} v(\xi') \frac{\sin[c(M + 1/2)(\xi - \xi')]}{\sin[c(\xi - \xi')/2]} d\xi' , \quad (3.78)$$

Equations (3.45) and (3.46) become [42]

$$u_n = \sqrt{\frac{c}{2\pi}} \int_{-1}^{1} F(\xi) \exp\{i \arg[Au](\xi) - icn\xi\} \, d\xi , \quad (3.79)$$

$$f(\xi) = \frac{c}{2\pi} \int_{-1}^{1} \frac{\sin[c(M + 1/2)(\xi - \xi')]}{\sin[c(\xi - \xi')/2]} F(\xi') \exp\left[i \arg f(\xi')\right] d\xi' . \quad (3.80)$$

3.1.6.3 The Hankel transform

The third example is related to the operator of the Hankel transform $A\colon L_2^{(w)}(0,c) \to L_2^{(w)}(0,\infty)$,

$$f(\xi) = [Au](\xi) = \int_0^c u(r) J_0(r\xi) r \cdot \mathrm{d}r\,, \tag{3.81}$$

where $L_2^{(w)}(0,c)$ is the Hilbert space with the weighted inner product

$$(u_1, u_2)_1 = \int_0^c u_1(r)\bar{u}(r) r \cdot \mathrm{d}r \tag{3.82}$$

and norm

$$\|u\|_1^2 = \int_0^c |u_1(r)|^2\, r \cdot \mathrm{d}r \tag{3.83}$$

associated with it. The scalar product $(u_1, u_2)_2$ and norm $\|u\|_2^2$ in the space $L_2^{(w)}(0,\infty)$ are defined in a similar way, J_0 is the Bessel function of zeroth order, $c > 0$ is a given real parameter. This operator appears in the two-dimensional Fourier transform of a compactly supported function in the case when the support of the transformed function is a circle and the function does not depend on the polar angle.

The functional to be minimized is

$$\sigma(u) = \int_0^\infty \{|[Au](\xi)| - F(\xi\}^2\, \xi \mathrm{d}\xi\,, \tag{3.84}$$

where the given function $F \in L_2^{(w)}(0,\infty)$ satisfies the condition

$$F(\xi) \begin{cases} \geq 0\,, & 0 \leq \xi \leq 1\,, \\ = 0\,, & \xi > 1\,. \end{cases} \tag{3.85}$$

It can be directly checked that the operator A defined by (3.75) is isometric. The adjoint operator $A^*: L_2^{(w)}(0,\infty) \to L_2^{(w)}(0,c)$ is defined as

$$\left[A^* v\right](r) = c \int_0^\infty v(\xi) J_0(r\xi)\, \xi \cdot \mathrm{d}\xi\,, \tag{3.86}$$

the operator $AA^*: L_2(0,\infty) \to L_2(0,\infty)$ is

$$\left[AA^* v\right](\xi) = c \int_0^\infty \frac{\xi J_0(c\xi') J_1(c\xi) - \xi' J_0(c\xi) J_1(c\xi')}{\xi^2 - (\xi')^2} v(\xi')\, \mathrm{d}\xi'\,. \tag{3.87}$$

Equations (3.45) and (3.46) take the forms (see, e.g. [25], p. 103)

$$u(r) = c \int_0^1 F(\xi) J_0(r\xi)\xi \cdot \mathrm{d}\xi\,, \tag{3.88}$$

$$f(\xi) = c \int_0^1 \frac{\xi J_0(c\xi') J_1(c\xi) - \xi' J_0(c\xi) J_1(c\xi')}{\xi^2 - (\xi')^2} \times F(\xi') \exp\left[i \arg f(\xi')\right] \mathrm{d}\xi'\,. \tag{3.89}$$

3.1.7 Branching of solutions

Here we investigate (3.46). If it has a solution $f(\xi)$, then this solution is nonunique. At least, the function $\alpha f(\xi)$ also solves (3.46) with α as a complex constant, $|\alpha| = 1$; we call this nonuniqueness *trivial*. If the operator $B = AA^*$ is real, that is, $B\bar{f} = \overline{Bf}$, then $\bar{f}(\xi)$ also solves (3.46).

If B depends on the real parameter c: $B = B(c)$, then the number of solutions may change as c varies. New solutions may appear at certain $c = c_k$ (*bifurcation points*). They can have a different nature depending on the existence of other solutions close to them in the neighborhood of c_k. We classified them using this property as:

a) *a branching point* (new solutions arise from an existing one continuously);
b) *an isolated appearance point* (there are no close solutions at $c < c_k$);
c) *an isolated disappearance point* (there are no close solutions at $c > c_k$).

Here we consider the first case when the new solutions continuously appear from existing ones. The last two occur more rarely; we will analyze them in the next chapter when considering concrete problems.

There are several books describing the methods for determining the branching points c_k and the structure of new solutions (see, e.g. [15]). For detailed results concerning the application of this technique for (3.74) see [47]. Here we show only a simplified scheme for determination of the points c_k where solutions may branch. For this purpose, we apply the perturbation method.

We assume that the parameter c is real and positive. At $c = c_0$, let the function f_0 solve the equation

$$f_0 = B_0 \left[F \exp\{i \arg(f_0)\} \right] . \tag{3.90}$$

The perturbation of the parameter $c = c_0 + \varepsilon$ leads to perturbations of the operator $B = B_0 + \varepsilon B_1$ and of the solution $f = f_0 + \varepsilon f_1$ to (3.46) (here we keep only the linear part of the perturbation with respect to ε). Before substituting the perturbed values in the equation, we calculate the perturbation of the function $\exp(i \arg(f))$, dropping the higher terms,

$$\begin{aligned} \exp[i \arg(f)] &= \sqrt{\frac{f}{\bar{f}}} = \sqrt{\frac{f_0 + \varepsilon f_1}{\bar{f}_0 + \varepsilon \bar{f}_1}} \\ &= \exp[i \arg(f_0)] \sqrt{1 + \varepsilon \frac{f_1}{f_0} - \varepsilon \frac{\bar{f}_1}{\bar{f}_0}} \\ &= \exp[i \arg(f_0)] \left[1 + i\varepsilon \operatorname{Im}\left(\frac{f_1}{f_0} \right) \right] . \end{aligned} \tag{3.91}$$

Substituting all perturbed values in the equation and equating the linear terms on both sides, we obtain

$$f_1 - i B_0 \left[F \exp\{i \arg(f_0)\} \operatorname{Im}\left(\frac{f_1}{f_0}\right)\right] = B_1 \left[F \exp\{i \arg(f_0)\}\right] . \qquad (3.92)$$

We consider the obtained equality as a nonhomogeneous equation with respect to f_1. This equation would have a unique solution only if the corresponding homogeneous system

$$f_1 - i B_0 \left[F \exp\{i \arg(f_0)\} \operatorname{Im}\left(\frac{f_1}{f_0}\right)\right] = 0 \qquad (3.93)$$

had no nontrivial solutions. In the other case f_1 is determined nonuniquely.

Introducing two real functions w, v instead of f_1 as

$$f_1 = (w + i v) f_0 , \qquad (3.94)$$

we rewrite (3.93) in the form

$$(v - i w) f_0 - B_0 \left[F \exp(i \arg(f_0)) \cdot v \right] = 0 . \qquad (3.95)$$

Since only the function v is under the operator in this equation, w can be eliminated from it and (3.95) can be reduced to one real equation with respect to v. For this it is sufficient to multiply the left-hand side of (3.95) by $\bar{f}_0$ and extract the real part. As a result, we obtain

$$v |f_0|^2 - \operatorname{Re}\left(\bar{f}_0 B_0 \left[F \exp(i \arg(f_0)) \cdot v \right]\right) = 0 . \qquad (3.96)$$

If v is found, then f_1 can be calculated from (3.93) as

$$f_1 = i B_0 \left[F \exp(i \arg(f_0)) \cdot v \right] . \qquad (3.97)$$

Equation (3.95) has the solution

$$w = 0 , \qquad v = 1 . \qquad (3.98)$$

Indeed, in this case, (3.95) transforms to (3.90). This fact was expected and relates to the trivial nonuniqueness, mentioned above. The point $c = c_0$ can be a branching point if the perturbation f_1 of the solution to (3.46) cannot be uniquely determined, that is, if there exists a pair $\{w, v\}$ different from (3.98), which solves (3.95).

Equation (3.96) can be considered as an eigenvalue problem with the spectral parameter c nonlinearly involved in the operator. It is convenient to reformulate this problem as a usual eigenvalue problem

$$\lambda v |f|^2 = \operatorname{Re}\left(\bar{f} B \left[F \exp(i \arg(f)) \cdot v \right]\right) \qquad (3.99)$$

(we drop the index 0 here) and to determine the branching points c_n as the points at which $\lambda = 1$ is the eigenvalue of multiplicity not less than two. The second (or larger) but different from (3.98), eigenfunction corresponding to $\lambda = 1$ indicates the direction in which the new solution branches from f_0.

Note that, in certain cases, the above points may also not be the branching ones, because the condition considered above for nonuniqueness f_1 is the only necessary one.

3.1.8 The case of a compact operator

3.1.8.1 Problem formulation

In contrast to the previous case of the isometric operator A, in which the minimization problem based on functional (3.6) is stable-formulated (up to the trivial nonuniqueness of the solution), it is not stable-formulated for the compact operator and must be complemented by a stabilization term. This term can be chosen as the squared norm of the required function, so that the functional to be minimized can be written in the form

$$\sigma_t(u) = \||Au| - F\|_2^2 + t\,\|u\|_1^2\,, \tag{3.100}$$

where $t > 0$ is the *regularization parameter*. This small parameter is used to reduce the influence of both the input data and calculation errors. It is chosen from the condition that the regularized solution u_t to the minimization problem for (3.100) tends towards the solution u to the problem for (3.6) if the errors tend to zero. An efficient way of achieving this is the *discrepancy principle* [44].

3.1.8.2 The Lagrange–Euler equation

The Lagrange–Euler equation for functional (3.100) can be obtained in a similar way to that for functional (3.6). The only difference consists in the additional term $2t\,\mathrm{Re}(\delta u, u)$ in (3.14) which is the first variation of the stabilization term in (3.100), so that

$$\delta\sigma_t(u, \delta u) = 2\,\mathrm{Re}\,\{(A[\delta u], Au)_2 - (A[\delta u]\cdot\exp[-i\arg(Au)], F)_2 \\ + t(\delta u, u)\}\,. \tag{3.101}$$

Repeating the considerations used for passing from (3.14) to (3.20) and (3.21), we obtain the required equations

$$A^*Au + tu = A^*\left[F\mathrm{e}^{i\arg Au}\right] \tag{3.102}$$

and

$$AA^*f + tf = AA^*\left[F\mathrm{e}^{i\arg f}\right]. \tag{3.103}$$

After the function f has been obtained from (3.103), the solution of (3.102) can be determined from the linear equation

$$A^*Au + tu = A^*\left[F\mathrm{e}^{i\arg f}\right]. \tag{3.104}$$

Note that (3.100) can be interpreted not only as a regularized form of the functional (3.6), but also has an independent meaning in other optimization problems. The same functional is obtained after applying the *Lagrange multiplier method* to the conventional optimization problem for functional (3.6) under the condition

$$\|u\|^2 \le N\,. \tag{3.105}$$

In this case t plays the role of the *Lagrange multiplier* and is determined from the condition that the function u_t solving the unconventional minimization problem for (3.100), satisfies (3.105). In this sense, functional (3.100) relates to Problems **C**, **K**, **P**.

The functional (3.100) appears as a particular case of the weighted functional in the two-criterion minimization problem for the functionals (3.6) and $\|u\|^2$. In this case t is the weight factor and, as a rule, it is determined in an empirical way by balancing the two above criteria.

The alternative functional similar to (3.48), is

$$\chi(v) = \frac{(|Av|, F)_2}{\|Av\|_2^2 + t\,\|v\|_1^2}\,. \tag{3.106}$$

The Lagrange–Euler equation for this functional has the form (3.103) with the indeterminate factor χ^{-1} on the right-hand side. This factor changes only the coefficient in v which is canceled when calculating χ by (3.106). The function u minimizing (3.100) is determined by v maximizing (3.106), as follows

$$u = \frac{(|Av|, F)_2}{\|Av\|_2^2}\,v\,. \tag{3.107}$$

3.1.8.3 The stability of minimizing sequences

In order to investigate the stability of the solution to the problem with compact operators A, we consider the functional

$$\sigma_{\Sigma\delta}\,(u) = \||A_\delta u| - F\|^2 + \delta\Sigma\,(u) \tag{3.108}$$

and introduce the quantity

$$\mu_{\Sigma\delta} = \inf_{u\in L_2(\Omega_1)} \sigma_{\Sigma\delta}\,(u) \tag{3.109}$$

which upper estimates the values μ_δ and μ.

Theorem 3.2

Let A and A_δ be compact continuous operators acting from H_1 into H_2, and the greatest lower bounds of the functionals $\sigma\,(u)$, $\sigma_\delta\,(u)$ and $\sigma_{\Sigma\delta}\,(u)$ be reachable. Then any numerical algorithm forming a minimizing sequence $\{u_n^{\Sigma\delta}\}$ for the functional (3.108) is stable, that is, it has the property $\sigma_{\Sigma\delta}\left(u_n^{\Sigma\delta}\right) \to \mu$ as $n \to \infty$, $\delta \to 0$.

Proof: We use the same scheme as in the previous theorem. First, we write the inequality

$$\left|\sigma_{\Sigma\delta}\left(u_n^{\delta}\right) - \mu\right| \le \left|\sigma_\delta\left(u_n^{\Sigma\delta}\right) - \mu_{\Sigma\delta}\right| + |\mu_{\Sigma\delta} - \mu|\,. \tag{3.110}$$

By theorem condition, the first addend on the right-hand side of (3.110) tends to zero as $n \to \infty$. It remains to prove that the second addend tends to zero as well.

Similar to Lemma 3.1, from the greatest lower boundedness of the functional $\sigma(u)$ it follows that for any $\varepsilon > 0$ there exists an element u_ε, such that

$$\||Au_\varepsilon| - F\|^2 \leq \mu + \varepsilon . \tag{3.111}$$

Since the values of the functional $\sigma_{\Sigma\delta}(u)$ estimate the values of $\sigma(u)$ from above, then

$$\mu \leq \mu_{\Sigma\delta} . \tag{3.112}$$

However, from (3.108), (3.32) and (3.112) we have

$$\begin{aligned}\mu_{\Sigma\delta} &\leq \||A_\delta u_\varepsilon| - F\|^2 + \delta\Sigma(u_\varepsilon) \\ &\leq \||Au_\varepsilon| - F\|^2 + 2\delta\Sigma(u_\varepsilon) \leq \mu + \varepsilon + 2\delta\Sigma(u_\varepsilon) .\end{aligned} \tag{3.113}$$

Inequalities (3.112) and (3.113) lead to

$$0 \leq \mu_{\Sigma\delta} - \mu \leq 2\delta\Sigma(u_\varepsilon) + \varepsilon . \tag{3.114}$$

Passing to the limit $\delta \to 0$ with taking into account (3.33), the arbitrariness of ε and the fixedness of u_ε, we have $\mu_{\Sigma\delta} \to \mu$. Finally, using (3.110), we arrive at $\sigma_{\Sigma\delta}\left(u_n^{\Sigma\delta}\right) \to \mu$, as $n \to \infty$, $\delta \to 0$. □

Note that the results of [26] allow us to claim that at $n \to \infty$ the minimizing sequence $\{u_n^{\Sigma\delta}\}$ tends weakly to the element $u_*^{\Sigma\delta}$ on which the lower bound of the functional $\sigma_{\Sigma\delta}$ is reached.

3.1.8.4 The iterative method

The iterative method similar to that described in Section 3.1.5.3 can be applied to equation (3.103). In each step of the method, the new approximation of the function f is determined by the previous one from the linear equation

$$AA^* f_{p+1} + t f_{p+1} = AA^*\left[F \exp\left(i \arg f_p\right)\right] , \qquad p = 1, 2, \ldots \tag{3.115}$$

To show the monotonicity of the method, we divide each step into two substeps with the half-step shift, as follows:

$$f_p = Au_p , \tag{3.116a}$$

$$A^* Au_{p+1} + tu_{p+1} = A^*\left[F \exp\left(i \arg f_p\right)\right] . \tag{3.116b}$$

We introduce the auxiliary functional

$$\sigma_t^{(p)}(u) = \left\|Au - F\exp\left(i\psi_p\right)\right\|_2^2 + t\,\|u\|_1^2 , \tag{3.117}$$

where, as before, $\psi_p = \arg f_p$. The function ψ_p is assumed to be given in (3.117). We transform this functional as follows

$$\sigma_t^{(p)}(u) = \|Au\|_2^2 - 2\,\mathrm{Re}(Au, F\exp\{i\psi_p\})_2 + \|F\|_2^2 + t\,\|u\|_1^2 . \tag{3.118}$$

The Lagrange–Euler equation for this functional has the form

$$A^* Au + tu = A^* \left[F \exp(i\psi_p) \right] \tag{3.119}$$

which is the linear equation for the function that minimizes $\sigma_t^{(p)}(u)$. Comparison of (3.119) with (3.116b) shows that this function coincides with u_{p+1}.

The same derivations as in (3.65) give

$$\sigma_t^{(p)}(u) - \sigma_t(u) \geq 0\,. \tag{3.120}$$

The equality is reached if and only if $\arg f$ coincides with ψ_p or differs from it by π. A chain of relations similar to (3.66) follows from the above results and leads to

$$\sigma_t(u_{p+1}) \leq \sigma_t(u_p)\,. \tag{3.121}$$

3.1.8.5 The particular case of the compact operator

As a particular case of the compact operator we consider the integral transform $A\colon L_2(S) \to L_2(0, 2\pi)$:

$$f(\varphi) = [Au]\,(\varphi) = \int_S u(s)\,K(s,\varphi)\mathrm{d}s\,, \tag{3.122}$$

where S is a curve (closed or not) in a plane,

$$K(s,\varphi) = \frac{k}{4}\sqrt{\frac{2}{\pi}}\exp\left(3\pi i/4\right)\exp\left[ikr\left(\theta\right)\cos\left(\varphi-\theta\right)\right]\,, \tag{3.123}$$

r, θ are the polar coordinates of the point $s \in S$ on the curve described by the equation $r = r(\theta)$. The functional to be minimized is

$$\sigma_t(u) = \int_0^{2\pi} \{|[Au](\varphi)| - F(\varphi)\}^2\,\mathrm{d}\varphi + t\int_S |u(s)|^2 \mathrm{d}S\,, \tag{3.124}$$

where $F \in L_2(0, 2\pi)$ is a given non-negative function.

The operator A^* is an integral one,

$$\left[A^* v\right](s) = \int_0^{2\pi} v(\varphi)\,\bar{K}(s,\varphi)\mathrm{d}\varphi\,, \tag{3.125}$$

so that (3.102) and (3.103) transform to

$$\begin{aligned} tu(s) &+ \int_S K_1\left(s,s'\right)u\left(s'\right)\mathrm{d}s' \\ &= \int_0^{2\pi} \bar{K}(s,\varphi)F(\varphi)\exp\left[i\arg\int_S K\left(s',\varphi\right)u\left(s'\right)\mathrm{d}s'\right]\mathrm{d}\varphi\,, \end{aligned} \tag{3.126}$$

where

$$K_1\left(\theta,\theta'\right) = \frac{k^2}{8\pi}\int_0^{2\pi} \mathrm{e}^{ik\left[r(\theta)\cos\left(\varphi-\theta'\right)-r(\theta')\cos(\varphi-\theta)\right]}\mathrm{d}\varphi\,, \tag{3.127}$$

and

$$t\,f(\varphi) + \int_0^{2\pi} K_2\left(\varphi, \varphi'\right) f\left(\varphi'\right) \mathrm{d}\varphi' = \int_0^{2\pi} K_2\left(\varphi, \varphi'\right) F(\varphi') \mathrm{e}^{i \arg f\left(\varphi'\right)} \mathrm{d}\varphi' \tag{3.128}$$

where

$$K_2\left(\varphi, \varphi'\right) = \frac{k^2}{8\pi} \int_S \mathrm{e}^{i k r(\theta)\left[\cos(\varphi-\theta)-\cos\left(\varphi'-\theta\right)\right]} \mathrm{d}s\,, \tag{3.129}$$

respectively. In particular, if S is the circle of the radius a, then (3.128) becomes (see [46])

$$t\,f\left(\varphi'\right) + \int_0^{2\pi} J_0\left(c \sin\left(\left(\varphi - \varphi'\right)/2\right)\right) f(\varphi) \mathrm{d}\varphi$$
$$= \int_0^{2\pi} J_0\left(c \sin\left(\left(\varphi - \varphi'\right)/2\right)\right) F(\varphi) \exp(i \arg f(\varphi)) \mathrm{d}\varphi'\,, \tag{3.130}$$

where $c = 2ka$. A constant factor dependent on a is included here in the parameter t.

3.1.8.6 Condition for equality of the norms

If the given function F is compactly supported (as in the above examples) such that $\operatorname{supp} F = \Delta \subset D_2$, where D_2 is the domain of function definition in H_2, then the functionals (3.6) and (3.100) can be supplied by the condition

$$\|Au\|_\Delta^2 = \|F\|_2^2 \tag{3.131}$$

where $\|Au\|_\Delta^2 = (Au, Au)_\Delta$; it differs from $\|Au\|_2$ in the definition domain of the inner product (Δ instead of D_2); note that $\|F\|_\Delta = \|F\|_2$ owing to the compact supporting of F. Sometimes it is expedient to introduce this condition, in order to avoid a decrease in the magnitude of $f = Au$ in comparison with the magnitude of F in its supporting domain.

If the operator A is isometric, then functional (3.6) can be rewritten in the form

$$\sigma(u) = \sigma_0(u) + \|u\|_1^2 - \|Au\|_\Delta^2\,, \tag{3.132}$$

where

$$\sigma_0(u) = \||Au| - F\|_\Delta^2\,. \tag{3.133}$$

By the Lagrange theorem, the minimization problem for functional (3.132) with condition (3.131) can be substituted by the problem for the functional

$$\sigma_\tau(u) = \||Au| - F\|_\Delta^2 + \|u\|_1^2 + (\tau - 1)\,\|Au\|_\Delta^2\,, \tag{3.134}$$

where τ is an indeterminate Lagrange multiplier. After simple derivations, $\sigma_t(u)$ reduces to

$$\sigma_t(u) = \|F\|_2^2 - 2(|Au|, F)_\Delta + \|u\|_1^2 + \tau\,\|Au\|_\Delta^2\,. \tag{3.135}$$

The Lagrange–Euler equation for functional (3.135) is obtained by the scheme which was applied to functionals (3.6) and (3.100). As a result, we obtain

$$u + \tau A^* Au = A^*[F \exp(i \arg(Au))] , \qquad (3.136)$$

or, after the action of the operator A and denoting $f = Au$,

$$f + \tau AA^* f = AA^*[F \exp(i \arg(f))] . \qquad (3.137)$$

The multiplier τ should be chosen such that condition (3.131) holds. In fact, (3.136) (or its equivalent (3.137)) together with condition (3.131) combine a nonlinear functional-transcendental system of equations for determining f and τ.

Equation (3.137) seems to be simpler for the numerical solution. If it is included in the system instead of (3.136), then, after its solution the function u is determined from (3.136) with substitution Au by f in its right-hand side.

The simplest iterative procedure for solving the above system can be a combination of the iterative scheme

$$f_\tau^{(p+1)} + \tau AA^* f_\tau^{(p+1)} = AA^*[F \exp(i \arg(f_\tau^{(p)}))] \qquad (3.138)$$

used for solving equation (3.137) with given τ, and some simple method (for instance, the chord method) for solving (3.131) as a transcendental equation with respect to τ. Scheme (3.138) describes the inner iterative loop in this procedure, whereas the transcendental equation solution is the outer loop. The monotonicity of the inner iterations is proved in a similar way to (3.115). The convergence of the chord method is provided by continuity of the discrepancy of (3.131) and by the existence of its solution. Sometimes the inner and outer steps may alternate, but this algorithm does not always converge. Probably, the best way is to apply methods of Newton type.

The statement of the above problem with the equality norm condition (3.131) can also be applied to the case of the compact operator A. In this case the initial functional can be chosen in the form

$$\sigma_t(u) = \| |Au| - F\|_\Delta^2 + t \|u\|_1^2 \qquad (3.139)$$

instead of (3.100), with the given regularizing parameter (or weight factor) t. The distinction is in the norm definition domain in the first addend. If the additional condition (3.131) is imposed onto the function u, then the method of Lagrange multipliers can be applied to the problem and the new functional to be minimized is

$$\begin{aligned} \sigma_{t\tau}(u) &= \| |Au| - F\|_\Delta^2 + t \|u\|_1^2 + \tau' \|Au\|_\Delta^2 \\ &= \|F\|_2^2 - 2(|Au|, F)_\Delta + t \|u\|_1^2 + \tau \|Au\|_\Delta^2 ; \end{aligned} \qquad (3.140)$$

here $\tau = (1 + \tau')$, τ' is the indeterminate Lagrange multiplier fulfiling the condition (3.131). The Lagrange–Euler equation of this functional has the form

$$tu + \tau A^* Au = A^*[F \exp(i \arg(Au))] , \qquad (3.141)$$

or, after the action of the operator A,

$$tf + \tau AA^* f = AA^*[F \exp(i \arg(f))] . \tag{3.142}$$

Similar to that above, one of these equation must be solved together with condition (3.131) considered as a transcendental equation with respect to τ. The simplest method for solving this equation system is the same as for the isometric operator.

The similarity of equations (3.136) (or (3.137)) and (3.141) (or (3.142)) is very interesting. It turns out that, in the case of an isometric operator A, the minimization problem for the functional $\sigma(u)$ with the additional condition (3.131) is fully equivalent to the problem for the functional (3.139) with $t = 1$ for the same operator A contracted in the image space to the support of function F (and, hence, no longer isometric). Note that no similar analogy is observed for the unconditional minimization problems (without the additional condition (3.131) imposed on the function u).

3.2
Variational problems with amplitude-phase restrictions on the required function

3.2.1
The phase optimization problem. The general case

In the phase optimization problems in which only arg u must be found, it is senseless to give the norm $\|u\|_1$ arbitrarily, independently of $\|F\|_2$. It is more expedient to write $|u|$ in the form $|u| = C\,|v|$ (see (3.36)) with the given $|v|$ and undetermined real positive C, and rewrite functional (3.6) in the form

$$\sigma(\psi, C) = \left\| C \cdot \left|A\left[|v| \exp(i\psi)\right]\right| - F \right\|_2^2 , \tag{3.143}$$

where $\psi = \arg u$. The problem is to find the real positive function ψ and constant C, which provides a minimum for $\sigma(\psi, C)$ with given real positive functions $|v|$ and F. Here we use the same notation σ for the functional as before, in order to emphasize its physical sense.

As follows from (3.37), the constant C depends on ψ explicitly (see (3.37))

$$C(\psi) = \frac{(|A|v| \exp(i\psi)|, F)_2}{\|A|v| \exp(i\psi)\|_2^2} . \tag{3.144}$$

The relation $\sigma(\psi, C) = \|F\|_2^2 - \chi^2(\psi)$ connects $\sigma(\psi, C)$ with the functional

$$\chi(\psi) = \frac{\left(\left|A\left[|v| \exp(i\psi)\right]\right|, F\right)_2}{\left\|A\left[|v| \exp(i\psi)\right]\right\|_2} , \tag{3.145}$$

which is independent of C (see (3.39)). The function ψ that maximizes $\chi(\psi)$ and the constant C determined by ψ using (3.144) minimize the functional $\sigma(\psi, C)$.

3.2.1.1 The Lagrange–Euler equation

We derive the Lagrange–Euler equation for the functional $\chi(\psi)$. For this end, we perturb the function ψ by $\delta\psi$, and calculate (in the first approximation)

$$\begin{aligned} v + \delta v &= |v| \exp\{i(\psi + \delta\psi)\} = |v| \exp(i\psi) \exp(i \cdot \delta\psi) \\ &= |v| \exp(i\psi)(1 + i\delta\psi) , \end{aligned} \tag{3.146}$$

so that

$$\delta v = i|v| \exp(i\psi)\delta\psi , \tag{3.147a}$$

$$\delta f = iA\left[|v| \exp(i\psi)\delta\psi\right] , \tag{3.147b}$$

where the notation

$$f = A[|v| \exp(i\psi)] \tag{3.148}$$

following from (3.9) is used. In order to calculate the first variation of functional (3.145), we can start from expression (3.40) for perturbation of analogous functional (3.39) and substitute in it δf by (3.147b). As the result, we obtain

$$\begin{aligned} \delta\chi(\psi, \delta\psi) &= \mathrm{Im}(A\left[|v| \exp(i\psi)\delta\psi\right], F \exp(i \arg f) - Cf)_2 \\ &= \mathrm{Im}(|v| \exp(i\psi)\delta\psi, A^*\left[F \exp(i \arg f) - Cf)\right])_1 \\ &= (\delta\psi, |v| \,\mathrm{Im}(\exp(-i\psi)A^*\left[F \exp(i \arg f) - Cf)\right]))_1 , \end{aligned} \tag{3.149}$$

where the constant C depends on ψ by (3.144). Consequently, for $\chi(\psi)$ to be maximal it is necessary that

$$\mathrm{Im}\left(\exp(-i\psi)A^*\left[F \exp(i \arg f) - Cf)\right]\right) = 0 . \tag{3.150}$$

This equality is the Lagrange-Euler equation for functional (3.145) at given F and $|v|$. Two simpler equations

$$\psi = \arg A^*\left[F \exp(i \arg f) - Cf)\right] \tag{3.151}$$

and

$$\psi = \arg A^*\left[Cf - F \exp(i \arg f)\right] , \tag{3.152}$$

follow from (3.150). Any solution to equation (3.152) (if exists) cannot provide the global minimum for (3.143). To prove this assertion, we transform $\sigma(\psi)$ from (3.143) to the form

$$\begin{aligned} \sigma(\psi) &= C(Av, Cf - 2F \exp(i \arg f))_2 + \|F\|_2^2 \\ &= C(v, A^*\left[Cf - F \exp(i \arg f)\right])_1 + \|F\|_2^2 - C(F, |f|)_2 \\ &= C(|v|, \exp(-i\psi)A^*\left[Cf - F \exp(i \arg f)\right])_1 + \|F\|_2^2 - C(F, |f|)_2 . \end{aligned} \tag{3.153}$$

The first summand in the last expression is positive for any solution ψ_1 to (3.152), and negative for corresponding solution $\psi_2 = \psi_1 + \pi$ to (3.151). The functions f obtained for the both solutions differ only in the sign, and the last summand in (3.153) is the same. Consequently, $\sigma(\psi_1) > \sigma(\psi_2)$, and only (3.151) can be considered as the Lagrange-Euler equation for functional (3.145), which contains the global minimum point among its solutions. The form (3.151) of the Lagrange-Euler equation is more convenient for numerical calculations, whereas (3.150) is more suitable for the theoretical investigation.

3.2.1.2 Branching of solutions

As the nonlinear equations considered in the previous section, (3.151) has nonunique solutions. If the operator A depends on parameter c: $A = A(c)$, then the solutions can branch with a variation of this parameter. To derive the homogeneous linear equation for the branching points we apply the same technique as in Section 3.1.7.

As above, we assume the parameter c to be real and positive. Let the function ψ_0 solve the equation

$$\operatorname{Im}\left\{\exp(-i\psi_0)A_0^*\left[C_0 f_0 - F\exp(i\arg A_0[|u|]\exp(i\psi_0)]\right\} = 0 \qquad (3.154)$$

at $c = c_0$. If the parameter c is perturbed, $c = c_0 + \varepsilon$, then the operator A and the solution ψ become (with accuracy to the terms of higher orders)

$$A = A_0 + \varepsilon A_1, \quad \psi = \psi_0 + \varepsilon\psi_1\,. \qquad (3.155)$$

Other variables participating in the derivations have the following perturbations:

$$\exp(\pm i\psi) = \exp(\pm i\psi_0)(1 \pm i\varepsilon\psi_1)\,, \qquad (3.156a)$$

$$u = u_0 + \varepsilon u_1\,, \qquad (3.156b)$$

$$f = f_0 + \varepsilon f_1\,, \qquad (3.156c)$$

where

$$u_1 = i u_0 \psi_1\,, \qquad (3.157a)$$

$$f_1 = A_0 u_1 + A_1 u_0\,. \qquad (3.157b)$$

Substituting the above expressions into (3.151) and keeping the linear term with respect to ε, we obtain

$$\operatorname{Im}\left\{\exp(-i\psi_0)\cdot\left(-i\psi_1 A_0^* d_0 + A_0^*\left[f_1 - iF\exp(i\arg f_0)\operatorname{Im}\frac{f_1}{f_0}\right] + A_1^* d_0\right)\right\} = 0\,, \qquad (3.158)$$

where

$$d_0 = f_0 - F\exp(i\arg f_0)\,. \qquad (3.159)$$

Taking into account (3.157) and moving all the terms independent of ψ_1 to the right-hand side of (3.158) leads to the nonhomogeneous linear equation

$$\begin{aligned}&\mathrm{Re}\left\{\exp(-i\psi_0)\left\{-\psi_1 A_0^* d_0 + A_0^* A_0[u_0\psi_1] - A_0^*\left[F\exp(i\arg f_0)\right.\right.\right.\\&\left.\left.\left.\times\,\mathrm{Re}\,\frac{A_0[u_0\psi_1]}{f_0}\right]\right\}\right\} = -\,\mathrm{Re}\{\exp(-i\psi_0)\left\{A_0^* A_1 u_0\right.\\&\left.\left.-A_0^*\left[F\exp(i\arg f_0)\,\mathrm{Im}\,\frac{A_1[u_0]}{f_0}\right]\right\}\right\} \qquad (3.160)\end{aligned}$$

with respect to ψ_1; here we use the fact that $\mathrm{Im}(iC) = \mathrm{Re}\,C$ for any complex C. This equation could have a unique solution if the respective homogeneous equation

$$\begin{aligned}&\mathrm{Re}\left(\exp(-i\psi_0)\left(-A_0^* d_0 + A_0^* A_0[u_0 v]\right.\right.\\&\qquad\left.\left.-A_0^*\left[F\exp(i\arg f_0)\,\mathrm{Re}\,\frac{A_0[u_0 v]}{f_0}\right]\right)\right) = 0 \qquad (3.161)\end{aligned}$$

only has the trivial one $v = 0$. Otherwise, ψ_1 is determined nonuniquely for any perturbation of the parameter c.

It can be easily checked that the left-hand side of (3.161) vanishes at $v = 1$. As in Section 3.1.7, this nonuniqueness is 'trivial'. It corresponds to the possibility of multiplication of the function u in (3.143) by the constant $\exp(i\beta)$ with arbitrary real β, which does not change the functional σ. Hence the point c_0 can be a branching point of (3.151) if $\lambda = 1$ is the multiple eigenvalue of the homogeneous equation

$$\begin{aligned}&v_n\,\mathrm{Re}\{\exp(-i\psi_0)\left(-A_0^* d_0\right)\} = -\lambda_n\,\mathrm{Re}\left\{\exp(-i\psi_0) A_0^* A_0[u_0 v_n]\right.\\&\qquad\left.-A_0^*\left[F\exp(i\arg f_0)\,\mathrm{Re}\,\frac{A_0[u_0 v_n]}{f_0}\right]\right\}. \qquad (3.162)\end{aligned}$$

In this case, at least two eigenfunctions $v_1 = 1$, $v_2 : (v_2, 1) = 0$, exist at $\lambda_{1,2} = 1$ and the solution of (3.160) is determined as $\psi_1 = \psi_1^0 + C_1 + C_2 v_2$, where ψ_1^0 is a partial solution to (3.161). The second eigenfunction v_2 points in the direction in which the new solution branches off from ψ_0. Since C_2 can have opposite signs, there are at least two solutions branching from ψ_0 in two opposite directions.

3.2.2
The phase optimization problem. A simplified alternative formulation

3.2.2.1 The Lagrange–Euler equations

In the phase optimization problem a simplified alternative formulation can be applied, which is based on the maximization of the functional

$$\chi_s(\psi) = (|A[|v|\exp(i\psi)]|, F)_2 \qquad (3.163)$$

at given $|u|$ and F. In the case of an isometric operator A when the norms $\|Av\|_2$ and $\|v\|_1$ are equal and independent of ψ, these norms may be chosen to be equal to unity. Then $\chi_s(\psi)$ coincides with $\chi(\psi)$ (3.145) and

$$\sigma(\psi) = \|u\|_1^2 + \|F\|_2^2 - 2\chi_s^2(\psi) . \tag{3.164}$$

In the general case, the functional (3.163) describes the *degree of similarity* in modulus of the obtained function $f = Av$ and the given function F. As before, the functions u and v are connected as $u = Cv$ with the constant positive C given by (3.144). Functional (3.163) and its generalization are used in the physical problems **Gp** (see (2.20) and (2.24)) and **Jp** (2.33) (after applying (3.38)).

In order to obtain the Lagrange–Euler equation for $\chi_s(u)$, we perturb the function ψ by $\delta\psi$. After using notations (3.148) and applying (3.13) and (3.147), we obtain

$$\begin{aligned}\delta\chi_s(\psi, \delta\psi) &= (\delta|f|, F)_2 = \mathrm{Re}(\{\delta f \cdot \exp(-i \arg f)\}, F)_2 \\ &= \mathrm{Im}\left(A\left[|v| \exp(i\psi)\delta\psi\right], F \exp(i \arg f)\right)_2 \\ &= \mathrm{Im}\left(|v| \exp(i\psi)\delta\psi, A^*\left[F \exp(i \arg f)\right]\right)_1 \\ &= \left(\delta\psi, |v| \,\mathrm{Im}\left\{\exp(-i\psi)A^*\left[F \exp(i \arg f)\right]\right\}\right)_1 .\end{aligned} \tag{3.165}$$

Consequently, for $\chi_s(\psi)$ to be maximal, it is necessary that

$$\mathrm{Im}\left\{\exp(-i\psi)A^*\left[F \exp(i \arg A[|v| \exp(i\psi)])\right]\right\} = 0 . \tag{3.166}$$

There are two equations which follow from (3.166). The first of these is

$$\psi = \arg\left(A^*\left[F \exp(i \arg f)\right]\right) . \tag{3.167}$$

The second one differs from it by the addend π on the right-hand side. It can be shown that the second equation has no solutions. Indeed, if

$$\psi = \arg\left(A^*\left[F \exp(i \arg f)\right]\right) + \pi , \tag{3.168}$$

then

$$\begin{aligned}\chi_s(v) &= (Av, F \exp(i \arg f))_2 = \left(v, A^*\left[F \exp(i \arg f)\right]\right)_1 \\ &= \left(|v|, \exp(-i\psi)A^*\left[F \exp(i \arg f)\right]\right)_1 < 0\end{aligned} \tag{3.169}$$

what contradicts (3.163) asserting that χ_s is always positive. Thus, (3.167) is the Lagrange–Euler equation for functional (3.163) at given F and $|v|$.

Another form of the Lagrange–Euler equation is obtained from (3.167) if we introduce the second auxiliary function

$$w = A^*\left[F \exp(i \arg f)\right] \tag{3.170a}$$

in addition to the function f given by (3.148). Then (3.167) can be rewritten as $\psi = \arg w$. After substituting it into (3.148) we obtain

$$f = A[|v| \exp(i \arg w)] . \tag{3.170b}$$

The two equalities (3.170) are equivalent to (3.167). They make up a system of nonlinear equations with respect to the functions w and f. After this system has been solved, the required function ψ can be calculated immediately as $\arg w$.

We emphasize the similarity of (3.170) to nonlinear equation (3.46), being the Lagrange–Euler equation for the functional (3.6) in the variational problem without amplitude-phase restrictions on the required functions for the case of the isometric operator. These two problems have the same form after rewriting system (3.170) in the matrix representation

$$\hat{f} = \hat{B}\left[\hat{F}\exp(i\arg\hat{f})\right], \tag{3.171}$$

where

$$\hat{f} = \begin{Bmatrix} f \\ w \end{Bmatrix}, \quad \hat{B} = \begin{Bmatrix} 0 & A \\ A^* & 0 \end{Bmatrix}, \quad \hat{F} = \begin{Bmatrix} F \\ |v| \end{Bmatrix}. \tag{3.172}$$

The only distinction is that in (3.46) the operator $B = AA^*$ is self-adjoint and positive, which is not true for the operator $\hat{B}$ in (3.171). Of course, this distinction is very important.

As before, nonlinear equation (3.167) (and its equivalent form (3.170)) has nonunique solutions. At least, if ψ solves (3.167), then $\psi + \beta$ also solves this equation with any real β. Indeed, from the definition of f, it follows that $\arg f$ obtains the addend β on the right-hand side of (3.167). It can be taken out from the operator and canceled with the same addend in the left-hand side. This solution describes the obvious invariance of the functional χ_s to the constant phase shift $\psi = \psi_0 + \beta$.

For certain types of operator A, several solutions may exist, coupled by explicit relations. Here we show a simple example of such an operator.

Definition 3.3 We consider the *operator* A to be *real on the complex function* v if $A\bar{v} = \overline{Av}$.

Lemma 3.2

Let the functions w and f solve system (3.170) and the operators A and A^* be real on the functions $|v|\exp(i\arg w)$ and $F\exp(i\arg f)$, respectively. Then the functions $\bar{w}$ and $\bar{f}$ solve (3.170), as well. Both solutions provide the same value of the functional (3.163).

Both assertions of the lemma can easily be checked with usage (3.170). The lemma will be used in the proof of the more complicated Theorem 3.3 related to the multi-operator phase optimization problem.

Keeping in mind that the operator A may depend on a real parameter c, below we will use the following form of system (3.170):

$$\Phi^{(1)}(w, f, c) = 0, \tag{3.173a}$$

$$\Phi^{(2)}(w, f, c) = 0, \tag{3.173b}$$

with

$$\Phi^{(1)}(w, f, c) = w - A^* \left[F \exp(i \arg f) \right] , \tag{3.174a}$$

$$\Phi^{(2)}(w, f, c) = f - A[|v| \exp(i \arg w)] . \tag{3.174b}$$

The formulation (3.173), (3.174) is convenient for the investigation of the branching of solutions to (3.170) and for the construction of numerical algorithms for its solution. The simplest iterative procedure for solving (3.167) is described by the formula

$$\psi^{(p+1)} = \arg \left(A^* \left[F \exp \left(i \arg f^{(p)} \right) \right] \right) , \tag{3.175}$$

where $f^{(p)}$ is calculated by $\psi^{(p)}$ using (3.148). To check the monotonicity of this procedure, we introduce the auxiliary functional

$$\begin{aligned} \chi_p(\psi) &= |(A[|v| \exp(i\psi)], F \exp(i\varphi_p))_2| \\ &\equiv \left(|v|, \exp(-i\psi) A^* \left[F \exp(i\varphi_p) \right] \right)_1 , \end{aligned} \tag{3.176}$$

where $\varphi_p = \arg f^{(p)}$. It is evident that this functional reaches its maximum at $\psi = \psi^{(p+1)}$. On the other hand, for any ψ we have

$$\begin{aligned} \chi_s(\psi) &= (|f|, F)_2 \geq |(|f| \exp(i(\arg f - \varphi_p)), F)_2| \\ &= |(f \exp(-i\varphi_p), F)_2| = |\chi_p(\psi)| ; \end{aligned} \tag{3.177}$$

the equality $\chi_s(\psi) = |\chi_p(\psi)|$ is reached if and only if $\arg(f)$ coincides with φ_p or differs from it by π. Consequently, the following chain of relations is valid:

$$\chi_s\left(\psi^{(p+1)}\right) \geq |\chi_{sp}\left(\psi^{(p+1)}\right)| \geq |\chi_{sp}\left(\psi^{(p)}\right)| = \chi_s\left(\psi^{(p)}\right) , \tag{3.178}$$

what means that the iteration procedure is monotonic.

Applying this algorithm to system (3.170), it transforms to

$$f^{(p)} = A \left[|v| \exp \left(i \arg w^{(p)} \right) \right] , \tag{3.179a}$$

$$w^{(p+1)} = A^* \left[F \exp \left(i \arg f^{(p)} \right) \right] . \tag{3.179b}$$

For a particular case of the operator A it was proposed in [48] and numerically realized in [49].

3.2.2.2 Branching of solutions

If the operator A depends on a real parameter c, the number of solutions may vary as this parameter varies. In order to obtain the equation for the branching points, we perturb the parameter c: $c = c_0 + \varepsilon$. Then the operators and unknown functions in (3.173) are perturbed as $A = A_0 + \varepsilon A_1$; $A^* = A_0^* + \varepsilon A_1^*$; $w = w_0 + \varepsilon w_1$; $f = f_0 + \varepsilon f_1$. Substituting these perturbations together with the first-order variations of the exponential functions

$$\exp(i \arg w) = \exp(i \arg w_0)\left[1 + i\varepsilon \operatorname{Im}\left(\frac{w_1}{w_0}\right)\right], \tag{3.180a}$$

$$\exp(i \arg f) = \exp(i \arg f_0)\left[1 + i\varepsilon \operatorname{Im}\left(\frac{f_1}{f_0}\right)\right] \tag{3.180b}$$

(see (3.91)) into (3.173) and retaining the linear terms with respect to ε, we obtain

$$\Phi_w^{(1)}(w_0, f_0, c_0, w_1) + \Phi_f^{(1)}(w_0, f_0, c_0, f_1) = -\Phi_c^{(1)}, \tag{3.181a}$$

$$\Phi_w^{(2)}(w_0, f_0, c_0, w_1) + \Phi_f^{(2)}(w_0, f_0, c_0, f_1) = -\Phi_c^{(2)}, \tag{3.181b}$$

where

$$\Phi_w^{(1)}(w_0, f_0, c_0, w_1) = w_1 \tag{3.182a}$$

$$\Phi_f^{(1)}(w_0, f_0, c_0, f_1) = -iA^*\left[F \exp(i \arg f_0) \operatorname{Im}\left(\frac{f_1}{f_0}\right)\right], \tag{3.182b}$$

$$\Phi_w^{(2)}(w_0, f_0, c_0, w_1) = -iA\left[|v| \exp(i \arg w_0) \operatorname{Im}\left(\frac{w_1}{w_0}\right)\right], \tag{3.182c}$$

$$\Phi_f^{(2)}(w_0, f_0, c_0, f_1) = f_1, \tag{3.182d}$$

$$\Phi_c^{(1)} = \left.\frac{\partial}{\partial c} A^*\left[F \exp(i \arg f_0)\right]\right|_{c=c_0}, \tag{3.183a}$$

$$\Phi_c^{(2)} = \left.\frac{\partial}{\partial c} A\left[|v| \exp(i \arg w_0)\right]\right|_{c=c_0}. \tag{3.183b}$$

Here the index '0' is omitted in the operators. The nonhomogeneous linear equation system (3.181) has the unique solution $\{\psi_1, \varphi_1\}$ at given ε, if the respective homogeneous system

$$w_1 - iA^*\left[F \exp(i \arg f_0) \operatorname{Im}\left(\frac{f_1}{f_0}\right)\right] = 0, \tag{3.184a}$$

$$f_1 - iA\left[|v| \exp(i \arg w_0) \operatorname{Im}\left(\frac{w_1}{w_0}\right)\right] = 0 \tag{3.184b}$$

has only trivial (zero) solution. In equations (3.186) the unknown functions under the operators are only in the form of imaginary parts of the functions w_1/w_0 and f_1/f_0. This fact allows us to simplify the equations by eliminating the real parts of these functions. For this, we denote $w_1/w_0 = \delta_1 + i\gamma_1$, $f_1/f_0 = \delta_2 + i\gamma_2$, multiply the both sides of equations (3.184a), (3.184b) by $\bar{w}_0$, $\bar{f}_0$, respectively, and separate the real parts in the obtained equalities. As a result, we obtain

$$|w_0|^2 \gamma_1 = \operatorname{Re}\left(\bar{w}_0 A^*\left[F \exp(i \arg f_0)\gamma_2\right]\right), \tag{3.185a}$$

$$|f_0|^2 \gamma_2 = \operatorname{Re}\left(\bar{f}_0 A\left[|v| \exp(i \arg w_0)\gamma_1\right]\right). \tag{3.185b}$$

This system can be considered as a nonlinear eigenvalue problem with the spectral parameter c. As before, it can be reformulated as the problem of finding the values of c, at which the linear problem

$$\lambda_n |w|^2 \gamma_1^{(n)} = \text{Re}\left(\bar{w} A^* \left[F \exp(i \arg f)\gamma_2^{(n)}\right]\right) , \tag{3.186a}$$

$$\lambda_n |f|^2 \gamma_2^{(n)} = \text{Re}\left(\bar{f} A \left[|v| \exp(i \arg w)\gamma_1^{(n)}\right]\right) \tag{3.186b}$$

has the eigenvalue $\lambda_n = 1$; here the index '0' in the functions f_0 and w_0 is dropped.

The equation system (3.186) has obvious ("trivial") solution $\lambda_0 = 1$, $\gamma_1^{(0)} = \gamma_2^{(0)} = 1$; then equations (3.186) become identities since the vector $\{w, f\}$ solves system (3.170). This solution describes the mentioned above arbitrary phase shift β. Therefore, the solutions can branch only if the eigenvalue $\lambda_n = 1$ in (3.186) is multiple (at least, double).

If one of the functions w or f has no zeros in its definition domain, then system (3.186) can be reduced to one equation with respect to one unknown function. For instance, if $f \neq 0$ everywhere, then from (3.186b) we have

$$\gamma_2^{(n)} = \frac{1}{\lambda_n |f|^2} \text{Re}\left(\bar{f} A \left[|v| \exp(i \arg w)\gamma_1^{(n)}\right]\right), \tag{3.187}$$

and, after substituting γ_2 into (3.186a),

$$\lambda_n^2 |w|^2 \gamma_1^{(n)} = \text{Re}\left(\bar{w} A^* \left[\frac{F}{|f|^2} \exp(i \arg f) \cdot \text{Re}\left(\bar{f} A \left[|v| \exp(i \arg w)\gamma_1^{(n)}\right]\right)\right]\right) . \tag{3.188}$$

Besides, if $w \neq 0$ at any point, then the both sides of (3.188) can be divided by w and the equation can be reduced to the usual eigenvalue problem.

3.2.2.3 The multi-operator phase optimization problem. The case of one-type operators

The optimization problem based on functional (3.163) can be generalized for the case when the operator A is a finite sequence of operators A_m, $m = 0, 1, \ldots, M-1$ of the same type. Let the functions $u_m \in H_m$, $m = 0, 1, \ldots, M$ be connected by the recurrent relations

$$u_{m+1} = A_m[u_m \exp(i\psi_m)], \quad m = 0, 1, \ldots, M-1 ; \tag{3.189}$$

H_m are the Hilbert spaces of L_2-type. The functional to be minimized has the form

$$\chi(\psi_0, \psi_1, \ldots, \psi_M) = |(u_M \exp(i\psi_M), v_M)_M| . \tag{3.190}$$

The functions $v_M \in H_M$ and u_0 are given, the real functions $\psi_0, \psi_1, \ldots, \psi_M$ should be found. In general, u_0 and v_M can be complex.

Instead of the standard way of deriving the Lagrange–Euler equations for the functional (3.190), here we use a simpler, more descriptive scheme. First of all, we introduce the auxiliary functions v_m, $m = 0, 1, \dots, M - 1$, by the following recurrent relations:

$$v_{m-1} = A^*_{m-1}[v_m \exp(-i\psi_m)], \quad m = M, M-1, \dots, 1\,, \tag{3.191}$$

and then consider the functionals

$$\chi_m(\psi_m) = |(u_m \exp(i\psi_m), v_m)_m|\,, \quad m = 0, 1, \dots, M\,. \tag{3.192}$$

The following lemma holds.

Lemma 3.3

If the functions u_m and v_m, $m = 0, 1, \dots, M$, satisfy the equalities (3.189) and (3.191), respectively, then

$$\chi_m(\psi_m) = \chi(\psi_0, \psi_1, \dots, \psi_M) \tag{3.193}$$

for any $m = 0, 1, \dots, M$.

Proof: We prove the lemma by the *induction method*. From (3.192) it can be seen that (3.193) holds for $m = M$. We should only prove that

$$\chi_{m-1}(\psi_{m-1}) = \chi_m(\psi_m) \tag{3.194}$$

for $m = M-1, M-2, \dots, 1$. Using definitions (3.189) and (3.191) for u_M, v_{M-1}, respectively, yields

$$\begin{aligned}\chi_M(\psi_M) &= |(u_M \exp(i\psi_M), v_M)_M| = |(u_M, v_M \exp(-i\psi_M))_M| \\ &= |(A_{M-1}[u_{M-1} \exp(i\psi_{M-1})], v_M \exp(-i\psi_M))_M| \\ &= \left|(u_{M-1} \exp(i\psi_{M-1}), A^*_{M-1}[v_M \exp(-i\psi_M)])_{M-1}\right| \\ &= |(u_{M-1} \exp(i\psi_{M-1}), v_{M-1})_{M-1}| = \chi_{M-1}(\psi_{M-1})\,.\end{aligned} \tag{3.195}$$

Assume that (3.194) holds for certain m. Then, in similar way, we obtain that (3.194) holds for $m - 1$, and the lemma is proved. □

From equalities (3.189) and (3.191) it can be seen that the functions u_m, v_m do not depend on ψ_m, and the functional χ_m depends on this function explicitly. Due to the above lemma, maximizing the functional χ_m is equivalent to maximizing $\chi(\psi_0, \psi_1, \dots, \psi_M)$ with respect to ψ_m under the condition that other arguments are fixed.

Maximization of $\chi(\psi_0, \psi_1, \dots, \psi_M)$ with respect to all its arguments can be substituted by the simultaneous maximization of the functionals $\chi_m(\psi_m)$, $m = 0, 1, \dots, M$ dependent only on one argument. The Lagrange–Euler equation for functionals $\chi_m(\psi_m)$ can be obtained from the following considerations. According to inequality (3.2), the modulus of the inner product in (3.192) is maximal if the arguments of both its complex multipliers are the same, which gives

$$\psi_m = \arg(v_m) - \arg(u_m), \quad m = 0, 1, \ldots, M . \tag{3.196}$$

Since the number $M + 1$ of these equalities coincides with the number of unknown functions, and each of them is valid at the maximum points of the functional $\chi(\psi_0, \psi_1, \ldots, \psi_M)$, the system (3.196) complemented by the relations (3.189) and (3.191) can be considered as the Lagrange–Euler equation system for this functional.

Equality (3.196) means that

$$\exp(i\psi_m) = \exp(i\arg(v_m)) \cdot \exp(-i\arg(u_m)) . \tag{3.197}$$

Substituting this expression into (3.189) and (3.191), we get another form of the Lagrange–Euler equations

$$v_{m-1} = A^*_{m-1}[|v_m|\exp(i\arg(u_m))] , \quad m = M, M-1, \ldots, 1 , \tag{3.198a}$$

$$u_{m+1} = A_m[|u_m|\exp(i\arg(v_m))] , \quad m = 0, 1, \ldots, M-1 . \tag{3.198b}$$

These equations can be rewritten in the matrix form generalizing (3.171), as follows

$$X = B\left[|D_1X|\exp(i\arg G_1X)\right] + H\left[D_2\begin{pmatrix}|u_0|\\|v_M|\end{pmatrix}\exp(i\arg G_2X)\right] , \tag{3.199}$$

where $X = (u_1, u_2, \ldots, u_M, v_{M-1}, v_{M-2}, \ldots, v_0)^T$ is the vector-column of $2M$ unknown functions,

$$D_1 = \begin{pmatrix} E & 0 & 0 & 0 \\ 0 & 0 & E & 0 \end{pmatrix} , \tag{3.200}$$

E is the $(M-1) \times (M-1)$ unit matrix, G_1 is the $M \times M$ skew diagonal unit matrix,

$$G_2 = \begin{pmatrix} 0 & E_M \\ E_M & 0 \end{pmatrix} , \tag{3.201}$$

$E_M = (0, \ldots, 0, 1)$ is the vector of dimension M, B is the $(2M-2) \times (2M-2)$ diagonal operator matrix with the diagonal $(A_1, \ldots, A_{M-1}, A^*_{M-2}, \ldots, A^*_0)$, D_2 is the 2×2 unit matrix.

Similar to the case of the simple phase optimization problem described in the previous subsection, the nonlinear equation system (3.198) has nonunique solutions. As before, an arbitrary common constant phase shift β can be added to all ψ_m, $m = 0, 1, \ldots, M$, connected with any solution of (3.198) by (3.197). Also, Lemma 3.2 can be generalized to the multi-operator case in the following way.

Theorem 3.3

Let the functions $u_0 \in H_0$, $v_M \in H_M$ be given and the functions u_m, $m = 1, 2, \ldots, M$, v_m, $m = 0, 1, \ldots, M-1$, solve the equation system (3.198). Also, for certain $m = n$, $n = 0, 1, \ldots, M-1$, let the operators A_n and A_n^* be real on the functions $|u_n| \exp(i \arg(v_n))$ and $|v_{n+1}| \exp(i \arg(u_{n+1}))$, respectively. Then the functions u_m^*, $m = 1, 2, \ldots, M$, v_m^*, $m = 0, 1, \ldots, M-1$, such that

$$u_m^* = u_m, \qquad m \neq n+1, \tag{3.202a}$$

$$u_{n+1}^* = \bar{u}_{n+1}; \tag{3.202b}$$

$$v_m^* = v_m, \qquad m \neq n, \tag{3.202c}$$

$$v_n^* = \bar{v}_n, \tag{3.202d}$$

also solve (3.198). The value of (3.190) on both solutions is the same.

Proof: If equalities (3.202a) and (3.202c) hold, then equations (3.198a) are satisfied by $\{u_m^*, v_m^*\}$ for all m except $m = n+1$, and (3.198b) are satisfied by $\{u_m^*, v_m^*\}$ for all m except $m = n$. Only the equations

$$v_n^* = A_n^* \left[\left| v_{n+1}^* \right| \exp\left(i \arg\left(u_{n+1}^* \right) \right) \right], \tag{3.203a}$$

$$u_{n+1}^* = A_n \left[\left| u_n^* \right| \exp\left(i \arg\left(v_n^* \right) \right) \right] \tag{3.203b}$$

should be checked. In these equations the functions $|v_{n+1}^*|$, $|u_n^*|$ are known, and v_n^*, u_{n+1}^* must be found. This problem is analogous to (3.170). According to Lemma 3.2, equations (3.203) are satisfied if equalities (3.202b) and (3.202d) hold.

Applying equalities (3.196) to the new solution gives

$$\psi_m^* = \arg(v_m) - \arg(u_m), \quad m \neq n, \quad m \neq n+1, \tag{3.204a}$$

$$\psi_n^* = -(\arg(v_n) + \arg(u_n)), \tag{3.204b}$$

$$\psi_{n+1}^* = \arg(v_{n+1}) + \arg(u_{n+1}). \tag{3.204c}$$

In order to check that the new solution provides the same value for the functional, we substitute (3.202a), (3.202d) and (3.204b) into (3.192) for $m = n$ and apply Lemma 3.3. Using property (3.1) for the functional space H_n yields

$$\begin{aligned} \chi(\psi_0^*, \psi_1^*, \ldots, \psi_M^*) &= \chi_n(\psi_n^*) \left| (u_n^* \exp(i\psi_n^*), v_n^*)_n \right| \\ &= \left| (u_n \exp(-i(\arg(v_n) + \arg(u_n))), \bar{v}_n)_n \right| \\ &= (|u_n|, |v_n|)_n = \chi_n(\psi_n) = \chi(\psi_0, \psi_1, \ldots, \psi_M), \end{aligned} \tag{3.205}$$

which completes the proof. □

Corollary Under the conditions of Theorem 3.3 functions u_m, $m = 1, 2, \ldots, M$, v_m, $m = 0, 1, \ldots, M-1$, that solve the equation system (3.198), make up an equivalent group of solutions to this system. All solutions in the group provide the same value for the functional χ. Any two solutions of this group can be obtained from each other by a finite number of substitutions u_{m+1}, v_m by $\bar{u}_{m+1}$, $\bar{v}_m$, respectively, for different m. The number of solutions in the group is 2^M.

The forms (3.189), (3.191) and (3.196) of the Lagrange–Euler system suggest the following simple iterative procedure for its solution:

$$u_{m+1}^{(p+1)} = A_m \left[u_m^{(p+1)} \exp\left(i \psi_m^{(p)} \right) \right], \quad m = 0, 1, \ldots, M-1, \tag{3.206a}$$

$$\psi_M^{(p+1)} = \arg(v_M) - \arg\left(u_M^{(p+1)} \right), \tag{3.206b}$$

$$v_{m-1}^{(p+1)} = A_{m-1}^* \left[v_m^{(p+1)} \exp\left(-i \psi_m^{(p+1)} \right) \right],$$

$$\psi_{m-1}^{(p+1)} = \arg\left(v_{m-1}^{(p+1)} \right) - \arg\left(u_{m-1}^{(p+1)} \right), \quad m = M, M-1, \ldots, 1. \tag{3.206c}$$

In these expressions the initial approximations $\psi_m^{(0)}$ for all $m = 0, 1, \ldots, M-1$ should be given, and $u_0^{(p)} = u_0$, $v_M^{(p)} = v_M$ should be chosen at each pth step. Remember, that the functions u_0 and v_M are given by conditions of the problem.

Substep (3.206b), as well as each second substep of (3.206c) increases the respective functional $\chi_m(\psi_m)$ at given $u_m^{(p+1)}$, $v_m^{(p+1)}$, which equals $\chi(\psi_0, \psi_1, \ldots, \psi_M)$ at all given ψ_t, $t \neq m$. All other substeps of (3.206) do not change the value of χ. Hence, the iterative procedure (3.206) generates a nondecreasing sequence of values of (3.190). Since this sequence is bounded from above owing to the boundedness of the operators A_m, $m = 0, 1, \ldots, M-1$, then it converges.

A more effective form of procedure (3.206) will be described in Section 5.3.1.2 in which numerical methods for solving the concrete problems will be analyzed.

In some particular cases, the above problem is equivalent to the problem formulated in Section 3.1.1. This equivalence takes place at $M = 2$, if the moduli of given functions $U_0(x_0)$ and $V_2(x_2)$ are identical as the functions of their arguments ($|U_0| \equiv |V_2|$), and the operators A_0, A_1 are connected by the relation

$$A_1^* u = \overline{A_0 \bar{u}}. \tag{3.207}$$

In order to prove this equivalence, we put $\psi_0 + \arg U_0 \equiv \psi_2 - \arg V_2 = \psi$ and write the functional to be minimized in the form (3.192) for $m = 1$

$$\chi_1(\psi_1) = |(u_1 \exp(i\psi_1), v_1)_1|, \tag{3.208}$$

where u_1, v_1 are calculated by (3.189) and (3.191), respectively, as follows:

$$u_1 = A_0[|U_0| \exp(i\psi)], \tag{3.209a}$$

$$v_1 = A_1^*[|V_2| \exp(-i\psi)]. \tag{3.209b}$$

According to (3.196), the function ψ_1 in (3.208) is chosen in such a way that both multipliers in the inner product are inphase. According to (3.1), the common phase is canceled out in the product. Due to the assumptions made, $|u_1| \equiv |v_1|$, so that

$$\begin{aligned}\chi_1 &= \|u_1\|^2 = (A_0[|U_0|\exp(i\psi)], A_0[|U_0|\exp(i\psi)])_1 \\ &= (U_0\exp(i\psi), A_0^* A_0[U_0\exp(i\psi)])_0 .\end{aligned} \tag{3.210}$$

According to inequality (3.2), the functional (3.210) is a maximum if both multipliers in the inner product are inphase, that is,

$$\psi = \arg\left(A_0^* A_0[|U_0|\exp(i\psi)\right) . \tag{3.211}$$

Denote $f = A_0^* A_0[|U_0|\exp(i\psi)]$. Then, according to (3.211), $\arg f = \psi$. Finally, (3.211) becomes

$$f = A_0^* A_0[|U_0|\exp(i\arg f)] . \tag{3.212}$$

It coincides with (3.46) using the notations $A = A_0^*$, $F = |U_0|$. Recall that (3.46) corresponds to the optimization problem with no restrictions on the required function for the case of an isometric operator A.

This result was first obtained in [27] for a particular case of the operator A and described in detail in [25].

3.2.2.4 **The multi-operator phase optimization problem. The case of double-type operators**

Here we consider a multi-operator phase optimization problem more complicated than that considered in the previous subsection (see Problems **S**, **T**, **U**). The problems differ in two key points. The first distinction is that each transforming element of the optimized system does not provide only the phase correction of the field but also divides the incoming field into two parts (sub-beams); the first part is forwarded farther along the line, the second one is separated from the entire field beam and added to similar sub-beams separated by other elements in order to create the field of the given structure. Both of these actions are accompanied by appropriate phase corrections. The exceptions are only the first and last elements: the first one forwards the entire field, and the last one adds the entire incoming field to the separated sub-beams. The linear operators that describe the propagation of the fields in the space are of two different types. The operators of the first type are similar to those considered in the previous subsection. They describe the sequential transmission of a wave beam along the line from one transforming element to the next one. The operators of the second type describe the propagation of the separated sub-beams.

The mathematical description of the system consists in the following. Let H_m, $m = 0, 1, \ldots, M$, and H_Ω be the Hilbert spaces of L_2-type, $(\cdot,\cdot)_m$, $m = 0, 1, \ldots, M$, and $(\cdot,\cdot)_\Omega$ be the inner products in these spaces, which generate the norms $\|\cdot\|_m$, $m = 0, 1, \ldots, M$, $\|\cdot\|_\Omega$, respectively. The linear bounded operators $A_m : H_m \to H_{m+1}$, $m = 0, 1, \ldots, M-1$, and $B_m : H_m \to H_\Omega$, $m = 1, 2, \ldots, M$ are given. The functions $u_m \in H_m$, $m = 0, 1, \dots, M$, $f_m \in H_\Omega$ are connected by the recurrent relations

$$u_{m+1} = A_m[u_m T_m], \quad m = 0, 1, \dots, M-1, \tag{3.213}$$

$$f_m = B_m[u_m R_m], \quad m = 1, \dots, M, \tag{3.214}$$

where $T_m, R_m \in H_m$,

$$|T_0| \equiv 1, \tag{3.215a}$$

$$|T_m|^2 + |R_m|^2 \equiv 1, \quad m = 1, 2, \dots, M-1, \tag{3.215b}$$

$$|R_M| \equiv 1. \tag{3.215c}$$

Depending on the physical demands, several optimization problems can be formulated. They are based on the maximization of different functionals and/or differ in the input data. We consider one of them in detail and briefly describe the others.

In the first problem the function u_0 (in general, complex) is given. The functional to be maximized has the form

$$\chi(T_0, T_1, \dots, T_{M-1}, R_1, R_2, \dots, R_M) = (|f|, F)_\Omega, \tag{3.216}$$

where

$$f = \sum_{m=1}^{M} f_m, \tag{3.217}$$

$F \in H_\Omega$ is a given (desired) real positive function.

The Lagrange–Euler equation system for (3.216) can be constructed by an artificial scheme similar to that applied in the previous section to the problem with one-type operators. First of all, we denote

$$\psi = \arg f \tag{3.218}$$

and add this function as an additional argument of the functional. Then (3.216) can be rewritten in the form

$$\chi(T_0, T_1, \dots, T_{M-1}, R_1, R_2, \dots, R_M, \psi) = |(f, \Phi)_\Omega| = \left|\sum_{m=1}^{M} (f_m, \Phi)_\Omega\right|, \tag{3.219}$$

where $\Phi = F\exp(i\psi)$. Equality (3.218) is one of the equations in the Lagrange–Euler system.

It can be seen from (3.214) that at any given T_m, $m = 0, 1, \dots, M-1$ and $|R_m|$, $m = 1, 2, \dots, M-1$, all summands in (3.219) can be made to be inphase (which maximally increases χ) by multiplication of each R_m by an appropriate constant phase factor $\exp(i\alpha_m)$. Without loss of generality, we assume that the addends are real and positive at the maximum points.

Introduce two sets of auxiliary functions. The functions $v_m \in H_m$, $m = 1, 2, \ldots, M$, are calculated from $F\exp(i\psi)$ as

$$v_m = B_m^*[F\exp(i\psi)] , \quad m = 1, 2, \ldots, M . \tag{3.220}$$

The functions $w_m \in H_m$, $m = M, M-1, \ldots, 0$, are calculated by the following recurrent relations

$$w_M \equiv 0 , \tag{3.221a}$$

$$w_{m-1} = A_{m-1}^*\left[v_m \bar{R}_m + w_m \bar{T}_m\right] , \quad m = M, M-1, \ldots, 1 . \tag{3.221b}$$

Denote

$$\chi_m(T_0, T_1, \ldots, T_{m-1}, R_1, R_2, \ldots, R_m, \psi) = \sum_{n=1}^{m} (f_n, F\exp(i\psi))_\Omega . \tag{3.222}$$

According to the above assumption, at the maximum points of the functional χ the values of χ_m are real and positive.

The following lemma holds.

Lemma 3.4

If the functions u_m, f_m, v_m, and w_m, $m = 1, 2, \ldots, M$, satisfy the equalities (3.213), (3.214), (3.220) and (3.221), respectively, then

$$\begin{aligned} \chi(T_0, \ldots, T_{M-1}, R_1, \ldots, R_M, \psi) = \big|\chi_{m-1}(T_0, \ldots, T_{m-2}, R_1, \ldots, R_{m-1}, \psi) \\ + \left(u_m, \left[v_m \bar{R}_m + w_m \bar{T}_m\right]\right)_m\big| \end{aligned} \tag{3.223}$$

for any $m = 0, 1, \ldots, M$; here $\chi_0 = \chi_{-1} = 0$, $v_0 = 0$ are assumed.

Proof: We prove the lemma by the *induction method* (for simplicity, we omit arguments of the functionals χ and χ_m). First, we show that (3.223) holds for $m = M$ and $m = M-1$. Using definitions (3.214) and (3.220) for f_M, v_M, respectively, and applying identity (3.1) for H_M, we obtain

$$\begin{aligned} \chi &= \left|\chi_{M-1} + (f_M, F\exp(i\psi))_\Omega\right| \\ &= \left|\chi_{M-1} + (B_M[u_M R_M], F\exp(i\psi))_\Omega\right| \\ &= \left|\chi_{M-1} + \left(u_M R_M, B_M^*\left[F\exp(i\psi)\right]\right)_M\right| \\ &= \left|\chi_{M-1} + \left(u_M, v_M \bar{R}_M\right)_M\right| . \end{aligned} \tag{3.224}$$

Taking into account (3.221a), we verify that (3.223) holds for $m = M$. Similarly, using (3.214), (3.220), (3.213) and (3.221) for f_{M-1}, v_{M-1}, u_M, and w_M, w_{M-1}, respectively, and (3.1) for H_{M-1}, we have

$$\begin{aligned}\chi &= \left|\chi_{M-2} + (f_{M-1}, F\exp(i\psi))_{\Omega} + \left(u_M, v_M \bar{R}_M\right)_M\right| \\ &= \left|\chi_{M-2} + \left(u_{M-1}R_{M-1}, B^*_{M-1}\left[F\exp(i\psi)\right]\right)_{M-1}\right. \\ &\quad \left. + \left(u_{M-1}T_{M-1}, A^*_{M-1}\left[v_M \bar{R}_M\right]\right)_{M-1}\right| \\ &= \left|\chi_{M-2} + \left(u_{M-1}, v_{M-1}\bar{R}_{M-1} + \bar{T}_{M-1}A^*_{M-1}\left[v_M \bar{R}_M\right]\right)_{M-1}\right| \\ &= \left|\chi_{M-2} + \left(u_{M-1}, v_{M-1}\bar{R}_{M-1} + w_{M-1}\bar{T}_{M-1}\right)_{M-1}\right| . \end{aligned} \tag{3.225}$$

Assume that (3.223) holds for certain $m > 1$. Then

$$\begin{aligned}\chi &= \left|\chi_{m-2} + (f_{m-1}, F\exp(i\psi))_{\Omega} + (u_m, v_m \bar{R}_m + w_m \bar{T}_m)_m\right| \\ &= \left|\chi_{m-2} + \left(u_{m-1}R_{m-1}, B^*_{m-1}\left[F\exp(i\psi)\right]\right)_{m-1}\right. \\ &\quad \left. + \left(u_{m-1}T_{m-1}, A^*_{m-1}\left[v_m \bar{R}_m + w_m \bar{T}_m\right]\right)_{m-1}\right] \\ &= \left|\chi_{m-2} + \left(u_{m-1}, v_{m-1}\bar{R}_{m-1} + w_{m-1}\bar{T}_{m-1}\right)_{m-1}\right| , \end{aligned} \tag{3.226}$$

that is, (3.223) holds for $m - 1$. It can be easily checked that (3.223) holds also for $m = 0$, which completes the proof. □

By the definitions, the function u_m, the functional χ_{m-1} and the functions v_m and w_m (if ψ is known) do not depend on R_m or T_m. The functional χ, written in the form (3.223), depends on these functions explicitly. Since χ_{m-1} is real and positive at the maximum points of χ, the second addend on the right-hand side of (3.223) should also be real and positive. Moreover, both terms in this addend become real and positive independently if we set

$$\arg R_m = \arg v_m - \arg u_m , \qquad m = 1, 2, \ldots , M , \tag{3.227a}$$

$$\arg T_m = \arg w_m - \arg u_m , \qquad m = 0, 1, \ldots , M - 1 . \tag{3.227b}$$

The functions $|R_m|$, $|T_m|$ can be found from the condition of maximality of the inner product

$$(|u_m|, |v_m|\,|R_m| + |w_m|\,|T_m|)_m \tag{3.228}$$

for $m = 1, 2, \ldots, M - 1$ under condition (3.215b). In the spaces considered here, the maximum is reached if the product

$$|u_m|(|v_m|\,|R_m| + |w_m|\,|T_m|) \tag{3.229}$$

is maximal at any point. According to (3.215b), this product is a function of a single variable (e.g. $|R_m|$), and it is maximal if

$$|R_m| = \frac{|v_m|^2}{|v_m|^2 + |w_m|^2} , \qquad m = 1, 2, \ldots , M - 1 , \tag{3.230a}$$

$$|T_m| = \frac{|w_m|^2}{|v_m|^2 + |w_m|^2} , \qquad m = 1, 2, \ldots , M - 1 . \tag{3.230b}$$

Equalities (3.213), (3.214), (3.217), (3.218), (3.227) and (3.230) together may be considered as the Lagrange–Euler equation system for the functional (3.216). All in all, we have $4M + 1$ complex and $4M - 1$ real equations, and the same number of unknown functions: $6M - 1$ complex, namely, u_m, f_m, v_m, $m = 1, 2, \ldots, M$; w_m, $m = 0, 1, \ldots, M-1$; f; R_m, T_m, $m = 1, 2, \ldots, M-1$; and three real: $\arg T_0$, $\arg R_M$ and ψ. These equations describe not only necessary, but also sufficient conditions for χ to be maximal. Indeed, except for (3.227) and (3.230), all other equations of the system describe the relations between unknown functions and do not concern the stationary points. Equations (3.227) and (3.230) are valid only at the maximal points of the functional χ.

A simple iterative method may be applied to solve the obtained system of equations. As in the previous subsection, the algorithm of the method has two levels. At the beginning of each exterior iteration, the previous approximations to the functions R_m, $m = 1, 2, \ldots, M$; T_m, $m = 0, 1, \ldots, M - 1$; u_m, $m = 1, 2, \ldots, M$, and ψ must be known. The initial approximations $R_m^{(0)}$, $m = 1, 2, \ldots, M$, $T_m^{(0)}$, $m = 0, 1, \ldots, M - 1$, for the functions R_m, T_m must be given at the beginning of the algorithm, and the functions $u_m^{(0)}$, $f_m^{(0)}$, $m = 1, 2, \ldots, M$, must be calculated by (3.213) and (3.214). These functions allow us to calculate $f^{(0)}$ and $\psi^{(0)}$ by (3.218) and (3.227). The functions $u_0^{(p)} = u_0$, $T_M^{(p)} \equiv 0$ and $w_M^{(p)} \equiv 0$ are set for all iterations.

If all results of the pth iteration are known, the next iteration is performed by the following algorithm (two loops of the interior iterations are separated by the semicolons):

$$v_M^{(p+1)} = B_M^*\left[F \exp\left(i\psi^{(p)}\right)\right], \tag{3.231a}$$

$$\arg R_M^{(p+1)} = \arg v_M^{(p)} - \arg u_M^{(p)}; \tag{3.231b}$$

$$v_m^{(p+1)} = B_m^*\left[F \exp\left(i\psi^{(p)}\right)\right], \tag{3.231c}$$

$$w_m^{(p+1)} = A_m^*\left[v_{m+1}^{(p+1)} \bar{R}_{m+1}^{(p+1)} + w_{m+1}^{(p+1)} \bar{T}_{m+1}^{(p+1)}\right], \tag{3.231d}$$

$$\arg R_m^{(p+1)} = \arg v_m^{(p+1)} - \arg u_m^{(p)}, \tag{3.231e}$$

$$\arg T_m^{(p+1)} = \arg w_m^{(p+1)} - \arg u_m^{(p)}, \tag{3.231f}$$

$$\left|R_m^{(p+1)}\right| = \frac{\left|v_m^{(p+1)}\right|^2}{\left|v_m^{(p+1)}\right|^2 + \left|w_m^{(p+1)}\right|^2}, \tag{3.231g}$$

$$\left|T_m^{(p+1)}\right| = \frac{\left|w_m^{(p+1)}\right|^2}{\left|v_m^{(p+1)}\right|^2 + \left|w_m^{(p+1)}\right|^2}, \quad m = M-1, M-2, \ldots, 1; \tag{3.231h}$$

$$w_0^{(p+1)} = A_1^*\left[w_1^{(p+1)} \bar{T}_1^{(p+1)}\right]; \tag{3.231i}$$

$$\arg T_0^{(p+1)} = \arg w_0^{(p+1)} - \arg u_0 \, ; \tag{3.231j}$$

$$u_m^{(p+1)} = A_{m-1}\left[u_{m-1}^{(p+1)} T_{m-1}^{(p+1)}\right] , \tag{3.231k}$$

$$f_m^{(p+1)} = B_m\left[u_m^{(p+1)} R_m^{(p+1)}\right] , \quad m = 1, \ldots, M \, . \tag{3.231l}$$

$$f^{(p+1)} = \sum_{m=1}^{M} f_m^{(p+1)} , \quad \psi^{(p+1)} = \arg f^{(p+1)} , \tag{3.231m}$$

$$\chi_{p+1} = \left(\left|f^{(p+1)}\right|, F\right)_\Omega \, . \tag{3.231n}$$

The value of functional χ is changed at two stages: at steps (3.231e)–(3.231h) and at step (3.231m). At both stages the value of χ is obviously increased. Since all the operators in the problem are bounded, the functional is bounded from above and the sequence $\{\chi_p\}$, generated by the algorithm, monotonously converges.

In Section 5.3.2 a more effective form of the algorithm will be described for the example of a concrete problem.

Below, we briefly consider other optimization problems for the system investigated. In the first the function u_0 is not given. The functional to be maximized is

$$\chi(u_0, T_1, \ldots, T_{M-1}, R_1, R_2, \ldots, R_M) = \frac{(|f|, F)_\Omega}{\|u_0\|_0} \, . \tag{3.232}$$

In contrast to (3.216), the complex function u_0 is considered as an optimization parameter instead of T_0 and an appropriate normalization is used. After reformulation, Lemma 3.4 remains valid for this functional for all m except $m = 0$. For $m = 0$, (3.223) must be rewritten as

$$\chi = \frac{|(u_0, w_0)_0|}{\|u_0\|_0} \, . \tag{3.233}$$

At a given w_0, the functional χ is maximal if

$$u_0 = w_0 \, . \tag{3.234}$$

This equation should substitute (3.227b) for $M = 0$ in the above Lagrange–Euler equation system (3.213), (3.214), (3.217), (3.218), (3.227) and (3.230).

The iterative algorithm (3.231) can be easily modified for the problem considered. Namely, an initial approximation $u_0^{(0)}$ to u_0 should be given and $T_0 = 1$ should be set for each iteration. Also, the operation (3.231j) should be substituted by

$$u_0^{(p+1)} = \frac{w_0^{(p+1)}}{\left\|w_0^{(p+1)}\right\|_0} \, . \tag{3.235}$$

The next two problems (with given and free u_0, respectively) are based on maximization of the functional

$$L(T_0, T_1, \ldots, T_{M-1}, R_1, R_2, \ldots, R_M) = \frac{\|f\|_\Omega^2}{\|u_0\|_0^2} \, . \tag{3.236}$$

For these problems, equation (3.218) should be excluded from the Lagrange–Euler systems, and the function $F\exp(i\psi)$ should be substituted by f everywhere within the systems. In the iterative algorithm, $F\exp(i\psi^{(p)})$ should be substituted by $f^{(p)}$ in (3.231a) and (3.231c), and (3.231n) should be replaced by

$$L_{p+1} = \frac{\|f^{(p+1)}\|_\Omega^2}{\|u_0\|_0^2} . \tag{3.237}$$

3.2.3 The amplitude optimization problem

3.2.3.1 The general case

Given F, $\psi = \arg u$. The problem is to find the real function v which minimizes the functional

$$\sigma(v) = \||A[v\exp(i\psi)]| - F\|_2^2 . \tag{3.238}$$

In this formulation the function v must only be real, but not necessarily positive. It can change the sign in its definition domain. If it is necessary to fix the points at which v changes sign, then this demand should be prescribed by the function ψ (it must have jumps of π at these points) and the function v should be found in the class of positive functions. We do not consider such a formulation here.

In order to calculate the first variation of $\sigma(v)$, in (3.238) we substitute the function v by $v_0 + \delta v$ with a real δv and keep the terms of first order with respect to δv. Similar to (3.10),

$$\sigma(v + \delta v) = \sigma(v) + 2\,\mathrm{Re}(\delta f, f)_2 - 2(\delta|f|, F)_2 , \tag{3.239}$$

where

$$f = A[v\exp(i\psi)] , \tag{3.240a}$$

$$\delta f = A[\delta v\exp(i\psi)] , \tag{3.240b}$$

$$\delta|f| = \mathrm{Re}\left[\delta f \cdot \exp(-i\arg f)\right] . \tag{3.240c}$$

Then,

$$\begin{aligned}\delta\sigma(v, \delta v) &= 2\,\mathrm{Re}\,(A[\delta v\exp(i\psi)], f)_2 \\ &\quad - 2\,\mathrm{Re}\,(A[\delta v\exp(i\psi)]\cdot\exp(-i\arg f), F)_2 \\ &= 2\,\mathrm{Re}\left\{(\delta v\exp(i\psi), A^* f)_2\right. \\ &\quad \left. - (\delta v\exp(i\psi), A^*\left[F\exp(-i\arg f)\right])_2\right\} \\ &= 2\left(\delta v, \mathrm{Re}\left\{\exp(-i\psi)A^* f\right.\right. \\ &\quad \left.\left. - \exp(-i\psi)A^*\left[F\exp(-i\arg f)\right]\right\}\right)_2 ,\end{aligned}$$

and the Lagrange–Euler equation for the functional (3.238) has the form

$$\operatorname{Re}\{\exp(-i\psi)A^*A[\nu\exp(i\psi)] \\ -\exp(-i\psi)A^*\left[F\exp(i\arg A[\nu\exp(i\psi)])\right]\}=0\,. \quad (3.241)$$

The iterative procedure applicable to (3.241) is based on the formula

$$\operatorname{Re}\left\{\exp(-i\psi)A^*A\left[\nu^{(p+1)}\exp(i\psi)\right]\right\} \\ =\operatorname{Re}\left\{\exp(-i\psi)A^*\left[F\exp\left(i\arg A\left[\nu^{(p)}\exp(i\psi)\right]\right)\right]\right\}\,. \quad (3.242)$$

In order to prove the monotonicity of this procedure, we introduce the auxiliary functional

$$\sigma_p(\nu)=\left\|A[\nu\exp(i\psi)]-F\exp(i\varphi_p)\right\|_2^2 \quad (3.243)$$

with $\varphi_p=\arg(f^{(p)})$, $f^{(p)}=A[\nu^{(p)}\exp(i\psi)]$. The Lagrange–Euler equation for this functional is the nonhomogeneous linear equation

$$\operatorname{Re}\{\exp(-i\psi)A^*A[\nu\exp(i\psi)]\}=\operatorname{Re}\{\exp(-i\psi)A^*\left[F\exp(i\varphi_p)\right]\} \quad (3.244)$$

and its solution coincides with $\nu^{(p+1)}$, that is, it provides the minimum to $\sigma_p(\nu)$.

Analogously to (3.65), it is proved that

$$\sigma_p(\nu)-\sigma(\nu)=2\left(|f|\left\{1-\cos\left(\arg f-\psi_p\right)\right\},F\right)_2\ge 0\,, \quad (3.245)$$

so that the following chain of relations holds:

$$\sigma(\nu_{p+1})\le\sigma_p(\nu_{p+1})\le\sigma_p(\nu_p)=\sigma(\nu_p)\,. \quad (3.246)$$

Similar to the previous problems, the nonlinear equation (3.242) has nonunique solutions which can branch as a parameter c of the problem varies. To derive the homogeneous equation for determining the branching points, we apply the perturbation method. The perturbation of the parameter $c=c_0+\varepsilon$ causes the perturbations of the operator

$$A=A_0+\varepsilon A_1 \quad (3.247)$$

and the solution

$$\nu=\nu_0+\varepsilon\nu_1\,. \quad (3.248)$$

After substituting all perturbed quantities into (3.241) and performing simple derivations we obtain the nonhomogeneous linear equation

$$\operatorname{Re}\left\{\exp(-i\psi)A_0^*[A_0[\nu_1\exp(i\psi)]-iF\exp(i\arg(f_0))\operatorname{Im}\frac{A_0[\nu_1\exp(i\psi)]}{f_0}\right\} \\ =-\operatorname{Re}\{\exp(-i\psi)A_1^*A_1[\nu_0\exp(i\psi)]\} \\ +\operatorname{Re}\{\exp(-i\psi)A_1^*\left[F\exp(\arg A_1[\nu_0\exp(i\psi)])\right]\}\,, \quad (3.249)$$

with respect to the perturbation v_1 of the solution; here

$$f_0 = A_0[v_0 \exp(i\psi)] . \tag{3.250}$$

The uniqueness condition for the solution v_1 to this equation is the absence of nontrivial solutions to the appropriate homogeneous equation

$$\operatorname{Re}\left\{\exp(-i\psi)A^*\left[A[v_1\exp(i\psi)] - iF\exp(i\arg(f_0))\operatorname{Im}\frac{A[v_1\exp(i\psi)]}{f_0}\right]\right\} = 0 ; \tag{3.251}$$

the index 0 in the operators is omitted. The solution v_0 to (3.241) can branch only at the values c at which (3.251) has a nontrivial solution. This is an eigenvalue problem, nonlinear with respect to the spectral parameter c. We reformulate it in the usual form of the linear eigenvalue problem

$$\lambda_n \operatorname{Re}\{\exp(-i\psi)A^*[A[w_n\exp(i\psi)]]\} = -\operatorname{Im}\left\{\exp(-i\psi)A^*\left[F\exp(i\arg(f_0))\operatorname{Im}\frac{A[w_n\exp(i\psi)]}{f_0}\right]\right\} . \tag{3.252}$$

For the point $c = c_0$ to be a branching point of the solution v_0 to (3.241) it is necessary that one of the eigenvalues λ_n (say, λ_1) is equal to unity. As before, the eigenfunction w_1 corresponding to the eigenvalue $\lambda_1 = 1$ points in the direction in which the new solution branches off from v_0, that is, in a small vicinity of c_0, the branched solutions have the form $v = v_0 \pm t(\varepsilon)w_1$. Of course, the initial solution v_0 can be smoothly continued by (3.248) if the nonhomogeneous equation (3.249) has a solution v_1 orthogonal to w_1.

Note that, in contrast to all previous cases when $\lambda = 1$ was the eigenvalue at all c and this value was multiple at the branching points, (3.241) has no solution of this type and $\lambda_1 = 1$ may be a simple eigenvalue at these points.

3.2.3.2 The case of a compact operator

In the case when A is a compact operator, the minimization problem for functional (3.238) is unstable, its solutions can be noticeably perturbed at small perturbations of the given function F and operator A. This functional must be completed by a stabilization term, for instance, as follows

$$\sigma_t(v) = \| |A[v\exp(i\psi)]| - F\|_2^2 + t\,\|v\|_1^2 , \tag{3.253}$$

where $t > 0$ is the regularization parameter. This term causes the additional summand $2t\operatorname{Re}(\delta v, v)_1$ in the first variation, so that the Lagrange–Euler equation for this functional has the form

$$\operatorname{Re}\left\{\exp(-i\psi)A^*A[v\exp(i\psi)] - \exp(-i\psi)A^*\left[F\exp(i\arg A[v\exp(i\psi)])\right]\right\} + tv = 0 . \tag{3.254}$$

The iterative procedure for this equation is changed in a respective way and is described by the relation

$$\begin{aligned}\mathrm{Re}\left\{\exp(-i\psi)A^*A[v^{(p+1)}\exp(i\psi)]\right\} + tv^{(p+1)} \\ = \mathrm{Re}\left\{\exp(-i\psi)A^*\left[F\exp(i\arg A[v^{(p)}\exp(i\psi)])\right]\right\}\,. \qquad (3.255)\end{aligned}$$

Its monotonicity is proved in a similar way to the procedure (3.242).

If the operator A depends on a real parameter c, which can vary, what causes the branching of solutions to (3.254), then the homogeneous linear equation for the possible branching points has the form

$$\begin{aligned}\lambda_n\left\{tw_n + \mathrm{Re}\{\exp(-i\psi)A^*A[w_n\exp(i\psi)]\}\right\} \\ = -\,\mathrm{Im}\left\{F\exp(i\arg(f_0))\,\mathrm{Im}\,\frac{A[w_n\exp(i\psi)]}{f_0}\right\}\,, \qquad (3.256)\end{aligned}$$

where f_0 is given by (3.250). As before, new solutions may branch off from v_0 at the values c, at which one of the eigenvalues λ_n of this equation equals the unity.

3.2.3.3 The case of an isometric operator

In the case of a isometric operator A, for which

$$\|A[v\exp(i\psi)]\|_2 = \|v\|_1\,, \qquad (3.257)$$

the functional is reduced to the form

$$\sigma(v) = \|v\|_1^2 - 2(|A[v\exp(i\psi)]|, F)_2 + \|F\|_2^2\,. \qquad (3.258)$$

The Lagrange–Euler equation for this functional has the form

$$v = \mathrm{Re}\left\{\exp(-i\psi)A^*\left[F\exp(i\arg A[v\exp(i\psi)])\right]\right\}\,. \qquad (3.259)$$

Its solutions can be calculated by the simple iterative procedure

$$v^{(p+1)} = \mathrm{Re}\left\{\exp\left(-i\psi\right)A^*\left[F\exp\left(i\arg A\left[v^{(p)}\exp\left(i\psi\right)\right]\right)\right]\right\}\,, \qquad (3.260)$$

whose monotonicity is proved similarly as for the procedure (3.242).

If the operator A depends on a real parameter c, then the branching points of solutions to (3.259) are the values c, at which the homogeneous linear equation

$$\lambda_n w_n = -\,\mathrm{Im}\left\{F\exp(i\arg(f_0))\,\mathrm{Im}\,\frac{A[w_n\exp(i\psi)]}{f_0}\right\} \qquad (3.261)$$

has the multiple eigenvalue $\lambda_1 = 1$. Here f_0 is given by (3.250).

3.3
Homogeneous optimization problems

There are physical problems in which no function is given, which the function $f = Au$ should approach in modulus. In these problems certain parameters must be optimized by choosing appropriate phase functions. Such problems include maximization of the transmission coefficient in the transmission line, minimization of losses of the eigenoscillation in open resonators or modes in the closed and open waveguides, etc. Most of them deal with the homogeneous equations, hence we call them *homogeneous optimization problems.*

3.3.1
Problems of energy concentration

The functional to be maximized is

$$\chi^2(u) = \frac{\|Au\|_2^2}{\|u\|_1^2} \,. \tag{3.262}$$

3.3.1.1 Amplitude-phase optimization

First, we consider the case when no restrictions are imposed on the function. In order to derive the Lagrange–Euler equation for this functional, we calculate its value on the perturbed function $u + \delta u$ keeping only the terms of first order with respect to δu:

$$\begin{aligned}
\chi^2(u + \delta u) &= \frac{(Au + A[\delta u], Au + A[\delta u])_2}{(u + \delta u, u + \delta u,)_1} \\
&= \frac{\|Au\|_2^2 + 2\,\mathrm{Re}(A[\delta u], Au)_2}{\|u\|_1^2 + 2\,\mathrm{Re}(\delta u, u)_1} \\
&= \frac{\|Au\|_2^2}{\|u\|_1^2}\left(1 + 2\,\mathrm{Re}\left\{\frac{(A[\delta u], Au)_2}{\|Au\|_2^2} - \frac{(\delta u, u)_1}{\|u\|_1^2}\right\}\right) \\
&= \chi^2(u) + 2\,\|u\|_1^{-2}\,\mathrm{Re}\left\{(A[\delta u], Au)_2 - \chi^2(u)(\delta u, u)_1\right\} \\
&= \chi^2(u) + 2\,\|u\|_1^{-2}\,\mathrm{Re}(\delta u, A^* Au - \chi^2(u)u)_1 \,.
\end{aligned} \tag{3.263}$$

With accuracy to the nonessential factor $2\,\|u\|_1^{-2}$, the first variation of $\chi^2(u)$ is

$$\delta\chi^2(u, \delta u) = \mathrm{Re}(\delta u, A^* Au - \chi^2(u)u)_1 \,. \tag{3.264}$$

Repeating the considerations from Section 3.1.2, based on the independent choice of both real and imaginary parts of δu, we obtain the Lagrange–Euler equation for functional (3.3.1) in the form

$$A^* Au - \chi^2(u)u = 0 \,. \tag{3.265}$$

This equation can be obtained in a simpler way by using the *Lagrange multipliers method* with the Lagrange multiplier χ^2. We have applied the above technique for methodical reasons.

Formally, (3.265) is the nonlinear equation with respect to u. However, it is obviously related to the linear homogeneous equation

$$A^*Aw_n - \lambda_n w_n = 0\,, \tag{3.266}$$

where the pair $\{\lambda_n, w_n\}$ is the *eigenvalue* and *eigenfunction* corresponding to it. Note that the operator in this equation is self-adjoint and all its eigenvalues are real and positive and the eigenfunctions are orthogonal and form a complete function set. It can be immediately checked that, if a function w_n satisfies this equation for a certain value of λ_n, then $\chi^2(w_n)$ calculated by (3.262) equals λ_n. Consequently, the maximum value of $\chi^2(u)$ is reached on the function $u = w_1$ corresponding to the maximal eigenvalue λ_1, and this value itself is λ_1. If the maximum eigenvalue is multiple, then the maximum of $\chi^2(u)$ is reached on any linear combination of the eigenfunctions corresponding to this eigenvalue.

There are various methods used for solving standard eigenvalue problems. The simplest of these intended for determining the single maximal eigenvalue and corresponding eigenfunction is the *power method*. It is based on the iterative procedure

$$w_1^{(p+1)} = A^*Aw_1^{(p)}\,, \tag{3.267a}$$

$$\lambda_1^{(p+1)} = \frac{\left\| w_1^{(p+1)} \right\|}{\left\| w_1^{(p)} \right\|}\,. \tag{3.267b}$$

If the eigenvalue λ_1 is single, that is, $\lambda_1 > \lambda_2 \geq \ldots$, and the initial approximation $w_1^{(0)}$ has nonzero coefficient C_1 in its expansion by the eigenfunctions w_n

$$w_1^{(0)} = \sum_{n=1}^{\infty} C_n w_n\,, \tag{3.268}$$

then this procedure converges as a geometric progression with the ratio (λ_2/λ_1), and the sequence $\{\chi^2(w_1^{(p)})\}$ is monotonically increasing. According to (3.267a),

$$\begin{aligned} w_1^{(p)} &= \sum_{n=1}^{\infty} C_n \lambda_n^p w_n \\ &= \lambda_1^p \left[C_1 w_1 + C_2 \left(\frac{\lambda_2}{\lambda_1}\right)^p w_2 + C_3 \left(\frac{\lambda_3}{\lambda_1}\right)^p w_3 + \ldots \right]. \end{aligned} \tag{3.269}$$

If λ_2 is close to λ_1, then the iteration procedure converges very slowly. In this case a modified iteration method [50] can be used, which is oriented to the problems having a group of eigenvalues close to λ_1 in modulus. Justification of this method is specified in [51].

3.3.1.2 Phase optimization

Let the modulus of the function u be given and the functional (3.262) be maximized by an appropriate choice of the function $\psi = \arg u$. Since the norm $\|u\|_1^2$ is independent of ψ, we can set $\|u\|_1^2 = 1$ and rewrite the functional in the form

$$\chi^2(\psi) = \|A[|u|\exp(i\psi)]\|_2^2 \,. \tag{3.270}$$

Using the notation

$$f = A\left[|u|\exp(i\psi)\right] \,, \tag{3.271}$$

we obtain

$$\chi^2(\psi + \delta\psi) = (f + \delta f, f + \delta f)_2 = \chi^2(\psi) + 2\,\mathrm{Re}(\delta f, f)_2 \,. \tag{3.272}$$

Similar to (3.147b),

$$\delta f = iA\left[|u|\exp(i\psi)\delta\psi\right] \,, \tag{3.273}$$

and

$$\begin{aligned} \delta\chi^2(\psi, \delta\psi) &= -2\,\mathrm{Im}\left(A\left[|u|\exp(i\psi)\delta\psi\right], A\left[|u|\exp(i\psi)\right]\right)_2 \\ &= -2\left(\delta\psi, |u|\,\mathrm{Im}\left\{\exp(-i\psi)\cdot A^* A\left[|u|\exp(i\psi)\right]\right\}\right)_1 \,. \end{aligned} \tag{3.274}$$

The Lagrange–Euler equation for functional (3.301) has the form

$$\mathrm{Im}\left\{\exp(-i\psi)\cdot A^* A\left[|u|\exp(i\psi)\right]\right\} = 0 \,. \tag{3.275}$$

The two equations

$$\psi = \arg\left\{A^* A\left[|u|\exp(i\psi)\right]\right\} \,, \tag{3.276a}$$

$$\psi = \arg\left\{A^* A\left[|u|\exp(i\psi)\right]\right\} + \pi \tag{3.276b}$$

follow from (3.275); however, only the first of them is solvable. If (3.276b) has a solution, then

$$\begin{aligned} \chi^2(\psi) &= (A[|u|\exp(i\psi)], A[|u|\exp(i\psi)])_2 \\ &= \left(|u|\exp(i\psi), A^* A[|u|\exp(i\psi)]\right)_1 < 0 \end{aligned} \tag{3.277}$$

which contradicts (3.270) that defines χ^2 as a positive functional.

Equation (3.276a) is analogous to (3.46) if the latter is rewritten as

$$\arg f = \arg\left\{AA^*[F\exp(i\arg f)]\right\} \,. \tag{3.278}$$

These two equations coincide after the following substitutions in (3.278):

$$\arg f \to \psi \,, \tag{3.279a}$$

$$A \to A^* \,, \tag{3.279b}$$

$$F \to |u| \,. \tag{3.279c}$$

This means that the problems of maximization of functionals (3.48) with respect to u and (3.270) with respect to ψ are equivalent. For instance, if the function u maximizes functional (3.48) at given F, then the function $\psi = \arg(Au)$ maximizes functional (3.270) with $|u| = F$ and the operator A coinciding with A^* from the first problem. The analog of the function f in (3.46) is $AA^* [|u| \exp\{i\psi\}]$.

Owing to this analogy, all the facts established for (3.46) (and, correspondingly, for functional (3.48)), are transferred to (3.276a). The iterative procedure (3.60) is rewritten as

$$\psi^{(p+1)} = \arg\left\{A^* A[|u| \exp(i\psi^{(p)})]\right\}, \qquad p = 0, 1, 2, \ldots \tag{3.280}$$

It is monotonic, that is, $\chi^2(\psi^{(p+1)}) \geq \chi^2(\psi^{(p)})$.

3.3.2
Maximization of the operator spectral radius

3.3.2.1 The general case

Let us consider the linear operator $B_\psi : H \to H$ acting in the Hilbert space H in the following way:

$$B_\psi u = A\left[u \exp(i\psi)\right], \tag{3.281}$$

where $A \colon H \to H$ is a linear operator, $\operatorname{Im} \psi = 0$. By definition, its spectral radius equals $|\lambda_1|$, where λ_1 is the first eigenvalue of the linear homogeneous equation

$$\lambda u = A\left[u \exp(i\psi)\right], \tag{3.282}$$

$|\lambda_1| \geq |\lambda_2| \geq |\lambda_3| \ldots$. The problem is to find the function ψ at which $|\lambda_1|$ is as large as possible.

A necessary condition for $|\lambda_1|$ to be maximum is equality to zero of the linear part of its perturbation with respect to the perturbation $\delta\psi$. Since

$$\delta|\lambda_1|^2 = (\lambda_1 + \delta\lambda_1)\left(\bar{\lambda}_1 + \bar{\delta}_1\right) - |\lambda_1|^2 = 2 \operatorname{Re}\left(\delta\lambda_1 \cdot \bar{\lambda}_1\right), \tag{3.283}$$

this condition is

$$\operatorname{Re}\left(\delta\lambda_1 \cdot \bar{\lambda}_1\right) = 0. \tag{3.284}$$

In order to calculate $\delta\lambda_1$, we introduce the adjoint equation

$$\bar{\lambda} v = \exp(-i\psi) A^* v \tag{3.285}$$

and consider the functional

$$\Lambda(u, v, \psi) = \frac{\left(A\left[u \exp(i\psi)\right], v\right)}{(u, v)}. \tag{3.286}$$

It can be seen that $\Lambda = \lambda$ if either u solves (3.282) or v solves (3.285). In particular, if $u = u_1$ corresponds to eigenvalues λ_1, or $v = v_1$ corresponds to $\bar{\lambda}_1$, then $\Lambda(u, v, \psi) = \lambda_1$. We find the first variation of this functional.

The perturbed value of $\Lambda(u, v, \psi)$ caused by the perturbations of its arguments equals

$$\Lambda(u+\delta u, v+\delta v, \psi+\delta\psi) = \frac{\left(A\left[(u+\delta u)\exp(i(\psi+\delta\psi))\right], v+\delta v\right)}{(u+\delta u, v+\delta v)}$$
$$= \frac{1}{(u,v)+(\delta u, v)+(u,\delta v)}\left\{\left(A\left[u\exp(i\psi)\right], v\right)+\left(A\left[\delta u\exp(i\psi)\right], v\right)\right.$$
$$\left. +i\left(A\left[u\exp(i\psi)\delta\psi\right], v\right)+\left(A\left[u\exp(i\psi)\right], \delta v\right)\right\}$$
$$= \Lambda(u,\psi)+\frac{1}{(u,v)}\left\{(\delta u, [\exp(-i\psi)A^*v-\bar{\Lambda}v])+\left(A\left[u\exp(i\psi)\right]\right.\right.$$
$$\left.\left. -\Lambda u, \delta v\right)+\left(\delta\psi, -i\bar{u}\exp(-i\psi)A^*v\right)\right\}. \quad (3.287)$$

The expression in the braces of the result is the required first variation of Λ. In order that the first two summands in (3.287) be zero, it is necessary that (3.285) and (3.282) are satisfied with $\lambda = \Lambda$. If the functions v, u solve these equations, then

$$\delta\Lambda(u, v, \psi, \delta u, \delta v, \delta\psi) = \frac{1}{(u,v)}\left(\delta\psi, -i\bar{u}\exp(-i\psi)A^*v\right). \quad (3.288)$$

Substituting this expression into (3.284) instead $\delta\lambda_1$ and dropping the index 1 yields

$$\mathrm{Re}\left(\frac{(\delta\psi, -i\bar{u}\exp(-i\psi)A^*v\cdot\lambda)}{(u,v)}\right) = \left(\delta\psi, |\lambda|^2\,\mathrm{Im}\left[\frac{(\bar{u}v)}{(u,v)}\right]\right) = 0. \quad (3.289)$$

Owing to the arbitrariness of $\delta\psi$, we have

$$\mathrm{Im}\left[\frac{\bar{u}v}{(u,v)}\right] = 0. \quad (3.290)$$

Arbitrary constant factors in u and v cancel out in (3.290), so that

$$\arg u = \arg v. \quad (3.291)$$

Equations (3.282), (3.285) and (3.291) make up the system for determining the functions ψ, u_1, v_1 which provide the maximum to the functional Λ. This is equivalent to the fact that the function ψ provides the maximal value of $|\lambda_1|$ as the spectral radius of the operator B_ψ.

3.3.2.2 Particular case. Optimization of the mirror shape in an open resonator

Consider the linear homogeneous integral equation

$$\lambda w(y) = \frac{1}{\sqrt{2\pi}}\int_{-c}^{c}\exp\left(\frac{i}{2}(x-y)^2\right)\exp\left(i[\varphi(x)+\varphi(y)]\right)w(x)\,dx \quad (3.292)$$

which arise in Problem **Y** about the minimization of loss of the main oscillation in the open quasi-optical resonator with identical mirrors, the shape of which is described by the function $\psi(x)$ (two-dimensional case). After denoting

$$u(x) = w(x)\exp\left(-i\left[\varphi(x) + \frac{x^2}{2}\right]\right), \tag{3.293a}$$

$$\psi(x) = x^2 + 2\varphi(x) \tag{3.293b}$$

this equation becomes

$$\lambda u(y) = \frac{1}{\sqrt{2\pi}} \int_{-c}^{c} \exp(-ixy)\exp\left(i\psi(x)\right) u(x)\,dx\,. \tag{3.294}$$

The adjoint equation is

$$\bar{\lambda} v(y) = \frac{1}{\sqrt{2\pi}} \exp\left(-i\psi(x)\right) \int_{-c}^{c} \exp\left(ixy\right) v(x)\,dx\,. \tag{3.295}$$

From a comparison of (3.294) with (3.295), it can be seen that the functions u and v are connected by the relation

$$v(x) = \bar{u}(x)\exp(-i\psi(x))\,. \tag{3.296}$$

Together with equality (3.291), this relation gives

$$\arg(u(x)) = -\arg(u(x)) - i\psi(x)\,, \tag{3.297}$$

that is,

$$\psi(x) = -2\arg(u(x))\,. \tag{3.298}$$

Substituting (3.298) in (3.294), we obtain the equation

$$\lambda u(y) = \frac{1}{\sqrt{2\pi}} \int_{-c}^{c} \exp(-ixy)\bar{u}(x)dx \tag{3.299}$$

which does not depend on $\psi(x)$. After renaming the independent arguments x, y in this equation as y, x, respectively, and complex conjugating both sides of the result, we obtain

$$\bar{u}(x) = \frac{1}{\bar{\lambda}\sqrt{2\pi}} \int_{-c}^{c} \exp\left(ixy'\right) u\left(y'\right) dy'\,. \tag{3.300}$$

Substituting this expression into the right-hand side of (3.299) and changing the integration order yields

$$|\lambda|^2 u(y) = \frac{1}{\pi} \int_{-c}^{c} \frac{\sin(y-x)}{y-x} u(x)dx\,. \tag{3.301}$$

This equation was investigated in detail, for example, in [78]). As an integral equation with a real symmetric kernel, it has real eigenvalues and eigenfunctions, so that, according to (3.298),

$$\psi(x) = 0\,. \tag{3.302}$$

Taking into account notation (3.293), we conclude that the function $\varphi(x)$ providing a maximal modulus of the eigenvalue λ_1 of (3.292) is

$$\varphi(x) = -\frac{x^2}{2}\,. \tag{3.303}$$

This result was obtained immediately in another way in [23]. The resonators with the mirrors which provide this phase correction are called *confocal resonators*. They have minimal loss of the main mode.

Another particular case of the problem related to the open resonators with shifted mirrors (Problem **Z**) will be considered in Section 5.5.2 where the nonlinear integral equation for the optimal mirror shape in such resonators will be obtained and numerically solved.

4
Analytical Solutions

We find that the nonlinear integral equations (3.74), (3.80) and (3.89), as particular cases of (3.46), have analytical solutions. They are defined by polynomials of finite degree. The roots of these polynomials are determined from finite-dimensional systems of transcendental equations.

Due to the similarity of the above equations, they can be generalized. In this chapter such a general equation is given and its properties are investigated. The theorems establishing the form of its solutions, boundedness of the polynomial degree, and existing equivalent groups of the solutions are proven. The transcendental equations for the branching points of the solutions are obtained. The results are detailed and numerically analyzed for the particular cases.

Similar results are also obtained for a general integral equation connected with (3.103).

The main results of this chapter are described in [28, 29] and [54–56].

4.1
Analytical solutions of a general class of nonlinear integral equation with free phase

Let us consider a nonlinear integral equation of Hammerstein type

$$f(\xi) = B[F\exp(i\arg f)] \equiv \int_a^b K(\xi,\xi')F(\xi')\exp(i\arg f(\xi'))\mathrm{d}\xi' \tag{4.1}$$

with the kernel

$$K(\xi,\xi') = \frac{s(\xi)q(\xi') - s(\xi')q(\xi)}{\tau(\xi) - \tau(\xi')} \tag{4.2}$$

generating the linear positive definite integral operator B acting from $L_2(a,b)$ to $L_2(a,b)$:

$$(Bg, g) > 0 \tag{4.3}$$

for any $g \in L_2(a,b)$; $s(\xi)$, $q(\xi), \tau(\xi)$ are real continuous functions such that the function sets $\{\tau^n(\xi)s(\xi)\}$, $\{\tau^n(\xi)q(\xi)\}(n = 0,1,\ldots)$ are linearly indepen-

Phase Optimization Problems. O. O. Bulatsyk, B. Z. Katsenelenbaum, Y. P. Topolyuk, and N. N. Voitovich

ISBN: 978-3-527-40799-6

dent; $F(\xi) \in L_2(a,b)$ is a given real positive function. This equation generalizes (3.74), (3.80) and (3.89) from the previous section, being the particular cases of the Lagrange–Euler equation for functional (3.6) with isometric operator A.

4.1.1 Finite-parametric representation of the solutions

We confine ourselves to the case when the solutions to (4.1) have no zeros at $\xi \in (a,b)$ and assume that they can be represented in the form

$$f(\xi) = \beta \hat{f}(\xi) P_N(\tau)\,, \tag{4.4}$$

where β is any complex constant with $|\beta| = 1$;

$$\hat{f}(\xi) = \frac{1}{|P_N(\tau)|}\left|\int_a^b K(\xi,\xi')F(\xi')\frac{P_N(\tau')}{|P_N(\tau')|}\mathrm{d}\xi'\right|\,, \tag{4.5}$$

$\tau = \tau(\xi)$, $\tau' = \tau(\xi')$;

$$P_N(\tau) = \prod_{k=1}^{N}(1 - \eta_{Nk}\tau) \tag{4.6}$$

is a polynomial of finite degree N with complex pairwise nonconjugated zeros η_{Nk}^{-1}:

$$\eta_{Nk} - \bar{\eta}_{Nm} \neq 0, \quad k, m = 1, 2, \ldots, N\,. \tag{4.7}$$

We call $P_N(\tau)$ the generating polynomial. It follows from (4.4) that

$$\exp(i \arg f(\xi)) = \beta \frac{P_N(\tau)}{|P_N(\tau)|}\,. \tag{4.8}$$

Introduce the symmetrical polynomial of two real variables

$$R_{N-1}(\tau,\tau') = \frac{2i\left[P_N(\tau')\bar{P}_N(\tau)\right]}{(\tau - \tau')} = \sum_{n,m=1}^{N} d_{nm}\tau^{n-1}(\tau')^{m-1} \tag{4.9}$$

and denote the matrix of its coefficients by $D = \{d_{nm}\}$. The determinant of D is

$$\det D = (-1)^{[N/2]} \prod_{k,m=1}^{N} (\bar{\eta}_{Nm} - \eta_{Nk})\,, \tag{4.10}$$

where the square brackets mean the integer part of the value. This fact follows from the condition 4^0 of the Bezudiant from [57] and is consistent with Theorem 7.8 from [58]. Its immediate proof is given in [29]. Due to condition (4.7), $\det D \neq 0$.

The conditions for the function $f(\xi)$ of the form (4.4) to be a solution to (4.1) are stated by the following theorem.

Theorem 4.1

Let a function $f(\xi)$ of the form (4.4) have no zeros at $\xi \in [a, b]$. In order for it to be a solution to (4.1), it is necessary and sufficient that the parameters η_{Nk} satisfy the following system of transcendental equations:

$$\int_a^b \frac{\tau^{n-1} s(\xi) F(\xi)}{|P_N(\tau)|} d\xi = 0\,, \quad n = 1, 2, \ldots, N\,; \tag{4.11a}$$

$$\int_a^b \frac{\tau^{n-1} q(\xi) F(\xi)}{|P_N(\tau)|} d\xi = 0, \quad n = 1, 2, \ldots, N\,. \tag{4.11b}$$

Proof: Before proving the theorem, we note that the solutions to (4.1) are determined to within the above arbitrary constant β, so that we can put $\beta = 1$ without loss of generality.

Necessity. Let function (4.4) be a solution to (4.1). Substituting (4.4) into (4.1), we have

$$\hat{f}(\xi) P_N(\tau) = \int_a^b \frac{\left[s(\xi)q(\xi') - s(\xi')q(\xi)\right] P_N(\tau')}{(\tau(\xi) - \tau(\xi'))\,|P_N(\tau')|} F(\xi') d\xi'\,. \tag{4.12}$$

Multiplying both sides of this equality by $\bar{P}_N(\tau)$ and taking the imaginary part leads to the identity

$$\int_a^b \frac{\left[s(\xi)q(\xi') - s(\xi')q(\xi)\right] R_{N-1}(\tau, \tau')}{|P_N(\tau')|} F(\xi') d\xi' \equiv 0\,. \tag{4.13}$$

Then, substituting (4.9) into (4.13) and interchanging the variables ξ and ξ', we have

$$\sum_{n,m=1}^{N} d_{nm} \left[q(\xi') \int_a^b \frac{\tau^{n-1} s(\xi) F(\xi)}{|P_N(\tau)|} d\xi \right. \\ \left. - s(\xi') \int_a^b \frac{\tau^{n-1} q(\xi) F(\xi)}{|P_N(\tau)|} d\xi \right] (\tau')^{m-1} \equiv 0\,. \tag{4.14}$$

Since the functions $\{\tau^n s\}$, $\{\tau^n q\}$, $n = 0, \ldots, N-1$, are linearly independent, (4.14) gives

$$\sum_{n=1}^{N} d_{nm} \int_a^b \frac{\tau^{n-1} s(\xi)\, F(\xi)}{|P_N(\tau)|} d\xi = 0, \quad n = 1, 2, \ldots, N\,; \tag{4.15a}$$

$$\sum_{n=1}^{N} d_{nm} \int_a^b \frac{\tau^{n-1} q(\xi)\, F(\xi)}{|P_N(\tau)|} d\xi = 0, \quad n = 1, 2, \ldots, N\,. \tag{4.15b}$$

Equalities (4.15) can be considered as the two independent systems of linear algebraic equations with respect to the unknown integrals. The determinant of their common matrix D does not equal zero owing to conditions (4.7), so that the systems have only zero solutions, that is, the transcendental equations (4.11) are satisfied.

Sufficiency. Let (4.11) hold at certain integer N and complex η_{Nk}, $k = 1, 2, \ldots, N$, satisfying conditions (4.7). Then, of course, equalities (4.15) are also satisfied and, hence the identities (4.14) and (4.13) also hold. This means that

$$\operatorname{Im}\left[\bar{P}_N(\tau)\int_a^b K(\xi,\xi')F(\xi')\frac{P_N(\tau')}{|P_N(\tau')|}\mathrm{d}\xi'\right] = 0\,. \tag{4.16}$$

From the theorem conditions, the expression in the square brackets has no zeros in $[a, b]$, that is, it is always either positive or negative. If the former is true, then

$$\arg P_N(\tau) = \arg\int_a^b K(\xi,\xi')F(\xi')\frac{P_N(\tau')}{|P_N(\tau')|}\mathrm{d}\xi' \tag{4.17}$$

and, due to (4.4) and (4.8),

$$\arg f(\xi) = \arg\int_a^b K(\xi,\xi')F(\xi')\exp(i\arg f(\xi'))\mathrm{d}\xi'\,. \tag{4.18}$$

Together with (4.5) this means that (4.1) holds.

If the left-hand side of (4.16) were negative, then the above considerations would lead to

$$f(\xi) = -\int_a^b K(\xi,\xi')F(\xi')\exp(i\arg f(\xi'))\mathrm{d}\xi'\,. \tag{4.19}$$

In order to check that this equation has no solution, we write it in the operator form

$$f = -B[F\exp(i\arg f)] \tag{4.20}$$

and calculate

$$(B[F\exp(i\arg f)], F\exp(i\arg f)) = -(|f|\exp(i\arg f), F\exp(i\arg f)) < 0\,. \tag{4.21}$$

This inequality contradicts condition (4.3). □

4.1.2
The main properties of the solutions

Solutions of the form (4.4) with different degrees of the generating polynomials P_N can exist simultaneously. The largest degree of the polynomials is associated with the number of zeros of the functions $s(\xi)$ and $q(\xi)$ in (a, b). This relation is established by the following theorem.

Theorem 4.2

Under the conditions of Theorem 4.1, function (4.4) can solve (4.1) only if the degree N of the generating polynomial $P_N(\tau)$ does not exceed the number of zeros of both the functions $q(\xi)$ and $s(\xi)$ in the interval (a, b).

Proof: The fulfilment of subsystem (4.11a) of transcendental equations means that the function

$$\Phi_s(\xi) = \frac{s(\xi)\,F(\xi)}{P_N(\tau)} \tag{4.22}$$

is orthogonal to any polynomial of degree less than N. This function should alternate its sign in the interval (a, b) not less than N times. In the opposite case a real polynomial of degree less than N could be constructed, the zeros of which would coincide with the zeros of $\Phi_s(\xi)$. However, the function $\Phi_s(\xi)$ cannot be orthogonal to such a polynomial, because the product is real and has a constant sign. On the other hand, the sign of $\Phi_s(\xi)$ can alternate only when alternating the sign of $s(\xi)$. Therefore, the function $s(\xi)$ should have not less than N zeros.

In an analogous way, the fulfilment of subsystem (4.11b) leads to the conclusion that the function $q(\xi)$ should also have not less than N zeros. □

Note that the above theorem does not establish the absence of solutions to (4.1) in a form different from (4.4) (in particular, with $N = \infty$), because it assumes that N is limited.

Theorem 4.3

If the function $f(\xi)$ of the form (4.4) with $\beta = 1$ solves (4.1), then the functions

$$f_k(\xi) = \hat{f}(\xi) P_N(\tau) \frac{1 - \bar{\eta}_{Nk}\tau}{1 - \eta_{Nk}\tau}, \quad k = 1, 2, \ldots, N, \tag{4.23}$$

also solve this equation.

Proof: For simplicity, we drop the index N in η_{Nk}.

Let $f(\xi) = \hat{f}(\xi) P_N(\tau)$ be a solution to (4.1). Then, according to Theorem 4.1, the transcendental equation system (4.11) is satisfied. Simultaneously, this system is satisfied after substituting the polynomial $P_N(\tau)$ by any other polynomial of degree N with the same modulus, in particular, by the polynomial

$$Q_N(\tau) = P_{N-1}(\tau)(1 - \bar{\eta}_k\tau), \tag{4.24}$$

where

$$P_{N-1}(\tau) = P_N(\tau)/(1 - \eta_k\tau), \tag{4.25}$$

that is, $f_k(\xi) = \hat{f}(\xi)Q_N(\tau)$. In these notations, $|P_N(\tau)| = |Q_N(\tau)|$, and system (4.11) has the form

$$\int_a^b \frac{\tau^{n-1}s(\xi)F(\xi)}{|Q_N(\tau)|}\mathrm{d}\xi = 0, \quad n = 1, 2, \ldots, N\,; \tag{4.26a}$$

$$\int_a^b \frac{\tau^{n-1}q(\xi)F(\xi)}{|Q_N(\tau)|}\mathrm{d}\xi = 0, \quad n = 1, 2, \ldots, N\,. \tag{4.26b}$$

Substitute $f_k(\xi)$ into the right-hand side of (4.1) instead of $f(\xi)$ and multiply and divide the result by $|Q_N(\tau)|^2$. Taking into account (4.24), we obtain the identity

$$\int_a^b K(\xi,\xi')\frac{Q_N(\tau')}{|Q_N(\tau')|}F(\xi')\mathrm{d}\xi' = \frac{(1-\bar{\eta}_k\tau)P_{N-1}(\tau)\bar{Q}_N(\tau)\int_a^b K(\xi,\xi')F(\xi')\frac{Q_N(\tau')}{|Q_N(\tau')|}\mathrm{d}\xi'}{|P_N(\tau)|^2}\,. \tag{4.27}$$

Since system (4.26) is satisfied, equality (4.16) holds by replacing P_N with Q_N, that is,

$$\mathrm{Im}\left[\bar{Q}_N(\tau)\int_a^b K(\xi,\xi')F(\xi')\frac{Q_N(\tau')}{|Q_N(\tau')|}\mathrm{d}\xi'\right] = 0\,. \tag{4.28}$$

After introducing the notation

$$\hat{f}_k(\xi) = \frac{\bar{Q}_N(\tau)}{|P_N(\tau)|^2}\int_a^b K(\xi,\xi')F(\xi')\frac{Q_N(\tau')}{|Q_N(\tau')|}\mathrm{d}\xi'\,, \tag{4.29}$$

identity (4.27) can be rewritten in the form

$$\int_a^b K(\xi,\xi')F(\xi')\frac{Q_N(\tau')}{|Q_N(\tau')|}\mathrm{d}\xi' = \hat{f}_k(\xi)P_{N-1}(\tau)(1-\bar{\eta}_k\tau)\,, \tag{4.30}$$

or, which is the same,

$$\hat{f}_k(\xi)P_{N-1}(\tau)(1-\bar{\eta}_k\tau) = \int_a^b K(\xi,\xi')F(\xi')\frac{P_{N-1}(\tau')(1-\bar{\eta}_k\tau')}{|P_N(\tau')|}\mathrm{d}\xi'\,. \tag{4.31}$$

It follows from (4.31) that the function $\hat{f}_k(\xi)P_{N-1}(\tau)(1-\bar{\eta}_k\tau)$ solves (4.1) if $\hat{f}_k(\xi)$ has no zeros in $[a, b]$.

In order to complete the proof, we must show that $\hat{f}_k(\xi) = \hat{f}(\xi)$. To this end, we first substitute the function $f(\xi) = \hat{f}(\xi)P_N(\tau)$ into (4.1) and obtain

$$\hat{f}(\xi)P_{N-1}(\tau)(1-\eta_k\tau) = \int_a^b K(\xi,\xi')F(\xi')\frac{P_{N-1}(\tau')(1-\eta_k\tau')}{|P_N(\tau')|}\mathrm{d}\xi'\,. \tag{4.32}$$

Then we multiply both sides of (4.31) and (4.32) by $(1-\eta_k\tau)$ and by $(1-\bar{\eta}_k\tau)$, respectively, and subtract the results termwise. Taking into account that

$$(1-\eta_k\tau)(1-\bar{\eta}_k\tau') - (1-\bar{\eta}_k\tau)(1-\eta_k\tau') = (\eta_k-\bar{\eta}_k)(\tau'-\tau)\,, \tag{4.33}$$

we obtain

$$(\hat{f}_k(\xi) - \hat{f}(\xi)) P_{N-1}(\tau)|1 - \eta_k \tau|^2 = (\eta_k - \bar{\eta}_k)$$
$$\times \left[q(\xi) \int_a^b s(\xi') F(\xi') \frac{P_{N-1}(\tau')}{|P_N(\tau')|} d\xi' - s(\xi) \int_a^b q(\xi') F(\xi') \frac{P_{N-1}(\tau')}{|P_N(\tau')|} d\xi' \right] . \tag{4.34}$$

Since $P_{N-1}(\tau)$ is a polynomial of degree $N - 1$, the integrals on the right-hand side of (4.34) equal zero, due to (4.11) and this side is zero on the whole. According to the theorem conditions, the functions $P_{N-1}(\tau)$ and $|1 - \eta_k \tau|^2$ on the left-hand side of the last equality do not equal zero. Hence, $\hat{f}_k(\xi) = \hat{f}(\xi)$. □

Corollary Solutions to the transcendental equation system (4.11) make up the equivalent groups, inside each of which the polynomials $P_N(\tau)$ differ only in the substitution of any number $s < N$ of the parameters η_k by their complex conjugates:

$$P_N^{(s)}(\tau) = \prod_{m=1}^{s} (1 - \eta_{k_m} \tau) \prod_{m=s+1}^{N} (1 - \bar{\eta}_{k_m} \tau) , \tag{4.35}$$

where $k_{m_1} \neq k_{m_2}$ if $m_1 \neq m_2$. Such polynomials generate the solutions to (4.1) with the same $|f(\xi)|$.

4.1.3 Branching of the solutions

Consider the case when the kernel $K(\xi, \xi')$ in (4.1) depends on a certain real parameter c. At any fixed value of this parameter, the analytical solutions to (4.1) are expressed in the form (4.4). The number of solutions depends on c and may vary as c varies (we do not consider the solutions caused by the arbitrariness of β).

Similar to the previous section, the necessary condition for the point c to be a branching point of solution (4.4) can be obtained using perturbation theory.

Theorem 4.4

Let the kernel $K(\xi, \xi'; c)$ in (4.1) depend on a real positive parameter c and the function $f(\xi; c)$ of form (4.4) solve this equation. In order that $c = c_0$ should be a branching point of this solution, it is necessary that the homogeneous integral equation

$$\lambda_n v_n(\xi) \int_a^b \frac{K(\xi, \xi'; c_0) F(\xi')}{|P_N(\tau')|} \operatorname{Re}\left[P_N(\tau') \bar{P}_N(\tau) \right] d\xi'$$
$$= \int_a^b \frac{K(\xi, \xi'; c_0) F(\xi')}{|P_N(\tau')|} \operatorname{Re}\left[P_N(\tau') \bar{P}_N(\tau) \right] v_n(\xi') d\xi' \tag{4.36}$$

has the eigenvalue $\lambda_1 = 1$ of multiplicity not less than two.

Proof: We use the equation in the operator form (4.20) with $B = B(c)$ and denote $B_0 = B(c_0)$, $f_0 = f(\xi, c_0)$. The perturbation $c = c_0 + \varepsilon$ of the parameter leads to the perturbations of the operator and solution, which to a first approximation are $B = B_0 + \varepsilon B_1$; $f = f_0 + \varepsilon f_1$. Branching is possible in the case when the function $f_1(\xi)$ is determined nonuniquely. The problem is reduced to a linear homogeneous equation having nontrivial solutions at the branching points.

After denoting $B = AA^*$, (4.20) coincides with the operator form of (3.46) from the previous section. Since this form of B was not used in obtaining the homogeneous equation (3.95), we start with the equation

$$\lambda(v - iw) f = B\left[F \exp\{i \arg(f)\} v\right], \tag{4.37}$$

where the notation

$$f_1 = (w + iv) f_0 \tag{4.38}$$

is used and the index 0 in B_0 and f_0 is dropped. It was established in Section 3.1.7 that this equation has the obvious solution $w = 0$, $v = 1$ at $\lambda = 1$ and any c. The existence of this solution is explained by the phase arbitrariness of the constant β in (4.4). At the branching points, nontrivial solutions different from that above should exist at $\lambda = 1$.

Since the function w does not occur under the operator on the right-hand side of (4.37), it can be eliminated from the equation. Denoting

$$f = f' + i f'', \tag{4.39a}$$

$$\exp(i \arg f) = \mathrm{e}' + i\mathrm{e}'' \tag{4.39b}$$

and separating the real and imaginary parts of (4.37) taking into account the reality of the operator B caused by the reality of the kernel $K(\xi, \xi'; c)$ and the eigenvalue λ, we reduce (4.37) to the real equation system

$$f' w - f'' v = -B[F\mathrm{e}'' v], \tag{4.40a}$$

$$f'' w + f' v = B[F\mathrm{e}' v] \tag{4.40b}$$

with respect to the functions w, v. In order to exclude w from these equations, we multiply (4.40a) and (4.40b) by f'' and f', respectively, and subtract them from one another. As a result, we obtain the equation

$$|f|^2 v = f' B[F\mathrm{e}'' v] + f'' B[F\mathrm{e}' v], \tag{4.41}$$

or, in the concrete form,

$$|f_0(\xi)|^2\, v(\xi) = \int_a^b K(\xi, \xi'; c_0) F(\xi') \left[\mathrm{e}'(\xi') f_0'(\xi) + \mathrm{e}''(\xi') f_0''(\xi)\right] v(\xi') \mathrm{d}\xi'. \tag{4.42}$$

Taking into account (4.8), we have

$$f_0(\xi) = |f_0(\xi)| \frac{P_N(\tau)}{|P_N(\tau)|} \tag{4.43}$$

or, after separating the real and imaginary parts,

$$f_0'(\xi) = |f_0(\xi)| \frac{\mathrm{Re}\, P_N(\tau)}{|P_N(\tau)|}, \tag{4.44a}$$

$$f_0''(\xi) = |f_0(\xi)| \frac{\mathrm{Im}\, P_N(\tau)}{|P_N(\tau)|}. \tag{4.44b}$$

Substituting these expressions into (4.42) and multiplying both sides by $|P_N(\tau)| \cdot |f_0(\xi)|^{-1}$, we obtain

$$\begin{aligned} |f_0(\xi)|\,|P_N(\tau)|\,\nu(\xi) = \int_a^b & \frac{K(\xi,\xi';c_0)F(\xi')}{|P_N(\tau')|} \\ & \times \left[\mathrm{Re}\, P_N(\tau')\,\mathrm{Re}\, P_N(\tau) + \mathrm{Im}\, P_N(\tau')\,\mathrm{Im}\, P_N(\tau)\right]\nu(\xi')\mathrm{d}\xi', \end{aligned} \tag{4.45}$$

which, together with (4.43), gives

$$\begin{aligned} f_0(\xi)\frac{|P_N(\tau)|}{P_N(\tau)}\,|P_N(\tau)|\,\nu(\xi) \\ = \int_a^b \frac{K(\xi,\xi';c_0)F(\xi')}{|P_N(\tau')|}\,\mathrm{Re}\left[P_N(\tau')\bar{P}_N(\tau)\right]\nu(\xi')\mathrm{d}\xi'. \end{aligned} \tag{4.46}$$

Since

$$f_0(\xi) = B_0\left[F(\xi)\frac{P_N(\tau)}{|P_N(\tau)|}\right], \tag{4.47}$$

(4.46) can be rewritten as

$$\begin{aligned} \int_a^b \frac{K(\xi,\xi';c_0)F(\xi')}{|P_N(\tau')|} & \{P_N(\tau')\bar{P}_N(\tau)\nu(\xi) \\ & - \mathrm{Re}\left[P_N(\tau')\bar{P}_N(\tau)\right]\nu(\xi')\}\,\mathrm{d}\xi' = 0. \end{aligned} \tag{4.48}$$

According to the transcendental equations (4.11), the identity (4.16) holds. We rewrite it in the form

$$\int_a^b \frac{K(\xi,\xi';c_0)F(\xi')}{|P_N(\tau')|}\,\mathrm{Im}[P_N(\tau')\bar{P}_N(\tau)]\mathrm{d}\xi' = 0. \tag{4.49}$$

Then equality (4.48) is reduced to

$$\int_a^b \frac{K(\xi,\xi';c_0)F(\xi')}{|P_N(\tau')|}\,\mathrm{Re}[P_N(\tau')\bar{P}_N(\tau)][\nu(\xi') - \nu(\xi)]\mathrm{d}\xi' = 0. \tag{4.50}$$

This equation can be considered as an eigenvalue problem, nonlinear with respect to the spectral parameter c. In the usual way, it can be rewritten in the form (4.36) as a problem of finding the values of c at which the eigenvalue λ of this linear problem equals one. Since, as mentioned before, the function $\nu(\xi) = const$ solves (4.50) at any c and describes the freedom of β in (4.4), the multiplicity of the eigenvalue $\lambda_n = 1$ in (4.36) must be two or more. □

There is another way to find the branching points of the solutions to (4.1). This is based on the immediate construction of the transcendental equations for the branching points. In the case when the solution of type (4.4) branches without changing the degree N of the generating polynomial P_N, the theory of implicit functions of several variables is applied for this purpose. If the polynomial degree is changed in the new solutions, the transcendental equation system is made up of equations of the type (4.11) written for the different values of N simultaneously.

First we consider the case when N does not change at the branching point, that is, new solutions are generated by the polynomials of the same degree as the initial one. We denote $\eta_{Nk} = \eta'_{Nk} + i\eta''_{Nk}$ and consider the real values η'_{Nk}, η''_{Nk} as functions of the parameter c. Then equations (4.11) are implicit forms of these functions:

$$\Phi_{Nn}(\eta'_{N1}, \ldots, \eta'_{NN}, \eta''_{N1}, \ldots, \eta''_{NN}; c) = 0, \quad n = 1, 2, \ldots, N\,; \tag{4.51a}$$

$$\Psi_{Nn}(\eta'_{N1}, \ldots, \eta'_{NN}, \eta''_{N1}, \ldots, \eta''_{NN}; c) = 0, \quad n = 1, 2, \ldots, N\,. \tag{4.51b}$$

We will use the notation

$$\mathcal{F} = \left\{\vec{\Phi}, \vec{\Psi}\right\}, \tag{4.52}$$

where $\vec{\Phi}$, $\vec{\Psi}$ are the vectors on the left-hand sides of (4.51).

According to the theory of implicit functions of several variables [60], Theorem 14.2), the points c, at which the solutions to system (4.11) are found, are determined from the condition

$$D(c) \equiv \det(\mathcal{F}') = 0\,, \tag{4.53}$$

where

$$\mathcal{F}' = \begin{pmatrix} \left\{\dfrac{\partial \Phi_{Nj}}{\partial \eta'_{Nk}}\right\}_{j,k=1}^{N} & \left\{\dfrac{\partial \Phi_{Nj}}{\partial \eta''_{Nk}}\right\}_{j,k=1}^{N} \\ \left\{\dfrac{\partial \Psi_{Nj}}{\partial \eta'_{Nk}}\right\}_{j,k=1}^{N} & \left\{\dfrac{\partial \Psi_{Nj}}{\partial \eta''_{Nk}}\right\}_{j,k=1}^{N} \end{pmatrix} \tag{4.54}$$

is the Jacobi matrix of these functions. Condition (4.53) is the transcendental equation with respect to c. It is the required equation for the branching points c. Of course, in order to calculate the left-hand side of this equation, we must solve system (4.11). After that the elements of determinant (4.54) become the integrals with explicit integrands.

The points at which the derivative $d\eta_{Nk}/dc$ equals infinity can be present among the roots of equation (4.53). These points can be the isolated bifurcation points, to the left or right of which the continuation of the new solution is impossible; they are the points of appearance or disappearance of the solution, respectively.

Consider the case when a solution $f_N(\xi)$ generated by the polynomial $P_N(\tau)$ branches on changing the polynomial degree. First, we assume that N increases by one at the branching point $c = c_j$. Since both the initial solution $f_N(\xi)$ and the branched one $f_{N+1}(\xi)$ solve (4.1), the necessary condition for the branching is $f_N(\xi) \equiv f_{N+1}(\xi)$ at $c = c_j$ or, after using (4.8),

$$\frac{P_N(\tau)}{|P_N(\tau)|} = \frac{P_{N+1}(\tau)}{|P_{N+1}(\tau)|} \,. \tag{4.55}$$

Substituting the polynomials $P_N(\tau)$ and $P_{N+1}(\tau)$ of the form (4.6) into (4.55), we obtain

$$\prod_{k=1}^{N} \frac{1-\eta_{Nk}\tau}{|1-\eta_{Nk}\tau|} = \prod_{k=1}^{N+1} \frac{1-\eta_{N+1,k}\tau}{|1-\eta_{N+1,k}\tau|} \,. \tag{4.56}$$

Equating the multipliers on both sides of (4.56) termwise gives

$$\eta_{Nk} = \eta_{N+1,k}, \qquad k = 1, 2, \ldots, N, \tag{4.57a}$$

$$\operatorname{Im} \eta_{N+1,N+1} = 0 \,. \tag{4.57b}$$

Hence, together with (4.11), the transcendental equation system

$$\int_a^b \frac{\tau^{n-1} s(\xi) F(\xi)}{|P_N(\tau)|} (1-\eta_{N+1,N+1}\tau) d\xi = 0, \quad n = 1, 2, \ldots, N+1 \,; \tag{4.58a}$$

$$\int_a^b \frac{\tau^{n-1} q(\xi) F(\xi)}{|P_N(\tau)|} (1-\eta_{N+1,N+1}\tau) d\xi = 0, \quad n = 1, 2, \ldots, N+1 \tag{4.58b}$$

with real $\eta_{N+1,N+1}$ should be satisfied at the branching point.

Since the identity

$$\frac{1}{1-\eta\tau} = 1 + \frac{\eta\tau}{1-\eta\tau} \tag{4.59}$$

is true, the nth equation of system (4.58) is a linear combination of the nth equation of system (4.11) and $(n+1)$th equation of system (4.58) for all $n = 1, 2, \ldots, N$. Hence, besides system (4.11), only two additional equations

$$\int_a^b \frac{\tau^{N} s(\xi) F(\xi)}{|P_N(\tau)|} (1-\eta_{N+1,N+1}\tau) d\xi = 0 \,, \tag{4.60a}$$

$$\int_a^b \frac{\tau^{N} q(\xi) F(\xi)}{|P_N(\tau)|} (1-\eta_{N+1,N+1}\tau) d\xi = 0 \tag{4.60b}$$

is satisfied at the branching point. As a result, we have $2N + 2$ real equations for determining $2N + 2$ real unknowns: N real and imaginary parts of the parameters η_{Nk}, $k = 1, 2, \ldots, N$, the real parameter $\eta_{N+1,N+1}$ and the branching point $c = c_j$.

In the particular case $N = 0$, system (4.11) is absent and only two equations

$$\int_a^b \frac{s(\xi)F(\xi)}{(1-\eta_{11}\tau)} d\xi = 0 , \tag{4.61a}$$

$$\int_a^b \frac{q(\xi)F(\xi)}{(1-\eta_{11}\tau)} d\xi = 0 \tag{4.61b}$$

remain to determine the two real unknowns η_{11} and c_0.

Now we consider the case when N increases by two at the branching point. The necessary condition for such branching is $f_N(\xi) \equiv f_{N+2}(\xi)$, that is

$$\prod_{k=1}^{N} \frac{1-\eta_{Nk}\tau}{|1-\eta_{Nk}\tau|} = \prod_{k=1}^{N+2} \frac{1-\eta_{N+2,k}\tau}{|1-\eta_{N+2,k}\tau|} . \tag{4.62}$$

at $c = c_j$. It can be seen that this equality holds only if

$$\eta_{Nk} = \eta_{N+2,k} , \quad k = 1, 2, \ldots, N , \tag{4.63}$$

and either

$$\eta_{N+2,N+1} = \bar{\eta}_{N+2,N+2} , \tag{4.64}$$

or

$$\mathrm{Im}\,\eta_{N+2,N+1} = \mathrm{Im}\,\eta_{N+2,N+2} = 0 . \tag{4.65}$$

Therefore, together with system (4.11) the following equations should be satisfied at the branching point:

$$\int_a^b \frac{\tau^{n-1}s(\xi)F(\xi)}{|P_N(\tau)|}(1-\eta_{N+2,N+1}\tau)(1-\eta_{N+2,N+2}\tau) d\xi = 0 , \quad n = 1, 2, \ldots, N+2 ; \tag{4.66a}$$

$$\int_a^b \frac{\tau^{n-1}q(\xi)F(\xi)}{|P_N(\tau)|}(1-\eta_{N+2,N+1}\tau)(1-\eta_{N+2,N+2}\tau) d\xi = 0 , \quad n = 1, 2, \ldots, N+2 , \tag{4.66b}$$

where the parameters $\eta_{N+2,N+1}$, $\eta_{N+2,N+2}$ are subject to conditions (4.63) and (4.64) or (4.63) and (4.65). Similar to the previous case, only four equations of system (4.66) are linearly independent of system (4.11), and four additional equations for $n = N+1, N+2$

$$\int_a^b \frac{\tau^{n-1}s(\xi)F(\xi)}{|P_N(\tau)|}(1-\eta_{N+2,N+1}\tau)(1-\eta_{N+2,N+2}\tau) d\xi = 0 , \quad n = N+1, N+2 ; \tag{4.67a}$$

$$\int_a^b \frac{\tau^{n-1} q(\xi) F(\xi)}{|P_N(\tau)|}(1-\eta_{N+2,N+1}\tau)(1-\eta_{N+2,N+2}\tau)\mathrm{d}\xi = 0\,,$$
$$n = N+1, N+2\,, \quad (4.67\mathrm{b})$$

is satisfied as well as system (4.11) at the branching points where the polynomial degree N increases by two. As a result, we have $2N+4$ real equations for determining $2N+3$ real unknowns: $2N$ real components of N complex parameters η_{Nk}, $n = 1, 2, \ldots, N$, real c_j and either one complex $\eta_{N+2,N+1}$ or two real $\eta_{N+2,N+1}$ and $\eta_{N+2,N+2}$. The particular cases related to the symmetrical data when solutions to such systems exist are considered in the next sections. In the general case without symmetry, the existence of such solutions is not very probable.

4.1.4 Strategy for numerical investigation

The aim of the numerical solution of nonlinear equations of type (4.1) dependent on a parameter c is to have their solutions either at a fixed $c = c_0$, or in a certain vicinity within its range. However, the first problem often turns out to be not much simpler, especially when we want to find all solutions at a fixed c. Then in a similar way to the general case, we should determine the total number of solutions, which is, as a rule, impossible without investigating the branching process in the entire range of the parameter variation. In most cases (at least, in all the problems considered here) the number of solutions and their initial approximations can be easily determined for small c. Due to this fact we will consider the problem of determining solutions to (4.1) for $0 < c \le c_{\max}$. We confine ourselves by the solutions arisen from the real one (with $N = 0$), as a result of the branching process.

As a result of the branching, the number of solutions to (4.1) rapidly increases. In order to begin determining each branch of solutions, we should specify the initial data consisting of the value c at which the branch starts, the degree N of the generating polynomial P_N and the initial values of the complex parameters η_{Nk} at the start point. During the calculations it is convenient to combine such data into a set, each (jth) element of which is the collection $\{N_j, c_j, \eta^j_{N_j k}, k = 1, ..N_j\}$.

There are two different schemes (strategies) for finding all the solutions in the chosen range of the parameter c. They differ in order of finding the branches of the solutions. In both cases the calculations start from the case $N = 0$. In the first scheme, the branches are found successively one by one, together with all the branching points on them. As a result of each (jth) step of this scheme, we obtain the solution f_j to (4.1) for all $c \in [c_j, c_{\max}]$ and the initial data $\{N_s, c_s, \eta^s_{N_s k}, k = 1, ..N_s\}$ for all (sth) branching points located on this branch. The graph from this scheme is illustrated in Figure 4.1a. The solid line shows the solutions obtained at the first step of the scheme, the dashed lines relate to the second step, the dotted lines to the third step and so on. All obtained initial data make up the data set in the order in which they are found.

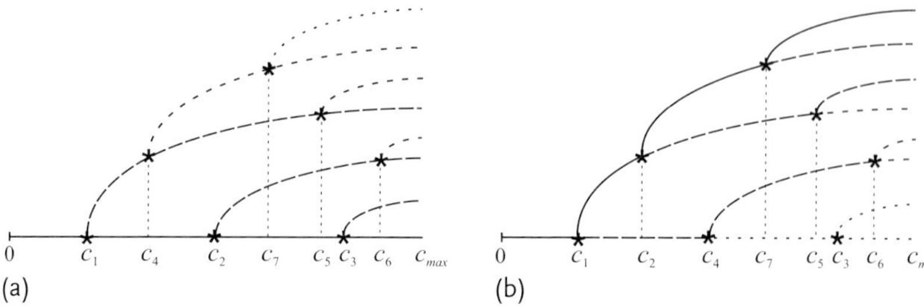

Figure 4.1 Illustration for calculation schemes.

In the second scheme, each (jth) branch is processed only to the next branching point $c = c_{j+1}$ on it. Then the collection $\{N_j, c_{j+1}, \eta^s_{N_j k}, k = 1, ..N_j\}$ is kept in the set for the next step of the scheme. The calculations continue with the initial data $\{N_{j+1}, c_{j+1}, \eta^s_{N_{j+1}k}, k = 1, ..N_{j+1}\}$, where N_{j+1} is the degree of the polynomial corresponding to the solution branching off at the found branching point, and $\eta^s_{N_{j+1}k}$ are the initial values of its parameters. If the branching points are no longer observed on the current branch up to value $c = c_{\max}$, we pass to the next step of the scheme. We return to the solution with the same or smaller value of N corresponding to the last branching point obtained at the previous step. Repeating this procedure we will find all nonisolated branches of the solutions of (4.1). Figure 4.1b illustrates the second scheme in the form of a graph.

In the first step of this scheme, all solutions generated by the polynomials of the maximum degree permissible for each value of c are determined. As will be seen from the numerical results obtained for the particular cases when (4.1) is the Lagrange–Euler equation for appropriate functionals, the minimum value of the functional is reached just on the solutions corresponding to this degree of the polynomial. If the problem consists only in minimization of the functional, then it is sufficient to proceed through the upper segment of the branches in the graph (solid lines in Figure 4.1b).

Obviously, the analysis of the equivalent solution groups having, as a rule, the same branching points (as will be shown later, there are exclusions from this rule which should be analyzed separately) allows us essentially to reduce the calculation expenses in both schemes.

Now we consider the first scheme in detail. Assume that the kernel $K(\xi, \xi')$ in (4.1) is real, and the function

$$f_0(\xi) = \int_a^b K(\xi, \xi')F(\xi')d\xi' \tag{4.68}$$

has no zeros in $[a, b]$ (as it takes place in the particular cases considered). Then, of course, $f_0(\xi)$ is a solution to (4.1) of the form (4.4) with $\beta = 1$, $\hat{f} = f_0(\xi)$, $N = 0$. This solution and all its subsequent branchings are the subject of our investigation. In the case when other solutions of the form (4.4) exist at small c (e.g. real solutions having zeros at $\xi \in (a, b)$), they can be investigated in a similar way.

The solution $f_0(\xi)$ (corresponding to $N = 0$) is calculated explicitly in the entire interval $0 < c < c_{\max}$. From this we determine the eigenvalues λ_n of (4.36), close to the value $\lambda = 1$ for all c and find the points $c = c_j$ (perhaps approximately) where the curves $\lambda_n(c)$ intersect this value. In addition to the possible branching points c_j, this equation allow us to determine the degree N_j of the polynomial P_N corresponding to the solutions branching off at these points. If the multiplicity of the eigenvalue $\lambda_n = 1$ equals two or three, then the degree N_j of a new solution either remains the same (impossible at $N = 0$) or increases by one or two ($N_j = 1$ or $N_j = 2$ at $N = 0$), respectively. To specify the value c_j and obtain the parameters $\eta_{N_j k}$ for the initial data of the new solution, we should solve the transcendental equation system (4.61) (for $N_j = 1$) or (4.67) (for $N_j = 2$, $P_0 = 1$). The values N_j, c_j, $\eta^j_{N_j k}$, $k = 1, ..N_j$ make up an element of the collection of initial data for processing in the next step of the scheme.

The next steps differ from the first one in the three positions. First, each of them starts from initial data taken from the completed set of the collections. Second, the solution $f_j(\xi)$ at each given c, $c_j < c \le c_{\max}$, is calculated by (4.4) with the polynomial P_N, $N = N_j$, in which the parameters η_{Nk} are determined from the equation system (4.11). Third, the system of transcendental equations for specifying a branching point c_s and determining the initial values of the parameters $\eta_{N_s k}$, consists of (4.11) and one of the systems (4.53), (4.60) or (4.67), used in the case when the polynomial degree N_s with new solutions is the same as in the initial one ($N_s = N_j$) or increased by one ($N_s = N_j + 1$) or by two ($N_s = N_j + 2$), respectively. Initial approximations for the deficient parameters in these sets can be found by minimizing the discrepancy of the additional equations.

In general, determination of every new solution on the branch enlarges the initial data set by a new collection corresponding to new branching point which exist on this branch. However, the total number of branching points in the bounded interval of values c is limited, and the calculation process is, of course, also limited. It should be kept in mind during the calculations that certain solution branches can be closed. Also, formally, each branching point belongs to both solution branches: initial and off-branched. Therefore, such points can occur in the initial data set two or more times. Particular examples of branching processes will be shown further in Figures 4.5 and 4.12.

The Newton method can be recommended for solving transcendental equation systems. The main peculiarity of its application to our problems will be detailed in Section 5.2.

4.2
A particular case: one-dimensional Fourier transformation

The simplest particular form of the general integral equation (4.1) is (3.74) which arises in the unconditional minimization problem for functional (3.6) with the isometric operator describing the one-dimensional Fourier transform of the finite function. Equation (3.74) is obtained from (4.1) with

$$[a, b] = [-1, 1]; \quad s(\xi) = \sin c\,\xi; \quad q(\xi) = \cos c\,\xi; \quad \tau(\xi) = \xi\,. \tag{4.69}$$

For convenience, we give this equation here:

$$f(\xi) = \frac{1}{\pi} \int_{-1}^{1} \frac{F(\xi') \sin c(\xi - \xi')}{(\xi - \xi')} \exp(i \arg f(\xi')) \mathrm{d}\xi'\,. \tag{4.70}$$

It has been partially analyzed in [56], where the concrete forms of the theorems were immediately proved and the branching process was investigated numerically for the case of symmetrical data (even function F) and their small perturbations. Here we only formulate the main theoretical results concerning this case and give new numerical results for nonsymmetrical $F(\xi)$. The only exception is Theorem 4.8 below which is not proved in the general case. Due to its key significance for optimization problems considered in this book, we give here its complete proof from [56] with some nonessential modifications.

4.2.1 Finite-parametric representation of solutions

Theorem 4.5

Let the function $f(\xi)$ having no zeros at $\xi \in [-1, 1]$ be expressed in the form

$$f(\xi) = \beta \hat{f}(\xi) P_N(\xi)\,, \tag{4.71}$$

with constant β, $|\beta| = 1$,

$$\hat{f}(\xi) = \frac{1}{|P_N(\xi)|} \left| \int_{-1}^{1} \frac{\sin c(\xi - \xi') F(\xi') P_N(\xi')}{(\xi - \xi') |P_N(\xi')|} \mathrm{d}\xi' \right|, \tag{4.72}$$

$$P_N(\xi) = \prod_{k=1}^{N} (1 - \eta_{Nk}\xi)\,, \tag{4.73}$$

where η_{Nk} are subject to conditions (4.7). For it to be a solution to (4.70), it is necessary and sufficient that the parameters η_{Nk} satisfy the following system of transcendental equations:

$$\Phi_{Nn} \equiv \int_{-1}^{1} \frac{\xi^{n-1} \sin(c\xi) F(\xi)}{|P_N(\xi)|} \mathrm{d}\xi = 0\,, \quad n = 1, 2, \ldots, N\,; \tag{4.74a}$$

$$\Psi_{Nn} \equiv \int_{-1}^{1} \frac{\xi^{n-1} \cos(c\xi) F(\xi)}{|P_N(\xi)|} \mathrm{d}\xi = 0\,, \quad n = 1, 2, \ldots, N\,. \tag{4.74b}$$

Note that the phase factor in (4.70) is now written in the form

$$\exp(i \arg f(\xi)) = \frac{P_N(\xi)}{|P_N(\xi)|}\,. \tag{4.75}$$

Theorem 4.6

Under the conditions of Theorem 4.5, function (4.71) can solve (4.70) only if the degree N of the generating polynomial $P_N(\xi)$ satisfies the condition

$$N < \frac{2c}{\pi} \,. \tag{4.76}$$

This result follows from Theorem 4.2 after taking into account that the number of zeros of the functions $\sin(c\xi)$ and $\cos(c\xi)$ in the interval $(-1, 1)$ is equal to $2[c/\pi]+1$, and $2[c/\pi]$, respectively. The latter, smaller of them leads to inequality (4.76).

Theorem 4.7

If the function $f(\xi)$ of form (4.71) with $\beta = 1$ solves (4.70), then the functions

$$f_n(\xi) = \hat{f}(\xi) P_N(\xi) \frac{1 - \bar{\eta}_{Nn}\xi}{1 - \eta_{Nn}\xi} \,, \quad n = 1, 2, \ldots, N \tag{4.77}$$

also solve this equation.

Corollary Any solution of equation set (4.74) generates an equivalent group of solutions which correspond to the polynomials $P_N(\xi)$ of the type

$$P_N^{(s)}(\xi) = \prod_{m=1}^{s} (1 - \eta_{n_m}\xi) \prod_{m=s+1}^{N} (1 - \bar{\eta}_{n_m}\xi),$$
$$s = 1, 2, \cdots, N-1, \, n_{m_1} \neq n_{m_2} \text{ if } m_1 \neq m_2 \,. \tag{4.78}$$

All of these solutions give the same value of $|f(\xi)|$ and, consequently, the same value of the functional $\sigma(u)$ (3.69) after substituting (3.73) with $Au = f$.

Note that all the proofs of the above theorems are based on the assumption that N is finite. The question of whether or not solutions of type (4.4) with $N = \infty$ exist, is still open. However, the following important assertion for equation (4.70) is true.

Theorem 4.8

Let a function $f_*(\xi)$ of the form (3.68) be a global minimum point of the functional $\sigma(u)$ (3.69) and have no zeros in $[-1, 1]$. Then this function has the form (4.71) with finite N.

Proof: It is known that any $f(\xi)$ of the form (3.68), not vanishing at $\xi = 0$, can be written as [61]

$$f(\xi) = f(0) \exp(it\xi) \prod_{n=1}^{\infty} (1 - \eta_{Nn}\xi), \quad \operatorname{Im} t = 0 \,, \tag{4.79}$$

with some complex (in general) η_{Nn}. Suppose that such a function is the global minimum point of the functional σ and that it has an infinite number of the complex nonconjugated pairwise zeros η_{Nn}^{-1}. Recall that according to the Paley–Wiener theorem [62], the function $f(\xi) \in L_2(-\infty; \infty)$ can be expressed in the form (3.68) with $u(x) \in L_2(-c; c)$ if and only if its analytical continuation into the complex plane is an entire function satisfying the condition

$$|f(\xi)| \leq B \exp(c|\xi|), \quad B < \infty . \tag{4.80}$$

The function $f_*(\xi)$ belongs to $L_2(-\infty; \infty)$ and it has the form (3.68); hence, it satisfies the condition (4.80). As an extreme point of the functional (3.69), it satisfies (4.70).

Let us consider the function

$$f_m(\xi) = f_*(\xi) \frac{1 - \bar{\eta}_{Nm}\xi}{1 - \eta_{Nm}\xi} , \tag{4.81}$$

where η_{Nm} is one of the parameters η_{Nn} satisfying (4.7). This function is entire, because it is expressed in the form (4.79) with the mth multiplier $1 - \eta_{Nm}\xi$ under the product symbol, replaced with $1 - \bar{\eta}_{Nm}\xi$. It satisfies condition (4.80), because

$$\lim_{|\xi| \to \infty} \left| \frac{1 - \bar{\eta}_{Nm}\xi}{1 - \eta_m \xi} \right| = 1 , \tag{4.82}$$

and, therefore, it can be expressed in the form (3.68). For real ξ, $|f_m(\xi)| = |f_*(\xi)|$. This means that $|f_m(\xi)|$ provides the same (global minimum) value of σ as does $f_*(\xi)$ and that is why $f_m(\xi)$ also satisfies (4.70). It can be easily seen that any function $f(\xi)$ satisfying (4.70) and having the same modulus as $|f_*(\xi)|$, solves the linear homogeneous integral equation

$$\nu f(\xi) = \frac{1}{\pi} \int_{-1}^{1} \frac{F(\xi)}{|f_*(\xi)|} \frac{\sin[c(\xi - \xi')]}{(\xi - \xi')} f(\xi) \mathrm{d}\xi \tag{4.83}$$

with $\nu = 1$. According to our assumption, the number of such functions is infinite.

Since $F(\xi)$ is integrable and $|f_*(\xi)|$ is bounded from below, the kernel

$$\sqrt{\frac{F(\xi)}{|f_*(\xi)|} \frac{F(\xi')}{|f_*(\xi')|}} \frac{\sin[c(\xi - \xi')]}{(\xi - \xi')} \tag{4.84}$$

of the appropriate symmetrized equation belongs to $L_2((-1; 1) \times (-1; 1))$. Its norm is finite and

$$\| K(\xi, \xi', c) \|^2 = \sum_{n=1}^{\infty} |\nu_n|^2 = \sum_{n=1}^{\infty} 1 + \sum_{n=1}^{\infty} \nu_n^2 \leq \infty \tag{4.85}$$

Consequently, (4.83) cannot have the eigenvalue $\nu = 1$ of infinite multiplicity. Therefore, the number of parameters η_{Nn} in expression (4.79) for $f_*(\xi)$, satisfying (4.7), is finite and we can write

$$f_*(\xi) = \beta \exp(it\xi)\hat{f}(\xi)P_N(\xi)\,, \tag{4.86}$$

where $\beta = \exp(i \arg f(0))$.

It remains to prove that $t = 0$. After substituting (4.86) into (4.70), we can calculate the asymptotic behavior of $f_*(\xi)$ as $\xi = \pm iz$, $\operatorname{Im} z = 0$, $z \to \infty$. Since $|\exp(i\beta\xi)| = 1$ with real β, ξ, we have

$$|f_*(\pm iz)| = \left| \int_{-1}^{1} \frac{\sin c(\xi - iz)F(\xi)}{(\xi - iz)} \frac{P_N(\xi)}{|P_N(\xi)|} \exp(i\beta\xi)\mathrm{d}\xi \right| . \tag{4.87}$$

Taking into account the asymptotic behavior

$$\sin c(\xi \pm iz) = i\mathrm{sh}z \cos c\xi \pm \mathrm{ch}z \sin c\xi \tag{4.88}$$

$$\cong i\frac{\exp z}{z}(\cos c\xi \mp i \sin c\xi) = i\frac{\exp z}{z} \exp(\mp ic\xi)\,, \tag{4.89}$$

we obtain

$$|f_*(\pm iz)| \cong \frac{\exp|z|}{2|z|} A_{\pm}^{(0)}\,, \tag{4.90}$$

where

$$A_{\pm}^{(0)} = \left| \int_{-1}^{1} \exp(i(\beta \pm c)\xi')F(\xi')\frac{P_N(\xi')}{|P_N(\xi')|}\mathrm{d}\xi' \right| . \tag{4.91}$$

If the coefficient $A_{\pm}^{(0)} = 0$, then it is necessary to include into the consideration the next terms of the expansion ξ'^{-1} by z^{-1}; in the general case we obtain

$$|f_*(\pm iz)| \simeq |A_{\pm}^{(m)}||z|^{-(m+1)}\exp c|z|\,, \tag{4.92}$$

where $A_{\pm}^{(m)}$ is the first nonzero coefficient

$$A_{\pm}^{(n)} = \frac{1}{2\pi} \int_{-1}^{1} \frac{F(\xi)P_N(\xi)\xi^n \exp(i(t \pm c)\xi)}{|P_N|}\mathrm{d}\xi\,,\, n = 1, 2, \ldots \tag{4.93}$$

All these coefficients are seen to be nozero simultaneously, except for the case $F(\xi) \equiv 0$. On the other hand, since $\hat{f}(\xi)$ has only real zeros and is real at real ξ, then, according to the Vieta theorem, the coefficients of its Taylor series are real, so that $\hat{f}(-iz) \simeq \bar{\hat{f}}(iz)$. Consequently, from (4.86) we have, that

$$|f_*(\pm iz)| \simeq |\hat{f}(iz)a_N z^N| \exp(\mp tz)\,, \tag{4.94}$$

where a_N is the higher coefficient of the polynomial P_N. This means that f_* has different exponential behavior on the upper and lower imaginary half-axes, which does not agree with (4.92). Consequently, $t = 0$, and the theorem is proved. □

4.2.2 Branching of the solutions

The next theorem is a particular case of the general Theorem 4.4.

Theorem 4.9

Let a function $f(\xi; c)$ of form (4.71) solve (4.70). In order that a value $c = c_0$ should be a branching point of this solution, it is necessary that, at this point, the homogeneous integral equation

$$\lambda_n \nu_n(\xi) \int_{-1}^{1} \frac{\sin c(\xi - \xi') F(\xi')}{(\xi - \xi')|P_N(\xi')|} \operatorname{Re}\left[P_N(\xi')\, \bar{P}_N(\xi)\right] \mathrm{d}\xi'$$
$$= \int_{-1}^{1} \frac{\sin c(\xi - \xi') F(\xi')}{(\xi - \xi')|P_N(\xi')|} \operatorname{Re}\left[P_N(\xi')\, \bar{P}_N(\xi)\right] \nu_n(\xi')\, \mathrm{d}\xi' \qquad (4.95)$$

has the eigenvalue $\lambda_1 = 1$ of multiplicity not less than two.

The branching of solutions to (4.70) can be one of the following three types: one found with the same polynomial degree N and two with N increasing by one or by two, respectively.

When determining the branching point $c = c_j$ in the case when the polynomial degree N does not change, the value c_j and parameters η_{Nk}, $k = 1, \ldots, N$ of both the initial and new polynomials are determined from the transcendental equation system (4.74) complemented by (4.53) with Φ_{Nn}, Ψ_{Nn} defined in (4.74).

At the branching points of solutions to (4.70) where the polynomial degree N changes by one, the parameters η_{Nk} of the initial polynomial $P_N(\xi)$ and parameters $\eta_{N+1,k}$ of the branched one $P_{N+1}(\xi)$ are connected by the equalities

$$\eta_{Nk} = \eta_{N+1,k}, \qquad k = 1, \ldots, N. \qquad (4.96)$$

In addition to (4.74), the two following equations

$$\int_{-1}^{1} \frac{\xi^N \sin(c\xi) F(\xi)}{|P_N(\xi)|\,(1 - \eta_{N+1,N+1}\xi)} \mathrm{d}\xi = 0, \qquad (4.97a)$$
$$\int_{-1}^{1} \frac{\xi^N \cos(c\xi) F(\xi)}{|P_N(\xi)|\,(1 - \eta_{N+1,N+1}\xi)} \mathrm{d}\xi = 0 \qquad (4.97b)$$

with $\operatorname{Im} \eta_{N+1,N+1} = 0$ also hold. On the whole, we have $2N + 2$ real equations to determine $2N + 2$ real unknowns: N real and imaginary parts of η_{Nk}, $k = 1, 2, \ldots, N$, and real $\eta_{N+1,N+1}$ and c_j.

At the branching points where the polynomial degree changes by two, the equalities

$$\eta_{Nk} = \eta_{N+2,k},\ k = 1, 2, \ldots, N. \qquad (4.98)$$

are valid. As well as (4.74), the four additional equations

$$\int_{-1}^{1} \frac{\xi^{n-1}\sin(c\xi)F(\xi)}{|P_N(\xi)|(1-\eta_{N+2,N+1}\xi)(1-\eta_{N+2,N+2}\xi)} d\xi = 0 , \quad n = N+1, N+2 ; \tag{4.99a}$$

$$\int_{-1}^{1} \frac{\xi^{n-1}\cos(c\xi)F(\xi)}{|P_N(\xi)|(1-\eta_{N+2,N+1}\xi)(1-\eta_{N+2,N+2}\xi)} d\xi = 0 , \quad n = N+1, N+2 , \tag{4.99b}$$

should be fulfiled with $\eta_{N+2,N+1}$, $\eta_{N+2,N+2}$ satisfying either conditions

$$\eta_{N+2,N+1} = \bar{\eta}_{N+2,N+2} \tag{4.100}$$

or

$$\operatorname{Im} \eta_{N+2,N+1} = \operatorname{Im} \eta_{N+2,N+2} = 0 . \tag{4.101}$$

Hence, we have $2N+4$ equations for $2N+3$ real unknowns: N complex η_{Nk}, $n = 1, 2, \ldots, N$, one real c_j, and one complex $\eta_{N+2,N+1}$ or two real $\eta_{N+2,N+1}$, $\eta_{N+2,N+2}$. As was mentioned in the previous section, the existence of solutions to such a system is unlikely in the general case. However, they may exist in the case when

$$F(\xi) = F(-\xi) . \tag{4.102}$$

Then those solutions to (4.74) are possible, which are generated by the polynomials with even modulus. The polynomial $P_{N+2}(\xi)$ should have this property, that is,

$$|P_{N+2}(\xi)| = |P_{N+2}(-\xi)| . \tag{4.103}$$

This equality decreases the number of unknowns twice: the parameters $\eta_{N+2,k}$ become imaginary or appear as pairs with opposite signs, whereas, according to (4.100), $\eta_{N+2,k}$, $k = N+1, N+2$, are always imaginary with opposite signs:

$$\operatorname{Re} \eta_{N+2,N+1} = \operatorname{Re} \eta_{N+2,N+2} = 0 , \tag{4.104a}$$

$$\eta_{N+2,N+1} = \bar{\eta}_{N+2,N+2} . \tag{4.104b}$$

On the other hand, conditions (4.102) and (4.103) also decrease the number of equations twice. N equations of system (4.74) and two additional equations (4.99a) become identities, because they have odd integrands on the left-hand side.

Finally, solution branching is possible by increasing the polynomial degree by two, if (4.102) is fulfilled and the following transcendental equation system is satisfied:

$$\int_{-1}^{1} \frac{\xi^{2n-1}\sin(c\xi)F(\xi)}{|P_N(\xi)|} d\xi = 0 , \quad n = 1, 2, \ldots, [N/2] , \tag{4.105a}$$

$$\int_{-1}^{1} \frac{\xi^{2n-2}\cos(c\xi)F(\xi)}{|P_N(\xi)|}\mathrm{d}\xi = 0\,, \quad n = 1, 2, \ldots, [(N+1)/2]\,, \tag{4.105b}$$

$$\int_{-1}^{1} \frac{\xi^{2[(N+2)/2]-1}\sin(c\xi)F(\xi)}{|P_N(\xi)|\,(1-\eta_{N+2,N+1}\xi)(1-\eta_{N+2,N+2}\xi)}\mathrm{d}\xi = 0\,, \tag{4.105c}$$

$$\int_{-1}^{1} \frac{\xi^{2[(N+1)/2]}\cos(c\xi)F(\xi)}{|P_N(\xi)|\,(1-\eta_{N+2,N+1}\xi)(1-\eta_{N+2,N+2}\xi)}\mathrm{d}\xi = 0\,, \tag{4.105d}$$

where η_{Nk}, $k = 1, \ldots, N$, are either imaginary or appear by pairs with opposite signs, and $\eta_{N+2,k}$, $k = N+1, N+2$ are subject to conditions (4.104). As a result, we have $N+2$ real equations with $N+2$ real unknowns.

4.2.3
Numerical results

The behavior of the solutions to (4.70) essentially depends on the evenness of the given function $F(\xi)$. One of the most important practical cases for even functions, the function $F(\xi) \equiv const$, as well as the perturbed one $F(\xi) = 1 + \beta\xi$ have been investigated in detail in [56]. In particular, the numerical results given there show that the number of solutions to (4.70) are fast increasing as the parameter c grows. In the case of even function $F(\xi)$, different types of bifurcations were observed: branching without changing the degree N of the polynomial P_N; changing N by one or by two; and the isolated points of appearance and disappearance of the solutions.

Small asymmetrical perturbations of the even $F(\xi)$ allowed investigation of the behavior of the solutions near their appearance and disappearance points, detailed analysis of the appearance of the branching points of the solutions to (4.70) and observation of the process of their disappearance themselves. Asymmetrical perturbations lead to transformation of the imaginary parameters η_{Nk} being the inverse zeros of the entire function $f(\xi)$ at certain values of c, into complex parameters with a nonzero real part. The parameters of the off-branched solutions start from real nonzero values.

For symmetrical (even) functions $F(\xi)$ the coupling of the branches of solutions was observed at the points where the polynomial degree changes by two. At small perturbations the branches disconnect and the branching of this type becomes impossible (as was predicted theoretically).

The mentioned results and those obtained for other functions $F(\xi)$ allow us to claim with a high probability that the minimal value of the functional σ (3.69) is attained at the solutions with the polynomial P_N of the highest degree permitted at a given value of c.

Here we investigate numerically the properties of the solutions to (4.70) with noneven function $F(\xi)$ for the case $F(\xi) = 1/(2+\xi)$ for $|\xi| \le 1$; $F(\xi) \equiv 0$ for $|\xi| > 1$. This function describes the so-called cosecant radiation pattern of the linear antenna, which is desired, for instance, for the uniform illumination of a lengthy flat domain near an elevated (e.g. TV) antenna (Figure 4.2). In this case $c = ka\sin\theta_0$, where k is the wave number, $2a$ is the antenna length, $\xi = \sin\theta / \sin\theta_0$

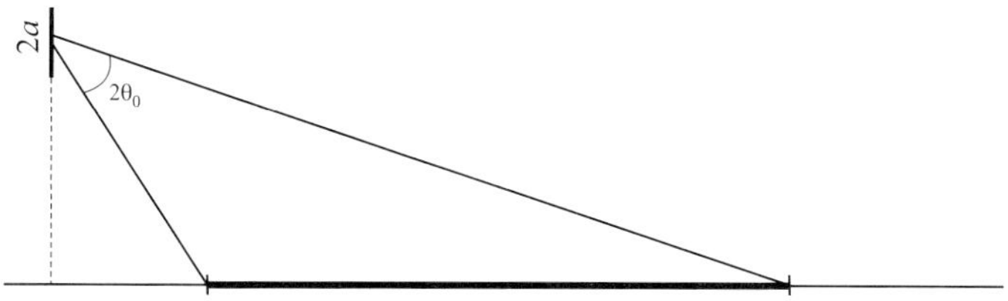

Figure 4.2 Illustration for the problem of a linear antenna.

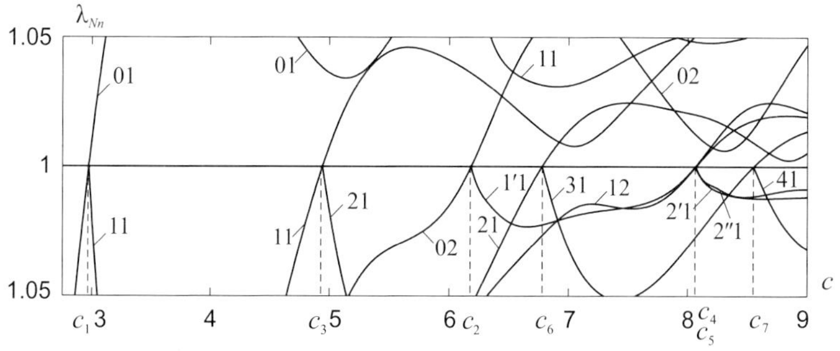

Figure 4.3 Eigenvalues of the homogeneous integral equation (4.95); $F(\xi) = 1/(2+\xi)$.

is the generalized angular coordinate; $2\theta_0$ is the angle in which the desired pattern differs from zero.

The calculations were carried out by the first strategy described above. The numerical results relating to the case considered are given in Figures 4.3–4.9. The eigenvalues λ_{Nn} of the homogeneous integral equation (4.95), close to $\lambda = 1$ are shown in Figure 4.3. The real and imaginary parts of the parameters η_{Nk}, corresponding to different solutions of (4.70), are shown in Figure 4.4. As a rule, only one representative (with $\operatorname{Im} \eta_{Nk} \geq 0$) of each equivalent group of solutions is shown (according to Theorem 4.9, the number of solutions in each group equals 2^N). The two-digit labels of curves correspond to the indices Nk; the curves describing different solutions related to the polynomials $P_N(\xi)$ with the same N are indicated as $N'k$, $N''k$, etc. Recall that we consider only the solutions having no zeros for $\xi \in [-1, 1]$.

At $N = 0$ we have $P_0(\xi) \equiv 1$ and the solution to (4.70) becomes real

$$f_0(\xi) = \frac{1}{\pi} \int_{-1}^{1} \frac{F(\xi') \sin c(\xi - \xi')}{(\xi - \xi')} d\xi' ; \tag{4.106}$$

such a solution exists for any $c > 0$ and arbitrary $F(\xi)$ (see [26], p. 51). Several eigenvalues λ_{0n} of (4.95), corresponding to this solution, are marked in Figure 4.3 by the labels $0n$. The eigenvalue $\lambda_0 = 1$, common for solutions with all N, is marked by 0.

The first branching point of the real solution $f_0(\xi)$ is the point $c = c_1$. At this point curve λ_{01} intersects the line $\lambda_0 = 1$ for the first time. Two solutions $f_1(\xi)$ with the degree $N = 1$ of the polynomials and complex conjugated parameters

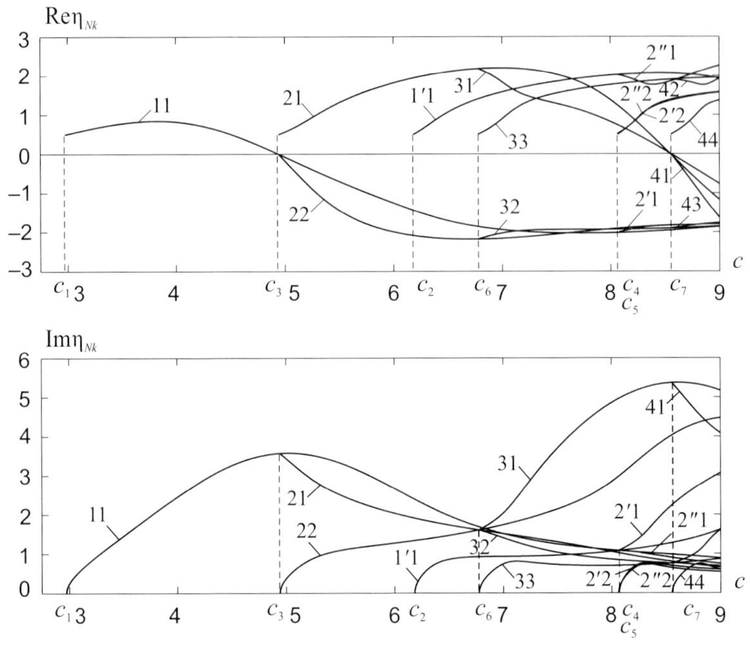

Figure 4.4 Real and imaginary parts of parameters η_{Nk}; $F(\xi) = 1/(2+\xi)$.

η_{11} arise (here and later, where this does not cause any misunderstanding, we use the index N in $f_N(\xi)$ to denote all solutions from the same equivalent group); at the branching point $c = c_1$ these parameters are real and coincide. At this point, the values of c_1 and real η_{11} are calculated from the transcendental equation system (4.61) having, in our case, the form

$$\int_{-1}^{1} \frac{F(\xi)\cos(c\xi)\mathrm{d}\xi}{(1-\eta_{11}\xi)} = 0\,, \tag{4.107a}$$

$$\int_{-1}^{1} \frac{F(\xi)\sin(c\xi)\mathrm{d}\xi}{(1-\eta_{11}\xi)} = 0\,. \tag{4.107b}$$

The point $c = c_1$ of intersection of the curve λ_{01} with $\lambda = 1$ in Figure 4.3, together with appropriate value of η_{11} can be chosen as initial approximations for solving this system. For this reason, c_1 may be calculated with small accuracy at this stage (this relates to all values $c = c_j$ given in Figure 4.3). The point $c = c_1$ is the first solution to (4.107).

The first eigenvalue λ_{11} of equation (4.95), corresponding to the solution $f_1(\xi)$ starts from $\lambda_0 = 1$ at $c = c_1$. This means that c_1 can be interpreted as a point at which the solution $f_0(\xi)$ branches off from $f_1(\xi)$ as c decreases.

Another eigenvalue λ_{02}, related to solution $f_0(\xi)$ with the polynomial $P_0 = 1$ (curve 02 in Figure 4.3) intersects the line $\lambda_0 = 1$ at the point $c = c_2$. This value belongs to the second solution of system (4.107). The second pair of complex-conjugated solutions with the polynomial $P_1(\xi)$ (curve $1'1$ in Figure 4.4) branches off from $f_0(\xi)$ at $c = c_2$; we denote it by $f_{1'}(\xi)$.

Since the real solution $f_0(\xi)$ has no other branching points in the given range of c, then, according to the chosen strategy, we begin to investigation of the first solution $f_1(\xi)$, which has branched off from $f_0(\xi)$ at $c = c_1$.

At the point $c = c_3$ where curve λ_{11} intersects the line $\lambda_0 = 1$, two solutions with $N = 2$ arise; we denote them by $f_2(\xi)$. Each of them has a pair complex parameters η_{21} and η_{22}. Each of these pairs are calculated separately (curves 21 and 22 in Figure 4.4). At the branching point, $\eta_{21} = \eta_{11}$ are complex and η_{22} is real; the values of these parameters are the same for both solutions. The values η_{11}, η_{22} and c_3 are determined from the transcendental equation system (4.74) (two equations at $N = 1$) complemented by additional equations (4.97). As a result, we have the following system of four equations:

$$\int_{-1}^{1} \frac{\sin(c\xi)F(\xi)}{|1-\eta_{11}\xi|}\mathrm{d}\xi = 0\,, \tag{4.108a}$$

$$\int_{-1}^{1} \frac{\cos(c\xi)F(\xi)}{|1-\eta_{11}\xi|}\mathrm{d}\xi = 0\,, \tag{4.108b}$$

$$\int_{-1}^{1} \frac{\xi\sin(c\xi)F(\xi)}{|1-\eta_{11}\xi|(1-\eta_{22}\xi)}\mathrm{d}\xi = 0\,, \tag{4.108c}$$

$$\int_{-1}^{1} \frac{\xi\cos(c\xi)F(\xi)}{|1-\eta_{11}\xi|(1-\eta_{22}\xi)}\mathrm{d}\xi = 0\,. \tag{4.108d}$$

The next eigenvalue λ_{12} relating to $f_1(\xi)$ intersects the line $\lambda = 1$ at $c = c_4$. At this point, the second pair of complex-conjugated solutions with the polynomial $N = 2$ branch off from $f_1(\xi)$; we denote one of them by $f_{2'}(\xi)$ (see curves $2'1$, $2'2$ in Figure 4.4). The value $c = c_4$ belongs to the second solution of the system (4.108).

The solution $f_1(\xi)$ has no more branching points, and we pass to the solution $f_{1'}(\xi)$ – the second solution with $N = 1$, which has branched off from $f_0(\xi)$ at $c = c_2$.

At $c = c_5$, the eigenvalue $\lambda_{1'1}$ corresponding to the solution $f_{1'}(\xi)$ intersects the line $\lambda = 1$. At this point two solutions with $N = 2$ arise once again; we denote one of them by $f_{2''}(\xi)$ (see curves $2''1$, $2''2$ in Figure 4.4). The values c_5 and $\eta_{2'1}, \eta_{2'2}$ are determined from the equation system analogous to (4.108).

Consider the solution $f_2(\xi)$, which has branched off from $f_1(\xi)$ at $c = c_3$. The eigenvalue λ_{21} of (4.95), corresponding to the solution $f_2(\xi)$, intersects the line $\lambda_0 = 1$ at $c = c_6$. Two solutions with $N = 3$ (we denote them as $f_3(\xi)$) arise at this point. Each of them has three complex parameters $\eta_{31}, \eta_{32}, \eta_{33}$ (curves 31, 32, 33 in Figure 4.4). At the branching point, $\eta_{31} = \eta_{21}$, $\eta_{32} = \eta_{22}$, and η_{33} is real. The parameters $\eta_{21}, \eta_{22}, \eta_{33}$, together with c_6 are determined from system (4.74) written for $N = 2$ and complemented by the additional equations (4.97). All they together make up the following system:

$$\int_{-1}^{1} \frac{\xi^{n-1}\sin(c\xi)F(\xi)}{|(1-\eta_{21}\xi)(1-\eta_{22}\xi)|}\mathrm{d}\xi = 0\,, \qquad n = 1, 2\,; \tag{4.109a}$$

$$\int_{-1}^{1} \frac{\xi^{n-1}\cos(c\xi)F(\xi)}{|(1-\eta_{21}\xi)(1-\eta_{22}\xi)|}\mathrm{d}\xi = 0\,, \qquad n = 1, 2\,; \tag{4.109b}$$

$$\int_{-1}^{1} \frac{\xi^{2}\sin(c\xi)F(\xi)}{|(1-\eta_{21}\xi)(1-\eta_{22}\xi)|\,(1-\eta_{33}\xi)}\mathrm{d}\xi = 0\,, \tag{4.109c}$$

$$\int_{-1}^{1} \frac{\xi^{2}\cos(c\xi)F(\xi)}{|(1-\eta_{21}\xi)(1-\eta_{22}\xi)|\,(1-\eta_{33}\xi)}\mathrm{d}\xi = 0\,. \tag{4.109d}$$

The solution $f_3(\xi)$ branches for the fist time at the point $c = c_7$. At this point, the eigenvalue λ_{31} intersects $\lambda = 1$. Solutions with polynomial degree $N = 4$ branch off from $f_3(\xi)$ (we denote one of them by $f_4(\xi)$). Each of these has four complex parameters (curves 41, 42, 43, 44 in Figure 4.4). At the branching point, $\eta_{41} = \eta_{31}$, $\eta_{42} = \eta_{32}$, $\eta_{43} = \eta_{33}$, and η_{44} is real. The values of these parameters and c_4 are determined from system (4.74) complemented by equations (4.97); the entire system is

$$\int_{-1}^{1} \frac{\xi^{n-1}\sin(c\xi)F(\xi)}{|\,(1-\eta_{31}\xi)(1-\eta_{32}\xi)(1-\eta_{33}\xi)\,|}\mathrm{d}\xi = 0\,, n = 1, 2, 3\,; \tag{4.110a}$$

$$\int_{-1}^{1} \frac{\xi^{n-1}\cos(c\xi)F(\xi)}{|\,(1-\eta_{31}\xi)(1-\eta_{32}\xi)(1-\eta_{33}\xi)\,|}\mathrm{d}\xi = 0\,, n = 1, 2, 3\,; \tag{4.110b}$$

$$\int_{-1}^{1} \frac{\xi^{3}\sin(c\xi)F(\xi)}{|\,(1-\eta_{31}\xi)(1-\eta_{32}\xi)(1-\eta_{33}\xi)\,|(1-\eta_{44}\xi)}\mathrm{d}\xi = 0\,, \tag{4.110c}$$

$$\int_{-1}^{1} \frac{\xi^{3}\cos(c\xi)F(\xi)}{|\,(1-\eta_{31}\xi)(1-\eta_{32}\xi)(1-\eta_{33}\xi)\,|(1-\eta_{44}\xi)}\mathrm{d}\xi = 0\,. \tag{4.110d}$$

From the above analysis, it can be seen that the number of solutions to (4.70) grows quickly as the parameter c increases. The whole branching process for $c \leq 9$ is shown in Figure 4.5 in graph form. The curve labels coincide with polynomial

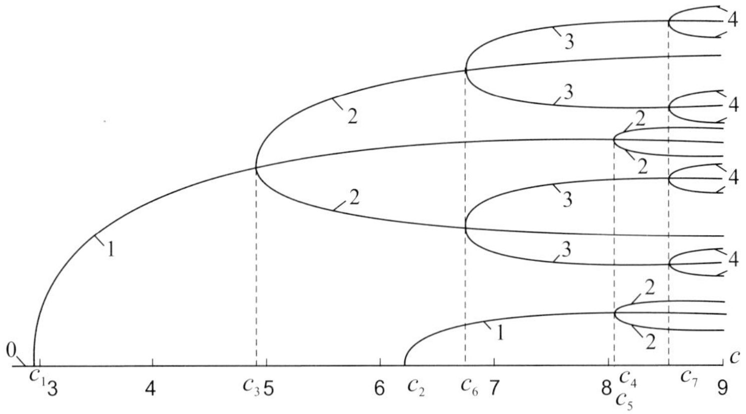

Figure 4.5 Graph of solutions to (4.70).

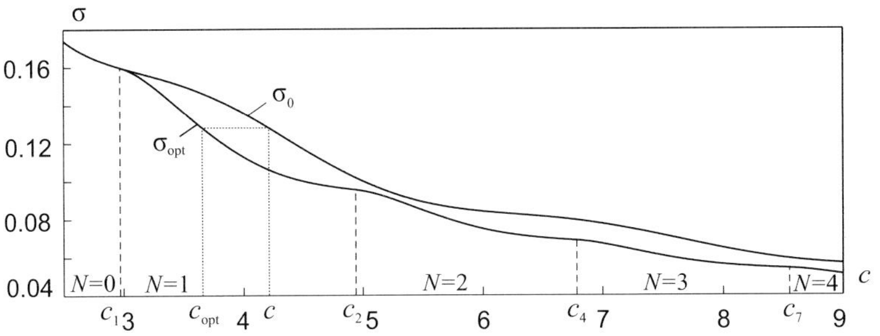

Figure 4.6 Comparison of values of the functional σ for different solutions to (4.70); $F(\xi) = 1/(2+\xi)$.

degree N. Only half of the solutions are shown here (to within the complex conjugate). So, at $c = 9$ the number of solutions is 41 (one real, 20 complex as shown in Figure 4.3, and the same number of conjugates to them).

Figure 4.6 demonstrates the comparison of the values of functional (3.69) obtained on the real solution $f_0(\xi)$ and on the optimal one $f_*(\xi)$ corresponding to the function u_* which provides the global minimum to the functional. The function $f_*(\xi)$ is compiled from functions $f_N(\xi)$ with different N on different segments of values c; these segments are separated by the dotted lines in the figure. The numerical results demonstrated here, as well as those obtained for many other functions $F(\xi)$ give a reason to assert that, at each fixed c, the optimal solutions $f_*(\xi)$ correspond to the largest possible value of $N = N_{\max}$ at this c. This fact is not yet justified theoretically.

The results given in Figure 4.6 allow us to calculate the efficiency factor $Q = (c - c_{\text{opt}})/c$ (see Figure 4.7), where c_{opt} and c are the values of c at the optimal and real solutions, respectively, corresponding to the same value of the functional $\sigma(u)$. This quantity shows how much the value of the parameter c can be decreased due to the free-phase formulation of the optimization problem in comparison with the case when $f(\xi)$ is found in the class of in-phase functions. If the antenna problem is considered, c is proportional to the antenna size and Q shows how much this size can be decreased (keeping the same value of σ) if the desired pattern is not given

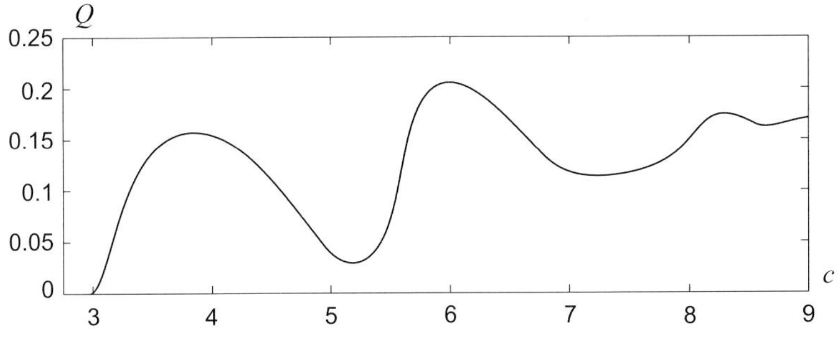

Figure 4.7 Efficiency of the free-phase choice; $F(\xi) = 1/(2+\xi)$.

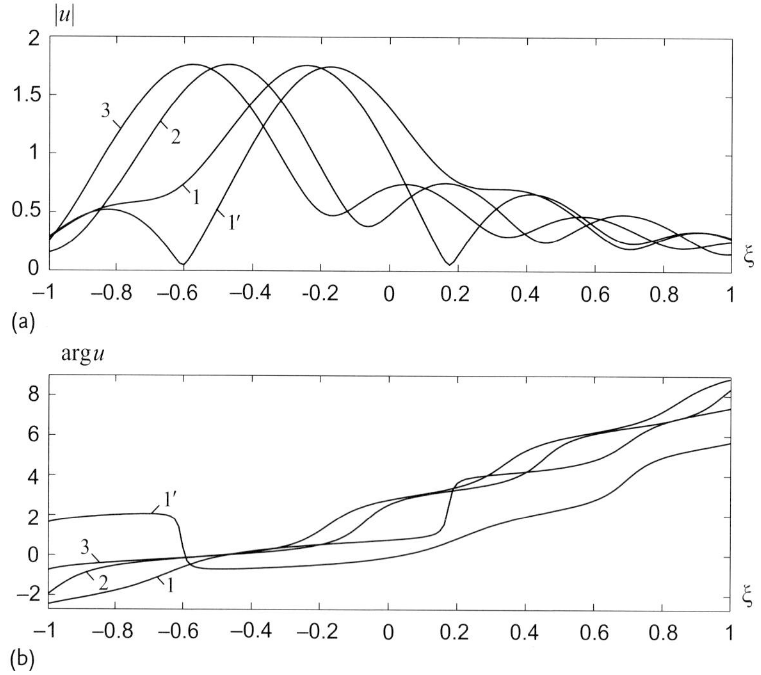

Figure 4.8 Comparison of solutions to (3.73) at $c = 7$; $F(\xi) = 1/(2 + \xi)$.

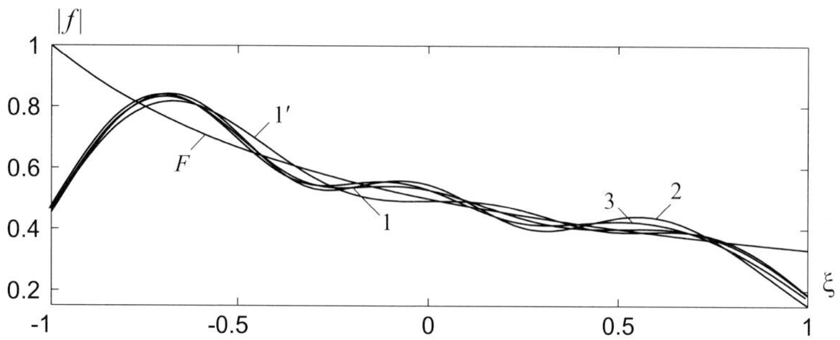

Figure 4.9 Comparison of solutions to (4.70) at $c = 7$; $F(\xi) = 1/(2 + \xi)$.

completely, but only its amplitude. It can be seen that the gain can reach about 20 percent.

Figures 4.8 and 4.9 show the behavior of the function $f(\xi)$ and, corresponding to it, $u(x)$ (see (3.73)) at $c = 7$. The labels on the curves correspond to the polynomial degree (similar to Figure 4.3).

4.3
A particular case: discrete Fourier transformation

The second particular case of general integral equation (4.1) is equation (3.80) which arises in the unconditional minimization problem for functional (3.76) where the isometric operator A describes the discrete Fourier transformation

$$f(\xi) = Au = \sqrt{\frac{\hat{c}}{2\pi}} \sum_{m=-M}^{M} u_m \exp(i\hat{c}m\xi) . \tag{4.111}$$

This operator acts from the T-dimensional ($T = 2M + 1$) space of complex vectors $u = \{u_m\}_{m=-M}^{M}$ into the space $L_2(-\pi/\hat{c}, \pi/\hat{c})$ of complex functions. Before establishing properties of the solutions to (3.80), we prove the following lemma.

Lemma 4.1

Any function $f(\xi)$ of the form (4.111) having no zeros in the interval $[-1, 1]$ can be expressed as

$$f(\xi) = \frac{\beta P_N(\tau) Q_{T-N-1}(\tau)}{(1+\tau^2)^M} , \tag{4.112}$$

where $|\beta| = 1$,

$$\tau = tg(\hat{c}\xi/2) , \tag{4.113}$$

$P_N(\tau)$ is a complex polynomial of degree N, containing all complex pairwise nonconjugated zeros of $f(\xi)$, $Q_{T-N-1}(\tau)$ is a polynomial of degree $T - N - 1$ ($T = 2M + 1$) with real coefficients, containing all real and pairwise conjugated complex zeros of $f(\xi)$.

Proof: It is easy to see that

$$\exp(\pm im\hat{c}\xi) = \frac{(1 \pm i\tau)^{2m}}{(1+\tau^2)^m} . \tag{4.114}$$

Substituting (4.114) into (4.111), we have

$$f(\xi) = \frac{1}{(1+\tau^2)^M} \left[u_0 + \sum_{m=1}^{M} (u_{-m}(1 - i\tau)^{2m} + u_m(1 + i\tau)^{2m})(1+\tau^2)^{M-m} \right] . \tag{4.115}$$

The square brackets in (4.115) contain a polynomial of τ of degree $2M$. In the general case it has the form

$$a_0 \prod_{n=1}^{2M} (1 - \eta_n \tau) , \tag{4.116}$$

where $a_0 = f(0)$, η_n^{-1} are zeros of this polynomial. Separating the factor $\beta = \exp(i \arg a_0)$ and dividing, in the obvious way, zeros of (4.116) into two groups; namely, collecting all complex pairwise nonconjugated zeros into the polynomial $P_N(\tau)$ and all real and complex pairwise conjugated ones into $Q_{T-N-1}(\tau)$ we obtain (4.112). □

Taking into account the obvious identity

$$\sin\frac{\hat{c}(\xi - \xi')}{2} = (\tau - \tau')\cos(\hat{c}\xi/2)\cos(\hat{c}\xi'/2)\,, \tag{4.117}$$

where $\tau = \tau(\xi)$ is defined by (4.113), we can rewrite the kernel in (3.80) in the form

$$K(\xi, \xi') = \frac{\hat{c}[\sin(T\hat{c}\xi/2)\cos(T\hat{c}\xi'/2) - \sin(T\hat{c}\xi'/2)\cos(T\hat{c}\xi/2)]}{2\pi[(\tau - \tau')\cos(\hat{c}\xi/2)\cos(\hat{c}\xi'/2)]}\,. \tag{4.118}$$

This expression coincides with (4.2) after introducing the notations

$$[a, b] = [-1, 1]\,; \quad s(\xi) = \frac{\sin(T\hat{c}\xi/2)}{\cos(\hat{c}\xi/2)}\,; \quad q(\xi) = \frac{\cos(T\hat{c}\xi/2)}{\cos(\hat{c}\xi/2)}\,. \tag{4.119}$$

4.3.1
Finite-parametric representation of solutions

Theorems 4.1 and 4.2 are reformulated for the case considered in the following forms.

Theorem 4.10

Let a function $f(\xi)$ having no zeros for $\xi \in [-1, 1]$ be expressed in the form

$$f(\xi) = \beta \hat{f}(\xi) P_N(\tau)\,, \tag{4.120}$$

with constant β, $|\beta| = 1$,

$$\hat{f}(\xi) = \frac{1}{|P_N(\tau)|}\left|\int_{-1}^{1} K(\xi, \xi')\frac{P_N(\tau')}{|P_N(\tau')|}\mathrm{d}\xi'\right| \tag{4.121}$$

$$P_N(\tau) = \prod_{k=1}^{N}(1 - \eta_{Nk}\tau)\,, \tag{4.122}$$

where η_{Nk} are subject to conditions (4.7), τ is defined by (4.113). For $f(\xi)$ to be a solution to (3.80), it is necessary and sufficient that the parameters η_{Nk} satisfy the following system of transcendental equations:

$$\Psi_{Nn} \equiv \int_{-1}^{1} \frac{\tau^{n-1}\sin(T\hat{c}\xi/2)F(\xi)}{\cos(\hat{c}\xi/2)|P_N(\tau)|}\mathrm{d}\xi = 0\,, \quad n = 1, 2, \ldots, N\,; \tag{4.123a}$$

$$\Psi_{Nn} \equiv \int_{-1}^{1} \frac{\tau^{n-1}\cos(T\hat{c}\xi/2)F(\xi)}{\cos(\hat{c}\xi/2)|P_N(\tau)|}\mathrm{d}\xi = 0\,, \quad n = 1, 2, \ldots, N\,. \tag{4.123b}$$

Theorem 4.11

The system of transcendental equations (4.123) can have solutions only when

$$N < \frac{T\hat{c}}{\pi} . \tag{4.124}$$

Theorems 4.3 and 4.4 remain the same, taking into account the concrete expressions (4.113) and (4.118) for $\tau(\xi)$ and $K(\xi, \xi')$, respectively. According to Lemma 4.1, the degree of $P_N(\tau)$ in (4.120) cannot exceed $T - 1$, and an analogy of Theorem 4.8 is not needed.

4.3.2
Branching of the solutions

According to Theorem 4.4, at the branching points $\hat{c} = \hat{c}_j$ the homogeneous integral equation

$$\begin{aligned} \lambda_n v_n(\xi) \int_{-1}^{1} & \frac{\sin(T\hat{c}(\xi - \xi')/2)}{\sin(\hat{c}(\xi - \xi')/2)|P_N(\tau)|} \left[P_N(\tau')\, \bar{P}_N(\tau)\right] \mathrm{d}\xi' \\ &= \int_{-1}^{1} \frac{\sin(T\hat{c}(\xi - \xi')/2)}{\sin(\hat{c}(\xi - \xi')/2)|P_N(\tau)|} \left[P_N(\tau')\, \bar{P}_N(\tau)\right] v_n(\xi')\, \mathrm{d}\xi' \end{aligned} \tag{4.125}$$

has multiple eigenvalues $\lambda_n = 1$.

In order to calculate the parameters $\hat{c}$ and η_{Nk}, $k = 1, 2, \ldots, N$ in the case when the solution branches, without changing the polynomial degree, it is necessary to complete system (4.123) by (4.53) using the definitions of Φ and Ψ given in (4.123).

At branching points where the polynomial $P_N(\tau)$ changes its degree by unity, the equalities

$$\eta_{Nk} = \eta_{N+1,k},\ k = 1, \ldots, N , \tag{4.126}$$

should hold, where $\eta_{N+1,k}$ are the parameters of the polynomial $P_{N+1}(\tau)$ in the branched solution. In addition to system (4.123), the two equations

$$\int_{-1}^{1} \frac{\tau^N \sin(T\hat{c}\xi/2) F(\xi)}{\cos(\hat{c}\xi/2)\, |P_N(\tau)|\, (1 - \eta_{N+1,N+1}\tau)} \mathrm{d}\xi = 0 , \tag{4.127a}$$

$$\int_{-1}^{1} \frac{\tau^N \cos(T\hat{c}\xi/2) F(\xi)}{\cos(\hat{c}\xi/2)\, |P_N(\tau)|\, (1 - \eta_{N+1,N+1}\tau)} \mathrm{d}\xi = 0 \tag{4.127b}$$

should also be fulfilled with real $\eta_{N+1,N+1}$.

At the branching points where N changes by two, the equalities

$$\eta_{Nk} = \eta_{N+2,k}, \qquad k = 1, 2, \ldots, N \tag{4.128}$$

should hold and the four additional equations:

$$\int_{-1}^{1} \frac{\tau^{n-1}\sin(T\hat{c}\xi/2)F(\xi)}{\cos(\hat{c}\xi/2)\,|P_N(\tau)|\,(1-\eta_{N+2,N+1}\tau)(1-\eta_{N+2,N+2}\tau)}\mathrm{d}\xi = 0\,; \quad n = N+1, N+2\,, \tag{4.129a}$$

$$\int_{-1}^{1} \frac{\tau^{n-1}\cos(T\hat{c}\xi/2)F(\xi)}{\cos(\hat{c}\xi/2)\,|P_N(\tau)|\,(1-\eta_{N+2,N+1}\tau)(1-\eta_{N+2,N+2}\tau)}\mathrm{d}\xi = 0\,; \quad n = N+1, N+2\,, \tag{4.129b}$$

should be fulfilled with parameters $\eta_{N+2,k}$, $k = N+1, N+2$ satisfying one of the conditions (4.100) or (4.101). As before, we have $2N+4$ equations with respect to $2N+3$ real unknowns. In the case when the function $F(\xi)$ is even, the solutions with even modulus can branch off, so that

$$|P_N(\tau)| = |P_N(-\tau)|\,, \qquad |P_{N+2}(\tau)| = |P_{N+2}(-\tau)|\,. \tag{4.130}$$

The situation is fully analogous to that considered in the preceding section. Under conditions (4.130), at the branching points where the polynomial degree changes by two, the following equation system:

$$\int_{-1}^{1} \frac{\tau^{2n-1}\sin(T\hat{c}\xi/2)F(\xi)}{\cos(\hat{c}\xi/2)\,|P_N(\tau)|}\mathrm{d}\xi = 0\,, \quad n = 1, 2, \ldots, [N/2]\,, \tag{4.131a}$$

$$\int_{-1}^{1} \frac{\tau^{2n-1}\cos(T\hat{c}\xi/2)F(\xi)}{\cos(\hat{c}\xi/2)\,|P_N(\tau)|}\mathrm{d}\xi = 0\,, \quad n = 1, 2, \ldots, [(N+1)/2]\,, \tag{4.131b}$$

$$\int_{-1}^{1} \frac{\tau^{2[(N+2)/2]-1}\sin(T\hat{c}\xi/2)F(\xi)}{\cos(\hat{c}\xi/2)\,|P_N(\tau)|\,(1-\eta_{N+2,N+1}\tau)(1-\eta_{N+2,N+2}\tau)}\mathrm{d}\xi = 0\,, \tag{4.131c}$$

$$\int_{-1}^{1} \frac{\tau^{2[(N+1)/2]}\cos(T\hat{c}\xi/2)F(\xi)}{\cos(\hat{c}\xi/2)\,|P_N(\tau)|\,(1-\eta_{N+2,N+1}\tau)(1-\eta_{N+2,N+2}\tau)}\mathrm{d}\xi = 0\,, \tag{4.131d}$$

should be satisfied, where the parameters η_{Nk}, $k = 1, 2, \ldots, N$, are either imaginary or appear as pairs with opposite signs, and $\eta_{N+2,k}$, $k = N+1, N+2$ are subject to conditions (4.104). Finally, we have $N+2$ real equations with respect to $N+2$ real unknowns. As before, the branching when changing the polynomial degree by more than two is not very probable.

4.3.3
Numerical results

4.3.3.1 The case of symmetrical data

We now analyze the numerical results obtained for (3.80) for a particularly important case of the given function and parameters. The results are represented in a way similar to that used for the case of the continuous Fourier transform (see Section 4.2.3).

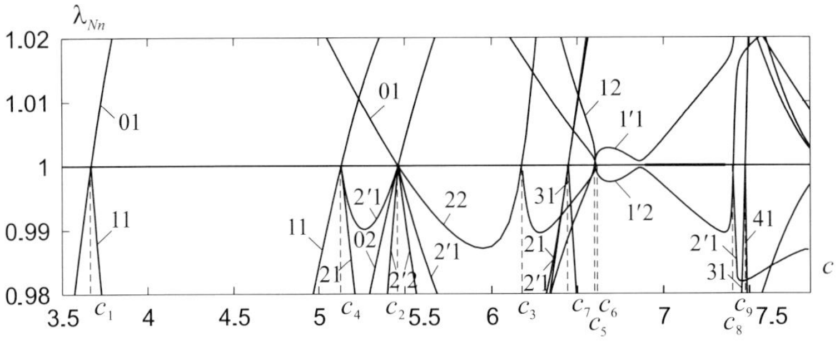

Figure 4.10 Eigenvalues of the homogeneous integral equation (4.125); $F(\xi) = \exp(-\xi^2)$.

In contrast to [56], here we consider another data for the discrete case, and extend the respective results from [56] shown there in only a fragmentary way, and analyze them much more carefully. Similar to [56] and [59], we consider the even function $F(\xi) = \exp(-\xi^2)$ and its asymmetrical perturbations $F(\xi) = \exp(-\xi^2) + \beta\xi$. The essential distinction of the discrete Fourier transform from the continuous one is the physical restriction $\hat{c} < \pi$ of the parameter $\hat{c}$.

Figure 4.10 shows the eigenvalues of the homogeneous equation (4.125) for $F(\xi) = \exp(-\xi^2)$, $T = 5$, versus the parameter $c = T\hat{c}/2$ for different possible values of degree of the polynomial $P_N(\tau)$. In this notation the above restriction is written as $c < T\pi/2$. The curves of the eigenvalues are similar to those for the case of the continuous Fourier transform (see Figure A.1 in [56]).

The parameters η_{Nk} of the polynomial $P_N(\tau)$ are determined from the transcendental equations system (4.123). As before, the system was solved by the Newton method. Calculations were provided by the first strategy. The results are represented in Figure 4.11 (as before, where this does not cause any misunderstanding we use the index N in $f_N(\xi)$ to denote all solutions from the same equivalent group).

At $N = 0$ the polynomial P_N in (4.120) has the form $P_0 = 1$ and the solution $f(\xi)$ becomes real; such a solution

$$f_0(\xi) = \frac{\hat{c}}{2\pi} \int_{-1}^{1} \frac{F(\xi')\sin(T\hat{c}(\xi - \xi')/2)}{\sin(\hat{c}(\xi - \xi')/2)} \mathrm{d}\xi' \tag{4.132}$$

exists at any $c < T\pi/2$ (which corresponds to $\hat{c} \in (0, \pi)$). The solution $f_0(\xi)$ has three branching points c_1, c_2, c_3.

At $c = c_1$, curve λ_{01} intersects the line $\lambda_0 = 1$ for the first time. Two solutions $f_1(\xi)$ with $N = 1$ and imaginary conjugated parameters η_{11} arise here; at the branching point $c = c_1$ these parameters are zero. The value of c_1 ($c_1 = T\hat{c}_1/2$) at this point is calculated from additional equations (4.127) which, in this case, become one equation:

$$\int_{-1}^{1} \frac{F(\xi)\cos(T\hat{c}\xi/2)}{\cos(\hat{c}\xi/2)} \mathrm{d}\xi = 0\,. \tag{4.133}$$

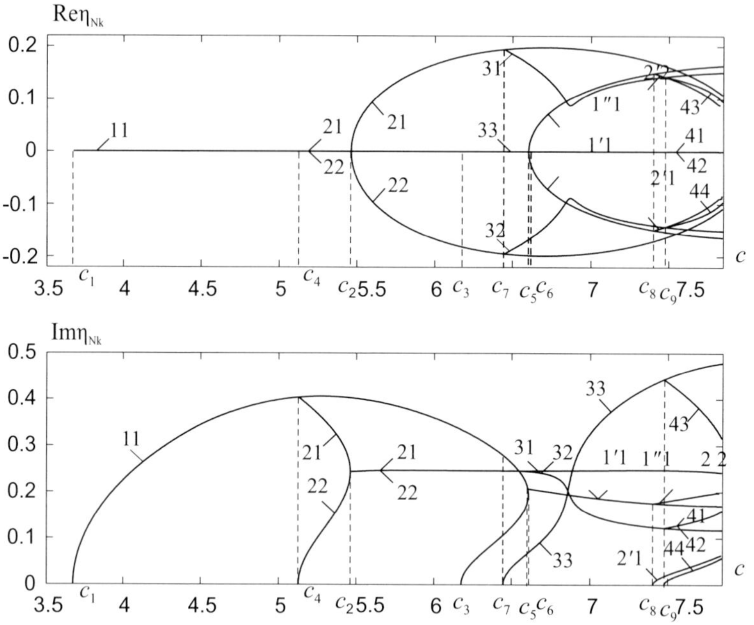

Figure 4.11 Real and imaginary parts of parameters η_{Nk}; $F(\xi) = \exp(-\xi^2)$.

The first eigenvalue λ_{11} of (4.125), corresponding to the solution $f_1(\xi)$ starts at $c = c_1$ from $\lambda_0 = 1$. This means that c_1 can be interpreted as a point at which the solution $f_0(\xi)$ branches off from $f_1(\xi)$ as c decreases.

At the point c_2 two eigenvalues λ_{01} and λ_{02} intersect the line $\lambda_0 = 1$ simultaneously. At this point the solution $f_0(\xi)$ branches; the polynomial degree is increased by two; we denote the new solutions by $f_2(\xi)$. Each of them has two complex parameters η_{21} and η_{22}. Each of these pairs are calculated separately (curves 21 in Figure 4.11). At the branching point the parameters $\eta_{21}$and η_{22} coincide and are imaginary. The values η_{21} and c_2 are determined from the transcendental equation system (4.131). Equation system (4.131) in this case becomes of the form:

$$\int_{-1}^{1} \frac{F(\xi)\cos(T\hat{c}\xi/2)d\xi}{\cos(\hat{c}\xi/2)(1-\eta_{21}^2\tau^2)} = 0\,, \tag{4.134a}$$

$$\int_{-1}^{1} \frac{\tau F(\xi)\sin(T\hat{c}\xi/2)d\xi}{\cos(\hat{c}\xi/2)(1-\eta_{21}^2\tau^2)} = 0\,. \tag{4.134b}$$

Note that this branching is double-sided, the off-branched solution exists for both $c < c_2$ and $c > c_2$. Formally, two new branches of solutions arise at this point, one at $c < c_2$ and the other at $c > c_2$. We will return to this point when analyzing other solutions with $N = 2$.

At $c > c_2$ the parameters η_{21}, η_{22} become complex with the same property $\eta_{21} = -\eta_{22}$ (curves 21 and 22 in Figure 4.11).

At the point $c = c_3$ the curve λ_{01} intersects the line $\lambda_0 = 1$ for a third time. This point is the second root of (4.133). However, any new solution does not branch off from $f_0(\xi)$ at this point; instead the solution $f_1(\xi)$ 'branches in' to $f_0(\xi)$ (see curve 11 in Figure 4.11). We will return to this point when analyzing solutions with $N = 1$.

The real solution $f_0(\xi)$ has no more branching points and we then investigate the first solution $f_1(\xi)$ which has branched off from it. At $c = c_4$ the curve λ_{11} intersects the line $\lambda_0 = 1$ for the first time. The solution $f_2(\xi)$ with $N = 2$ branches off from the solution $f_1(\xi)$. The system for the value c_4, imaginary parameters $\eta_{21} = \eta_{11}$, and η_{22} has the form:

$$\int_{-1}^{1} \frac{F(\xi)\cos(T\hat{c}_2\xi/2)\mathrm{d}\xi}{\cos(\hat{c}_2\xi/2)(1-\eta_{11}^2\tau^2)^{1/2}} = 0\,, \tag{4.135a}$$

$$\int_{-1}^{1} \frac{\tau F(\xi)\sin(T\hat{c}\xi/2)\mathrm{d}\xi}{\cos(\hat{c}\xi/2)(1-\eta_{11}^2\tau^2)^{1/2}(1-\eta_{22}\tau)} = 0\,, \tag{4.135b}$$

$$\int_{-1}^{1} \frac{\tau F(\xi)\cos(T\hat{c}\xi/2)\mathrm{d}\xi}{\cos(\hat{c}\xi/2)(1-\eta_{11}^2\tau^2)^{1/2}(1-\eta_{22}\tau)} = 0\,. \tag{4.135c}$$

Two solutions of this type (with different signs for η_{22}), branching off from $f_1(\xi)$ and $\bar{f}_1(\xi)$, make up (together with their complex-conjugates) the equivalent group of four solutions.

The equivalent group generated by this solution contains two appreciably different functions: $f_2(\xi)$ (with same signs as η_{21}, η_{22}) and $f_{2'}(\xi)$ (with opposite signs from η_{21}, η_{22}). The corresponding curves of the eigenvalues λ_{21}, $\lambda_{2'1}$ start at $c = c_4$ from $\lambda_0 = 1$, but their behavior is different. The eigenvalue λ_{21} varies smoothly up to the point c_7, where it intersects the line $\lambda = 1$ for a second time, whereas $\lambda_{2'1}$ returns to this line and only touches it at $c = c_2$. At this point the imaginary parameters η_{21}, η_{22} become the complex-conjugates and the solution becomes real, coinciding with $f_0(\xi)$. Therefore, the point c_2 can be treated either as the branching point of the solution $f_0(\xi)$ increasing the polynomial degree by two, or as the branching point of the solution $f_{2'}(\xi)$ decreasing the polynomial degree by two.

At the point $c = c_5$ the solution $f_1(\xi)$ branches without changing the polynomial degree and two new solutions $f_{1'}(\xi)$ and $f_{1''}(\xi)$ with $N = 1$ branch off from it. They have complex parameters $\bar{\eta}_{1'1} = -\eta_{1''1}$ (curves $1'1$, $1''1$ in Figure 4.11). The equation system for the branching point c_5 and imaginary (at this point) $\eta_{11} = \eta'_{11}$ is obtained from (4.53)

$$\int_{-1}^{1} \frac{F(\xi)\cos(T\hat{c}\xi/2)\mathrm{d}\xi}{\cos(\hat{c}\xi/2)(1-\eta_{11}^2\tau^2)^{1/2}} = 0\,, \tag{4.136a}$$

$$\eta_{11}\int_{-1}^{1} \frac{\tau F(\xi)\sin(T\hat{c}\xi/2)\mathrm{d}\xi}{\cos(\hat{c}\xi/2)(1-\eta_{11}^2\tau^2)^{3/2}}\int_{-1}^{1} \frac{\tau^2 F(\xi)\cos(T\hat{c}\xi/2)\mathrm{d}\xi}{\cos(\hat{c}\xi/2)(1-\eta_{11}^2\tau^2)^{3/2}} = 0\,. \tag{4.136b}$$

The point $c = c_5$ belongs to the first root of this system; here the first integral in (4.136b) becomes zero. In contrast to the case of the continuous Fourier transformation for the same given function $F(\xi) = const$ (see Figure A.2 in [56]), in the discrete transformation case the point c_5 does not coincide with c_4.

The next specific point is $c = c_6$ where the solutions $f_1(\xi)$ and $f_{1'}(\xi)$ coincide; the values η_{11} (equal to $\eta_{1'1}$) and c_6 are calculated from system (4.136); at this point the second integral in (4.136b) is zero. The point $(\eta_{11}; c_6)$ can be considered as an isolated point of disappearance of the solution, because there is no solution in its neighborhood at $c > c_6$. The distance between the points c_5, c_6 depends on the given function $F(\xi)$, if this function is varied, these points can move away or draw together, until they coincide.

The solution $f_1(\xi)$ has no more branching points and we can investigate the first solution $f_2(\xi)$, branching off from it at $c = c_4$. At the point $c = c_7$, the curve λ_{21}, corresponding to $f_2(\xi)$, intersects the line $\lambda = 1$; the two solutions with $N = 3$ branch off from each solution of the considered equivalent group (we denote one by $f_3(\xi)$). Each of the off-branched solutions is generated by a polynomial P_3 with two complex parameters η_{31}, $\eta_{32} = -\bar{\eta}_{31}$ and one imaginary parameter $\pm\eta_{33}$. At this point $\eta_{31} = \eta_{21}$, and the parameters η_{21}, η_{33} and $c = c_7$ are calculated from system (4.123) complemented by (4.127) in this case having the form:

$$\int_{-1}^{1} \frac{F(\xi)\cos(T\hat{c}\xi/2)\mathrm{d}\xi}{\cos(\hat{c}\xi/2)\left|1-\eta_{21}^2\tau^2\right|} = 0\,, \tag{4.137a}$$

$$\int_{-1}^{1} \frac{\tau F(\xi)\sin(T\hat{c}\xi/2)\mathrm{d}\xi}{\cos(\hat{c}\xi/2)\left|1-\eta_{21}^2\tau^2\right|} = 0\,, \tag{4.137b}$$

$$\int_{-1}^{1} \frac{\tau^2 F(\xi)\cos(T\hat{c}\xi/2)\mathrm{d}\xi}{\cos(\hat{c}\xi/2)\left|1-\eta_{21}^2\tau^2\right|(1-\eta_{33}\tau)} = 0\,, \tag{4.137c}$$

$$\int_{-1}^{1} \frac{\tau^2 F(\xi)\sin(T\hat{c}\xi/2)\mathrm{d}\xi}{\cos(\hat{c}\xi/2)\left|1-\eta_{21}^2\tau^2\right|(1-\eta_{33}\tau)} = 0\,. \tag{4.137d}$$

The point c_7 is the unique branching point of the solution $f_2(\xi)$. The next solution which must be investigated in the chosen strategy, is $f_{1'}(\xi)$, branched off from $f_1(\xi)$ at $c = c_5$. At $c = c_8$ the curve $\lambda_{1'1}$ corresponding to this solution intersects the line $\lambda_0 = 1$ for the first time. The solution $f_2(\xi)$ with $N = 2$ branches off from the solution $f_{1'}(\xi)$. The system for the value c_8, complex parameter $\eta_{2'1} = \eta_{1'1}$ and $\eta_{2'2}$ has the form:

$$\int_{-1}^{1} \frac{F(\xi)\cos(T\hat{c}_2\xi/2)\mathrm{d}\xi}{\cos(\hat{c}_2\xi/2)\left|1-\eta_{1'1}\tau\right|} = 0\,, \tag{4.138a}$$

$$\int_{-1}^{1} \frac{F(\xi)\sin(T\hat{c}_2\xi/2)\mathrm{d}\xi}{\cos(\hat{c}_2\xi/2)\left|1-\eta_{1'1}\tau\right|} = 0\,, \tag{4.138b}$$

$$\int_{-1}^{1} \frac{\tau F(\xi)\sin(T\hat{c}\xi/2)\mathrm{d}\xi}{\cos(\hat{c}\xi/2)\left|1-\eta_{1'1}\tau\right|(1-\eta_{2'2}\tau)} = 0\,, \tag{4.138c}$$

$$\int_{-1}^{1} \frac{\tau F(\xi)\cos(T\hat{c}\xi/2)\mathrm{d}\xi}{\cos(\hat{c}\xi/2)\left|1-\eta_{1'1}\tau\right|(1-\eta_{2'2}\tau)} = 0\,. \tag{4.138d}$$

Since $f_{1'}(\xi)$ has no more branching points, we consider the solution $f_3(\xi)$ branched off from $f_2(\xi)$ at $c = c_7$. At the point $c = c_9$ two solutions with $N = 4$ branch off from $f_3(\xi)$, each of these has two complex parameters η_{41}, $\eta_{42} = -\bar{\eta}_{41}$, and two imaginary ones η_{43}, $\pm\eta_{44}$. At the branching point we have

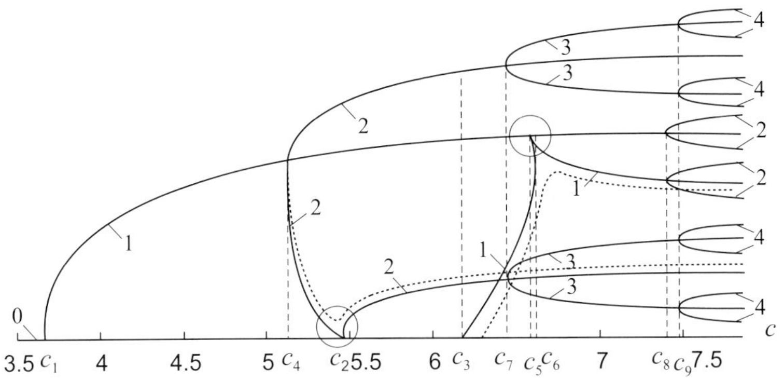

Figure 4.12 Graph of solutions to (3.80).

the two complex parameters $\eta_{41} = \eta_{31}$, $\eta_{42} = -\bar{\eta}_{31}$, one imaginary $\eta_{43} = \eta_{33}$, and one real η_{44}. The system for unknown η_{31}, η_{33}, η_{44}, and c_9 are obtained from the system (4.123) and additional equations (4.127). As a result, we have five real equations for five real parameters:

$$\int_{-1}^{1} \frac{F(\xi)\cos(T\hat{c}\xi/2)\mathrm{d}\xi}{\cos(\hat{c}\xi/2)\left|1-\eta_{31}^2\tau^2\right|(1-\eta_{33}^2\tau^2)} = 0\,, \tag{4.139a}$$

$$\int_{-1}^{1} \frac{\tau F(\xi)\sin(T\hat{c}\xi/2)\mathrm{d}\xi}{\cos(\hat{c}\xi/2)\left|1-\eta_{31}^2\tau^2\right|(1-\eta_{33}^2\tau^2)} = 0\,, \tag{4.139b}$$

$$\int_{-1}^{1} \frac{\tau^2 F(\xi)\cos(T\hat{c}\xi/2)\mathrm{d}\xi}{\cos(\hat{c}\xi/2)\left|1-\eta_{31}^2\tau^2\right|(1-\eta_{33}^2\tau^2)} = 0\,, \tag{4.139c}$$

$$\int_{-1}^{1} \frac{\tau^3 F(\xi)\cos(T\hat{c}\xi/2)}{\cos(\hat{c}\xi/2)\left|1-\eta_{31}^2\tau^2\right|(1-\eta_{33}^2\tau^2)(1-\eta_{44}\tau)}\mathrm{d}\xi = 0\,, \tag{4.139d}$$

$$\int_{-1}^{1} \frac{\tau^3 F(\xi)\sin(T\hat{c}\xi/2)}{\cos(\hat{c}\xi/2)\left|1-\eta_{31}^2\tau^2\right|(1-\eta_{33}^2\tau^2)(1-\eta_{44}\tau)}\mathrm{d}\xi = 0\,. \tag{4.139e}$$

The branching process for the problem considered is thus completed. It is illustrated by the solid line in Figure 4.12 as a graph. As before, the labels of the branches correspond to the polynomial degree N.

For comparison, in Figure 4.13 we give the values of the functional σ (3.76) obtained on the real solution f_0 (line σ_0) and on the complex ones f_N providing minimum σ (line σ_{opt}) for $F(\xi) = \exp(-\xi^2)$, $T = 5$. The segments with different values of optimal polynomial degree N are separated by vertical lines. As in the continuous case, the minimal values of σ are reached at maximal values of N permitted at current c.

The efficiency factor Q obtained on the optimal solutions of (3.80) is given in Figure 4.14. As in previous example (see Figure 4.7), this factor depends essentially on c. In particular, in a vicinity of $c \approx 5.5$ it is approximately zero, whereas at $c \approx 5.6$ it increases stepwise.

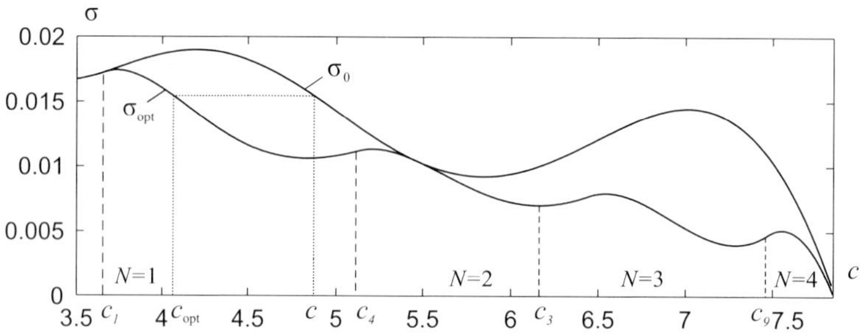

Figure 4.13 Comparison of values of the functional σ for different solutions to (3.80); $F(\xi) = \exp(-\xi^2)$, $T = 5$.

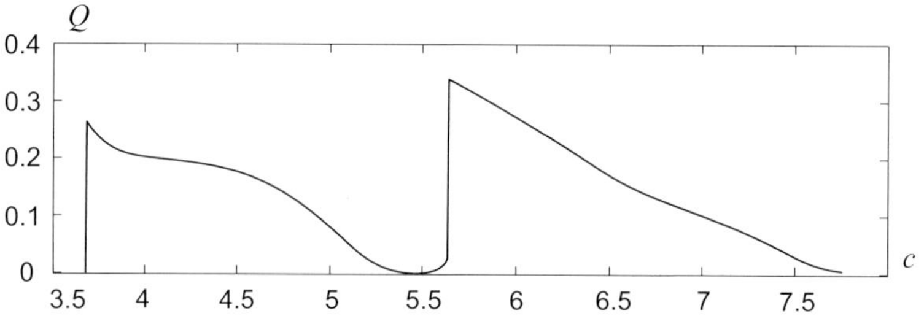

Figure 4.14 Efficiency of the free-phase choice; $F(\xi) = \exp(-\xi^2)$, $T = 5$.

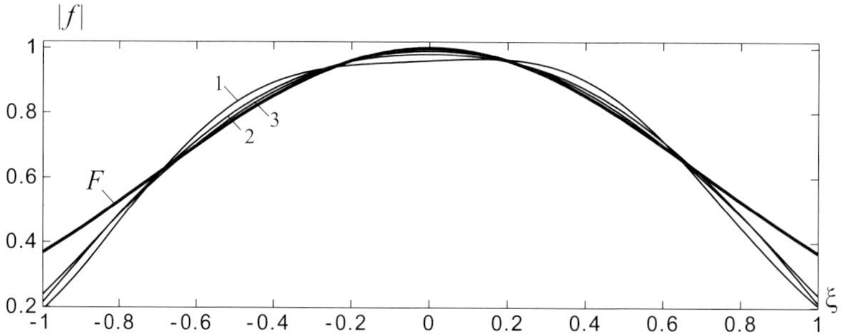

Figure 4.15 Comparison of solutions to (3.80) at $c = 7$; $F(\xi) = \exp(-\xi^2)$, $T = 5$.

As an example, the amplitude and phase distributions of different solutions at $c = 7$ are given in Figure 4.15. The labels on the curves correspond to the polynomial degree.

4.3.3.2 **Asymmetrical perturbed data**

The existence of the appearance and disappearance points observed in the case of even functions $F(\xi)$ causes us to analyze the behavior of the solutions to (3.80) with nonsymmetrical perturbations of the input data. We make such an analysis for the example of the given function $F(\xi) = \exp(-\xi^2) + \beta\xi$, for small β at $T = 5$.

In contrast to the symmetrical case, for nonsymmetrical $F(\xi)$, all parameters η_{Nk} for all $N \neq 0$ are complex (with nonzero real parts). At the branching points new parameters η_{Nk} in off-branched solutions start from certain real nonzero values. This difference is seen most clearly near the specific values of c, indicated in Figure 4.12 by small circles. The dotted curves in Figure 4.12 show deformations of the branching process at small asymmetrical perturbations. We consider some essential distinctions between the symmetrical and asymmetrical cases.

At the point c_2 in the symmetrical case two different solutions of the same equivalent group at $N = 2$, corresponding to imaginary parameters η_{2k}, have transformed into the complex ones (with nonzero real parts) in different ways. One of them had the coinciding parameters $\eta_{22} = \eta_{21}$ and those obtained the real parts with opposite signs. The second one had the conjugated ones $\eta_{22} = -\eta_{21}$ giving the real function $f_2(\xi)$ coinciding with $f_0(\xi)$. The point $c = c_2$ was the branching point with N changing by two. In the nonsymmetrical case the curves of the eigenvalues for $N = 2$ (which intersect at $\beta = 0$ (see Figure 4.12)) disconnect, they do not intersect the value $\lambda = 1$. The branching point $c = c_2$ does not exist.

A specific situation arises around the points c_3, c_5, c_6. Figure 4.16 shows the comparative behavior of the eigenvalues of (4.125) with $N = 0$ and $N = 1$ for the

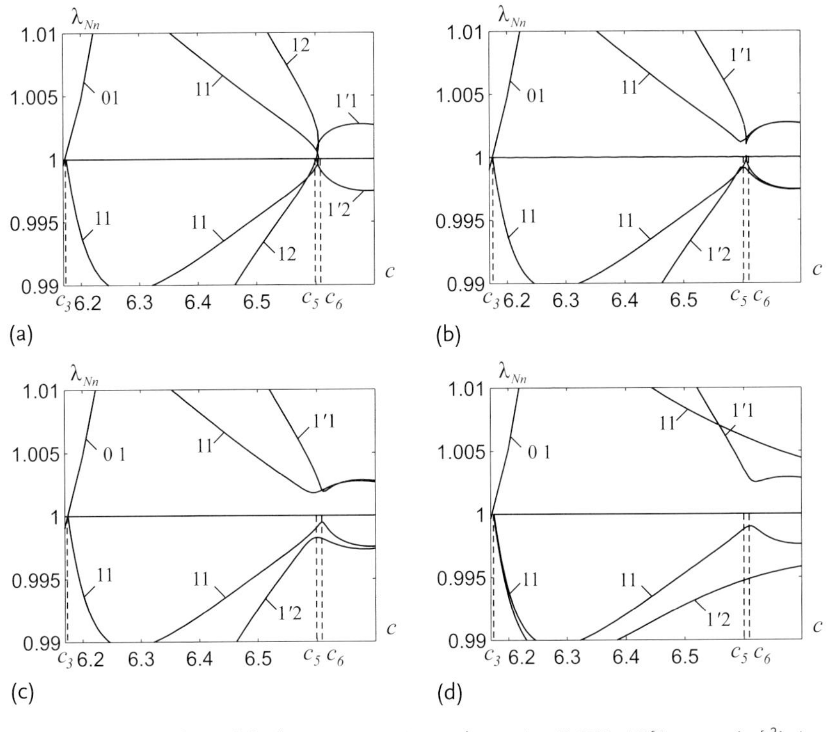

Figure 4.16 Eigenvalues of the homogeneous integral equation (4.125); $F(\xi) = \exp(-\xi^2) + \beta\xi$, $T = 5$.

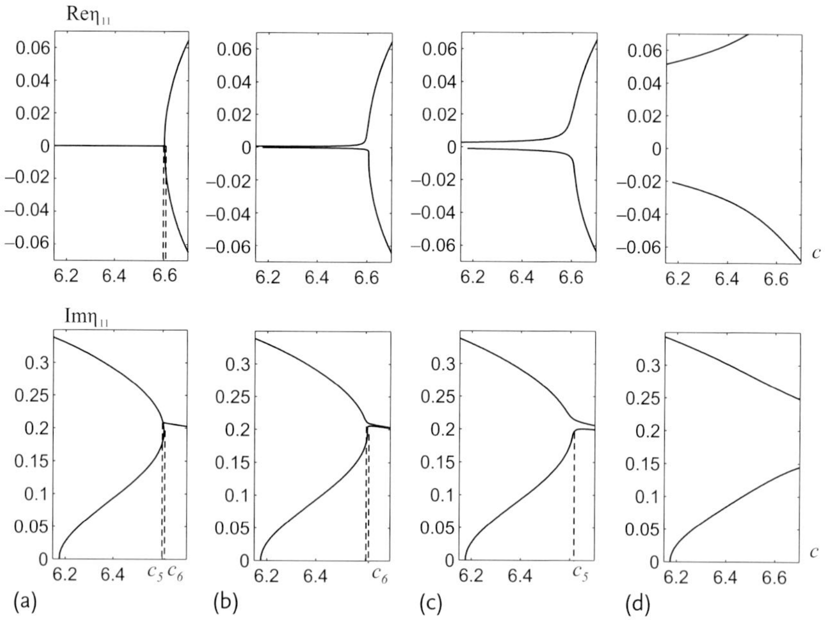

Figure 4.17 Real and imaginas parts of parametrs η_{Nk}; $F(\xi) = \exp(-\xi^2) + \beta\xi$, $T = 5$.

symmetrical ($\beta = 0$; Figure 4.16a) and nonsymmetrical ($\beta \neq 0$; Figure 4.16b–d) cases. Corresponding dependencies of the parameter η_{11} on c in different solutions are given in Figure 4.17a–d. The values: $\beta = 0$ (a); 0.005 (b); 0.01 (c); 0.1 (d) are chosen such that all typical stages of solution transformation are shown. Since the topology of the eigenvalue curves changes during this transformation, we must change the indication of some curves in comparison with Figures 4.10 and 4.11.

In the case of a symmetrical function $F(\xi)$ ($\beta = 0$) we had two branching points $c = c_3$, $c = c_5$, of the solutions $f_0(\xi)$, $f_1(\xi)$ and the disappearance of points for the solution $f_1(\xi)$: $c = c_6$. At small perturbations of the given function $F(\xi)$ the curves of the eigenvalues for different solutions with the same N disconnect at the point of their multiplicity, making acute angles (see Figures 4.16a–d). The point c_3 remains the branching point. Here the solution $f_{1'}(\xi)$ branches off from $f_0(\xi)$. For very small β and $c_5 < c < c_6$ there are four solutions with $N = 1$. At $\beta \approx 0.01$ the points c_5, c_6 coincide and only two solutions remain. This effect is more demonstrative for other $F(\xi)$ (see Section A.2.5 in [56]).

4.4
A particular case: the Hankel transform

The third particular case of the general integral equation (4.1) is (3.89) which arises in the unconditional minimization problem for functional (3.6) with the isometric operator describing the Hankel transform. The following notations are applied:

$$a = 0, \quad b = 1, \quad s(\xi) = \xi J_1(c\xi), \quad q(\xi) = J_0(c\xi), \quad \tau(\xi) = \xi^2 \tag{4.140}$$

to reduce (3.89) to the form (4.1). For convenience, we rewrite this equation

$$f(\xi) = c \int_0^1 \frac{\xi J_0(c\xi') J_1(c\xi) - \xi' J_0(c\xi) J_1(c\xi')}{\xi^2 - (\xi')^2} \cdot F(\xi') \exp[i \arg f(\xi')] d\xi' . \tag{4.141}$$

Equation (4.141) is obtained in the problem of minimization of the functional

$$\sigma(u) = \int_0^{2\pi} \int_0^1 [F(\xi,\varphi) - |f(\xi,\varphi)|]^2 \xi d\xi d\varphi + \int_0^{2\pi} \int_1^{\infty} |f(\xi,\varphi)|^2 \xi d\xi d\varphi , \tag{4.142}$$

where

$$F(\xi,\varphi) \geq 0 \tag{4.143}$$

is a given function. The Lagrange–Euler equation for this functional has the form:

$$f(\xi,\varphi) = c \int_0^{2\pi} \int_0^1 \frac{J_1(c\,|\vec{\xi} - \vec{\xi}'|)}{|\vec{\xi} - \vec{\xi}'|} F(\xi',\varphi') \exp(i \arg f(\xi',\varphi'))\xi' d\xi' d\varphi' , \tag{4.144}$$

where

$$\left|\vec{\xi} - \vec{\xi}'\right| = \sqrt{\xi^2 + \xi'^2 - 2\xi\xi' \cos(\varphi - \varphi')} . \tag{4.145}$$

In the case when the given function F does not depend on φ and we are interested in the same type of solution $f(\xi)$, then (4.144) adopts the form (4.141). The equation (4.144) arises in Problem **B** for the transmitting line with circular apertures. Then (4.141) describes the solutions $f(\xi,\varphi)$ to (4.144), which do not depend on the angular coordinate φ. The solutions depending on φ, can be found straightforwardly from (4.144) by the iterative method described in Section 3.1.5.3.

Solutions to (4.141) are found in the form (4.4), and Theorems 4.1–4.4 apply here. We reformulate two of them, which gives the set of transcendental equations for determining the parameters η_{Nk} and the limitation of the polynomial degree; the last two remain without modification.

Theorem 4.12

Let a function $f(\xi)$ having no zeros at $\xi \in [0,1]$ be expressed in the form (4.4)–(4.6) with conditions (4.7). For $f(\xi)$ to be a solution to (4.141), it is necessary and

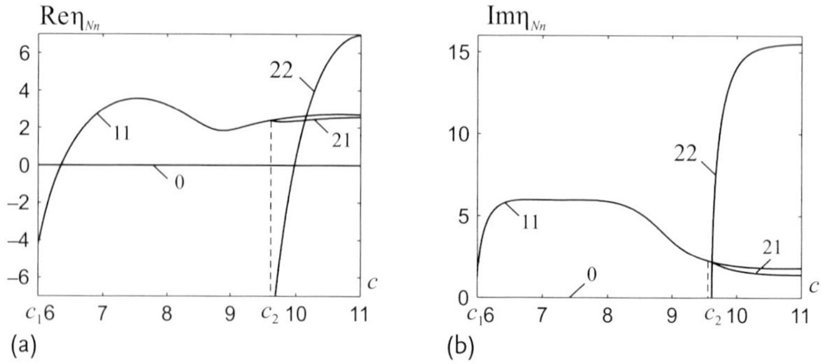

Figure 4.18 Real and imaginary parts of parametrs η_{Nk}; $F(\xi) = const$.

sufficient that the parameters η_{Nk} satisfy the following system of transcendental equations:

$$\int_0^1 \frac{(\xi^2)^{n-1}\xi J_1(c\xi)F(\xi)}{|P_N(\tau)|}\mathrm{d}\xi = 0\,, n = 1, 2, \ldots, N\,; \tag{4.146a}$$

$$\int_0^1 \frac{(\xi^2)^{n-1} J_0(c\xi)F(\xi)}{|P_N(\tau)|}\mathrm{d}\xi = 0\,, \quad n = 1, 2, \ldots, N\,. \tag{4.146b}$$

Theorem 4.13

Under the conditions of Theorem 4.12 function (4.4) can solve (4.141) only if the degree N of the generating polynomial $P_N(\tau)$ does not exceed the number of zeros of the functions $J_0(c\xi)$ in the interval $\xi \in (0, 1)$.

The function $\xi J_1(c\xi)$ does not participate in this theorem because it has more zeros than $J_0(c\xi)$.

The investigation of the branching process for this case is analogous to the previous ones and is not given here.

Below, the numerical results are presented for the case $F(\xi) = const$. The real and imaginary parts of the parameters η_{Nk} are shown in Figure 4.18. Figure 4.19 demonstrates the values of functional (3.84) versus c for the trivial solution $f_0(\xi)$ (dashed line) and for the solutions $f_N(\xi)$, which provide a minimum to the functional (solid lines). The curves are labeled by the number N, denoting the polynomial degree.

Branching with a change in the polynomial degree by a unit is observed at $c = c_1$. The solution $f_1(\xi)$ branches off from $f_0(\xi)$ at this point. At $N = 0$ system (4.146) is absent and only two additional equations (4.60) for the two real unknowns η_{11} and c_1 remain. In our case these equations are

$$\int_0^1 \frac{\xi J_1(c\xi)F(\xi)}{(1 - \eta_{11}\tau)}\mathrm{d}\xi = 0\,, \tag{4.147a}$$

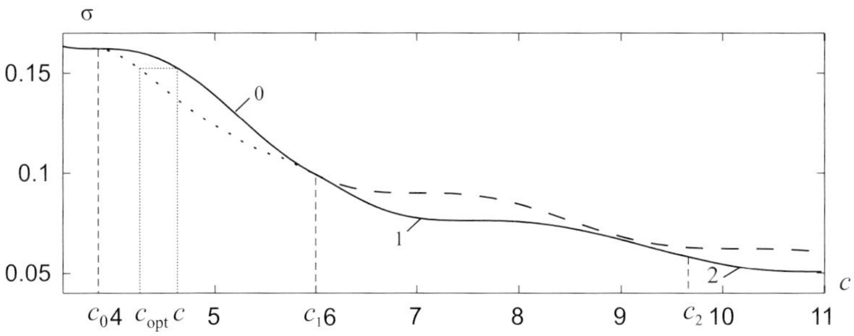

Figure 4.19 Comparison of values of the functional σ for different solutions to (3.80); $F(\xi) = const.$

$$\int_0^1 \frac{J_0(c\xi)F(\xi)}{(1-\eta_{11}\tau)}\mathrm{d}\xi = 0\,. \tag{4.147b}$$

At $c = c_2$ two solutions with $N = 2$ branch off from $f_1(\xi)$. The equation system

$$\int_{-1}^1 \frac{F(\xi)\xi J_1(c\xi)\mathrm{d}\xi}{|1-\eta_{11}\tau|} = 0\,, \tag{4.148a}$$

$$\int_{-1}^1 \frac{F(\xi)J_0(c\xi)\mathrm{d}\xi}{|1-\eta_{11}\tau|} = 0\,, \tag{4.148b}$$

$$\int_{-1}^1 \frac{\tau F(\xi)\xi J_0(c\xi)\mathrm{d}\xi}{|1-\eta_{11}\tau|\,(1-\eta_{22}\tau)} = 0\,, \tag{4.148c}$$

$$\int_{-1}^1 \frac{\tau F(\xi)J_0(c\xi)\mathrm{d}\xi}{|1-\eta_{11}\tau|\,(1-\eta_{22}\tau)} = 0\,, \tag{4.148d}$$

for the complex η_{11}, real η_{22} and $c = c_2$ is obtained from system (4.146) and additional equations (4.60).

The first of these branches off from $f_0(\xi)$ at $c = c_0$. It has an odd phase with respect to φ: arg $f(\xi,-\varphi) = -$ arg $f(\xi,-\varphi)$. The value of σ (3.84) for this solution is shown in Figure 4.19 by the dotted line. As was expected, it is smaller than that reached for $f_0(\xi)$. Similar to the one-dimensional problem (4.141), the solution with the odd phase appears earlier than the symmetrical one and it is optimal in the segment $c_0 < c < c_1$. The optimal distribution of the field $u(r,\varphi)$ on the antenna for this solution is shown by its isolines in Figure 4.20 for $c = 4.5$. The isolines corresponding to the symmetrical solution f_0 are given by the dotted lines.

The quantity $Q = (c - c_{opt})/c$ introduced in Section 4.2.3, is shown in Figure 4.21. This quantity characterizes the gain obtained by using the free phase in the desired field F.

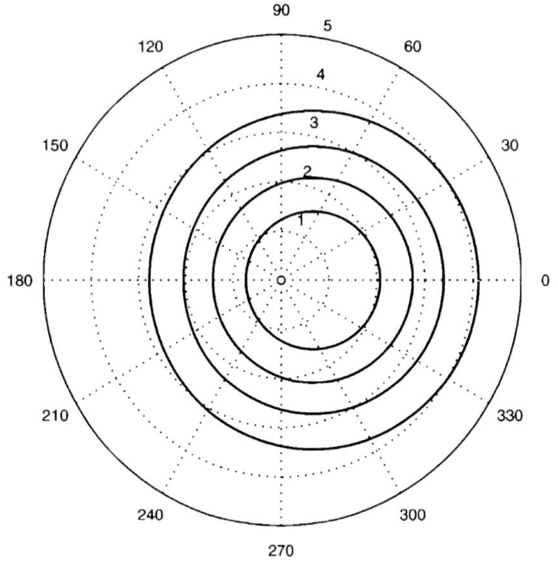

Figure 4.20 Isolines of optimal $u(r, \varphi)$ for asymmetrical solutions to (4.144).

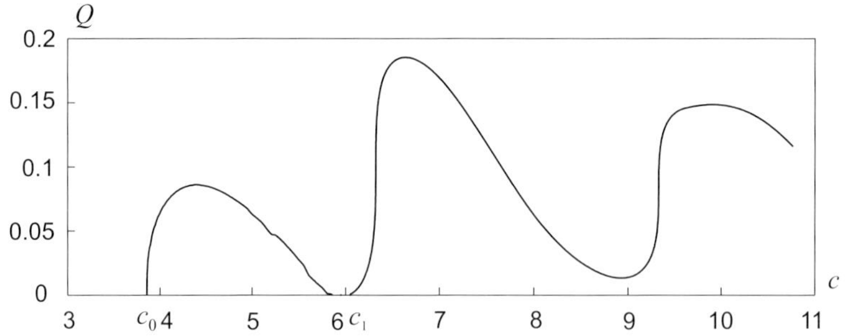

Figure 4.21 Efficiency of the free-phase choice; $F(\xi) = const$.

4.5
Generalized nonlinear integral equation for the compact operator

A generalized equation analogous to (4.1) can also be constructed for the case considered in Section 3.1.8, when the operator $B = AA^*$ in the equation

$$\alpha f + Bf = B[F \exp(i \arg f)] \tag{4.149}$$

is compact. Here $\alpha > 0$ and $F(\xi) \geq 0$ are a given constant and function, respectively. This generalized equation has the form:

$$\alpha f(\xi) + \int_a^b K(\xi, \xi', c) f(\xi') d\xi' = \int_a^b K(\xi, \xi', c) F(\xi') \exp(i \arg f(\xi')) d\xi', \tag{4.150}$$

with the same notation as (4.2) and properties of the functions the same as in the previous section. A more general equation is considered in [99].

4.5.1 Finite-parametric representation of solutions

Similar to the solutions for (4.1), we will find the solutions to (4.150) having no zeros at $\xi \in [a, b]$ in the same form as (4.4):

$$f(\xi) = \beta \hat{f}(\xi) P_N(\tau) . \tag{4.151}$$

As before, $\hat{f}(\xi)$ is a real positive function having no zeros at $\xi \in [a, b]$; β is a complex constant with $|\beta| = 1$;

$$P_N(\tau) = \prod_{k=1}^{N} (1 - \eta_{Nk}\tau) \tag{4.152}$$

is a polynomial of finite degree N with complex nonconjugated pairwise zeros η_{Nk}^{-1}:

$$\eta_{Nk} - \bar{\eta}_{Nm} \neq 0, \quad k, m = 1, 2, \ldots, N . \tag{4.153}$$

From (4.151) it follows that

$$\exp(i \arg f(\xi)) = \frac{P_N(\tau)}{|P_N(\tau)|} . \tag{4.154}$$

Theorem 4.14

A function $f(\xi)$ of form (4.4) having no zeros at $\xi \in [a, b]$ is a solution to (4.150) if and only if the real function $|f(\xi)|$ and complex parameters η_{Nk} with condition (4.153) satisfy the following integral equation:

$$\alpha |f(\xi)| + \frac{1}{|P_N(\tau)|} \int_a^b |f(\xi')| K(\xi, \xi') \frac{\mathrm{Re}\left[\bar{P}_N(\tau) P_N(\tau')\right]}{|P_N(\tau')|} \mathrm{d}\xi'$$
$$= \frac{1}{|P_N(\tau)|} \int_a^b F(\xi') K(\xi, \xi') \frac{\mathrm{Re}\left[\bar{P}_N(\tau) P_N(\tau')\right]}{|P_N(\tau')|} \mathrm{d}\xi' \tag{4.155}$$

and the system of transcendental equations:

$$\Phi_{Nn}(\eta_{N1}, \eta_{N2,\ldots} \eta_{NN}) = 0 , \quad n = 1, 2, \ldots, N , \tag{4.156a}$$

$$\Psi_{Nn}(\eta_{N1}, \eta_{N2,\ldots} \eta_{NN}) = 0 , \quad n = 1, 2, \ldots, N , \tag{4.156b}$$

where

$$\Phi_{Nn} = \int_a^b \tau^{k-1} s(\xi) \frac{F(\xi) - |f(\xi)|}{|P_N(\tau)|} \mathrm{d}\xi , \tag{4.157a}$$

$$\Psi_{Nn} = \int_a^b \tau^{k-1} q(\xi) \frac{F(\xi) - |f(\xi)|}{|P_N(\tau)|} \mathrm{d}\xi . \tag{4.157b}$$

Proof: *Necessity*: Let function (4.151) be a solution to (4.150). It can be seen that we can put $\beta = 1$ without loss of generality. Substituting (4.151) and (4.154) into (4.150), multiplying both its left- and right-hand sides by $\bar{P}_N(\xi)$, and taking the real part of the result, we obtain

$$\alpha \hat{f}(\xi)\,|P_N(\tau)|^2 + \int_a^b \hat{f}(\xi')K(\xi,\xi')\,\mathrm{Re}\left[\bar{P}_N(\tau)P_N(\tau')\right]\mathrm{d}\xi' = \int_a^b F(\xi')K(\xi,\xi')\frac{\mathrm{Re}\left[\bar{P}_N(\tau)P_N(\tau')\right]}{|P_N(\tau')|}\mathrm{d}\xi' . \quad (4.158)$$

This equation takes the form (4.155) after dividing both sides by $|P_N(\tau)|$ and substituting

$$\hat{f}(\xi)\,|P_N(\tau)| = |f(\xi)| . \quad (4.159)$$

On the other hand, after taking the imaginary part of the same result, we have

$$\int_a^b K(\xi,\xi')\hat{f}(\xi')R_{N-1}(\tau,\tau')\mathrm{d}\xi' = \int_a^b K(\xi,\xi')\frac{F(\xi')}{|P_N(\tau')|}R_{N-1}(\tau,\tau')\mathrm{d}\xi' , \quad (4.160)$$

where

$$R_{N-1}(\tau,\tau') = \frac{\mathrm{Im}\left[P_N(\tau')\bar{P}_N(\tau)\right]}{\tau-\tau'} = \sum_{k,m=1}^{N} d_{km}\tau^{k-1}(\tau')^{m-1} \quad (4.161)$$

is a symmetrical polynomial of two variables with the matrix of coefficients $D = \{d_{km}\}$. Writing (4.160) in the form

$$\int_a^b K(\xi,\xi')\frac{F(\xi')-|f(\xi')|}{|P_N(\tau')|}R_{N-1}(\tau,\tau')\mathrm{d}\xi' \equiv 0 , \quad (4.162)$$

and using (4.161), we have

$$\sum_{k,m=1}^{N} d_{km}\left[q(\xi')\int_a^b \tau^{k-1}s(\xi)\frac{F(\xi')-|f(\xi')|}{|P_N(\tau')|}\mathrm{d}\xi - s(\xi')\int_a^b \tau^{k-1}q(\xi)\frac{F(\xi')-|f(\xi')|}{|P_N(\tau')|}\mathrm{d}\xi\right](\tau')^{m-1} \equiv 0 . \quad (4.163)$$

Due to the linear independence of the functions $\{\tau^k s(\xi)\}$ and $\{\tau^k q(\xi)\}$ at $k = 0, 1, \ldots, N-1$, identity (4.163) leads to

$$\sum_{k=1}^{N} d_{km}\Phi_{Nn} = 0 , \quad n = 1, 2, \ldots, N , \quad (4.164a)$$

$$\sum_{k=1}^{N} d_{km}\Psi_{Nn} = 0 , \quad n = 1, 2, \ldots, N , \quad (4.164b)$$

where Φ_{Nn}, Ψ_{Nn} are defined in (4.157). If the coefficients d_{km} are to be found, these equalities could be considered as a collection of two independent systems of linear algebraic equations with respect to unknowns Φ_{Nn} and Ψ_{Nn}, respectively, with the same matrix D. Since, according to (4.10), $\det D \neq 0$, these systems can have only zero solutions, which means that the transcendental equation system (4.156) holds and the proof is completed.

Sufficiency. Let the integral equation (4.155), together with the transcendental equation system (4.156) be satisfied at certain integer N and certain complex η_{Nk}, $k = 1, 2, \dots, N$, under condition (4.153). We want to show that function (4.151) is a solution to (4.150).

It follows from (4.156) that equalities (4.164) are fulfilled and, after using (4.157), identities (4.163) and (4.162) also hold. With the aid of (4.161), we obtain from (4.162):

$$\operatorname{Im}\left[\bar{P}_N(\tau)\int_a^b K(\xi,\xi')\frac{F(\xi')-|f(\xi')|}{|P_N(\tau')|}P_N(\tau')\mathrm{d}\xi'\right]=0\,, \tag{4.165}$$

or, after adding the real function $\alpha\hat{f}(\xi)\,|P_N(\tau)|^2$ under the symbol of imaginary part,

$$\operatorname{Im}\left[\alpha\hat{f}(\xi)\,|P_N(\tau)|^2+\bar{P}_N(\tau)\int_a^b K(\xi,\xi')\frac{F(\xi')-|f(\xi')|}{|P_N(\tau')|}P_N(\tau')\mathrm{d}\xi'\right]=0\,. \tag{4.166}$$

Taking into account (4.159) and dividing both sides of (4.166) by the real positive function $|P_N(\tau)|$, we obtain

$$\operatorname{Im}\left[\alpha|f(\xi)|+\frac{\bar{P}_N(\tau)}{|P_N(\tau)|}\int_a^b K(\xi,\xi')\frac{F(\xi')-|f(\xi')|}{|P_N(\tau')|}P_N(\tau')\mathrm{d}\xi'\right]=0\,. \tag{4.167}$$

On the other hand, the integral equation (4.155) can be written in the form

$$\operatorname{Re}\left[\alpha\,|f(\xi)|+\frac{\bar{P}_N(\tau)}{|P_N(\tau)|}\int_a^b K(\xi,\xi')\frac{F(\xi')-|f(\xi')|}{|P_N(\tau')|}P_N(\tau')\mathrm{d}\xi'\right]=0\,. \tag{4.168}$$

Equalities (4.167) and (4.168) together imply that the expression in their square brackets equals zero, which after multiplying by (4.154) shows that, function (4.151) solves the integral equation (4.150).

□

Theorem 4.15

If the function $f(\xi)$ of form (4.151) with $\beta = 1$ solves equation (4.150), then the functions

$$f_n(\xi) = \hat{f}(\xi) P_N(\tau) \frac{1 - \bar{\eta}_n \tau}{1 - \eta_n \tau}, \quad n = 1, 2, \ldots, N, \tag{4.169}$$

also solve this equation.

For simplicity, we drop here the index N in η_{Nk}.

Proof: We denote

$$Q_N(\tau) = P_{N-1}(\tau)(1 - \bar{\eta}_n \tau), \tag{4.170}$$

where

$$P_{N-1}(\tau) = P_N(\tau)/(1 - \eta_n \tau). \tag{4.171}$$

In this notation, $f_n(\xi) = \hat{f}(\xi) Q_N(\tau)$. It can be seen that

$$|P_N(\tau)| = |Q_N(\tau)|. \tag{4.172}$$

Let the function $f(\xi)$ of the form (4.151) solve (4.150). Then, according to Theorem 4.14, the transcendental equation system (4.156) is fulfilled. At the same time, this system is fulfilled after replacing the polynomial $P_N(\tau)$ within it by any other polynomial of degree N with the same modulus, in particular, by $Q_N(\tau)$, which is

$$\int_a^b \tau^{n-1} s(\xi) \frac{F(\xi) - |f(\xi)|}{|Q_N(\tau)|} \mathrm{d}\xi = 0, \qquad n = 1, 2, \ldots, N. \tag{4.173a}$$

$$\int_a^b \tau^{n-1} q(\xi) \frac{F(\xi) - |f(\xi)|}{|Q_N(\tau)|} \mathrm{d}\xi = 0, \quad n = 1, 2, \ldots, N. \tag{4.173b}$$

We rewrite (4.150) in the form

$$\alpha f(\xi) = \int_a^b K(\xi, \xi') \left[F(\xi') \exp(i \arg f(\xi')) - f(\xi') \right] \mathrm{d}\xi'. \tag{4.174}$$

In order to show that $f_n(\xi)$ satisfies this equation, we substitute it into the right-hand side instead of $f(\xi)$. After multiplying and dividing this side by $|Q_N(\tau)|^2$ and using (4.170), (4.172), we obtain

$$\begin{aligned} &\int_a^b K(\xi, \xi') \left(F(\xi') \exp(i \arg f_n(\xi')) - f_n(\xi') \right) \mathrm{d}\xi' \\ &= \int_a^b K(\xi, \xi') \frac{F(\xi') - |f(\xi')|}{|Q_N(\tau')|} Q_N(\tau') \mathrm{d}\xi' \\ &= \frac{(1 - \bar{\eta}_n \tau) P_{N-1}(\xi) \bar{Q}_N(\xi) \int_a^b K(\xi, \xi') \frac{F(\xi') - |f(\xi')|}{|Q_N(\tau')|} Q_N(\tau') \mathrm{d}\xi'}{|Q_N(\tau)|^2}. \end{aligned} \tag{4.175}$$

Since equation system (4.173) holds, then equality (4.165) after substitution $P_N(\xi)$ by $Q_N(\xi)$ also holds, that is,

$$\operatorname{Im}\left\{\bar{Q}_N(\tau)\int_a^b K(\xi,\xi')\frac{F(\xi')-|f(\xi')|}{|Q_N(\tau')|}Q_N(\tau')\mathrm{d}\xi'\right\}=0\,. \tag{4.176}$$

We introduce the function

$$\hat{f}_n(\xi)=\frac{\bar{Q}_N(\tau)\int_a^b K(\xi,\xi')\frac{F(\xi')-|f(\xi')|}{|Q_N(\tau')|}Q_N(\tau')\mathrm{d}\xi'}{\alpha\,|Q_N(\tau)|^2}\,. \tag{4.177}$$

According to (4.176), this function is real. With this notation, identity (4.175) can be rewritten in the form:

$$\alpha\,\hat{f}_n(\xi)(1-\bar{\eta}_n\tau)P_{N-1}(\tau)=\int_a^b K(\xi,\xi')\frac{F(\xi')-|f(\xi')|}{|Q_N(\tau')|}Q_N(\tau')\mathrm{d}\xi'\,. \tag{4.178}$$

With the aid of (4.170), we have

$$\begin{aligned}\alpha\,\hat{f}_n(\xi)&(1-\bar{\eta}_n\tau)P_{N-1}(\tau)\\&=\int_a^b K(\xi,\xi')\frac{F(\xi)-|f(\xi)|}{|P_N(\tau')|}(1-\bar{\eta}_n\tau')P_{N-1}(\tau')\mathrm{d}\xi'\,.\end{aligned} \tag{4.179}$$

This equality implies that, if $\hat{f}_n(\xi)$ has no zeros on $[a,b]$, then the function $\hat{f}_n(\xi)P_{N-1}(\tau)(1-\bar{\eta}_n\tau)$ solves (4.150).

In order to complete the proof, it remains to show that $\hat{f}_n(\xi)=\hat{f}(\xi)$. For this reason, we substitute the function $f(\xi)=\hat{f}(\xi)P_N(\xi)$ into (4.174) and use (4.172):

$$\begin{aligned}\alpha\,\hat{f}(\xi)&(1-\eta_n\tau)P_{N-1}(\tau)\\&=\int_a^b K(\xi,\xi')\frac{F(\xi')-|f(\xi')|}{|Q_N(\tau')|}(1-\eta_n\tau')P_{N-1}(\tau')\mathrm{d}\xi'\,.\end{aligned} \tag{4.180}$$

Now we multiply both sides of (4.179) by $(1-\eta_n\tau)$, and both sides of (4.180) by $(1-\bar{\eta}_n\tau)$, and subtract them from one other. Using the identity

$$(1-\eta_n\tau)(1-\bar{\eta}_n\tau')-(1-\bar{\eta}_n\tau)(1-\eta_n\tau')=(\eta_n-\bar{\eta}_n)(\tau'-\tau)\,, \tag{4.181}$$

we obtain

$$\begin{aligned}&(\hat{f}_n(\xi)-\hat{f}(\xi))P_{N-1}(\tau)\,|1-\eta_n\tau|^2=(\eta_n-\bar{\eta}_n)\\&\times\left[q(\xi)\int_a^b s(\xi')\frac{F(\xi')-|f(\xi')|}{|Q_N(\tau')|}P_{N-1}(\tau')\mathrm{d}\xi'\right.\\&\left.-s(\xi)\int_a^b q(\xi')\frac{F(\xi')-|f(\xi')|}{|Q_N(\tau')|}P_{N-1}(\tau')\mathrm{d}\xi'\right].\end{aligned} \tag{4.182}$$

Since $P_{N-1}(\tau)$ is a polynomial of degree $N-1$, the integrals on the right-hand side of (4.182) equal zero owing to (4.156). Under the theorem condition, the functions $P_{N-1}(\tau)$ and $|1-\eta_n\tau|^2$ on the left-hand side of (4.182) have no zeros on $[a,b]$, hence $\hat{f}_n(\xi)=\hat{f}(\xi)$ and the proof is complete. □

Corollary The solutions to equation systems (4.155) and (4.156) make up the equivalent groups inside which the function $\hat{f}(\xi)$ is the same and the polynomials $P_N(\tau)$ differ only by substitution of any number $s<N$ of the parameters η_n with the complex conjugates:

$$P_N^{(s)}(\tau)=\prod_{m=1}^{s}(1-\eta_{n_m}\tau)\prod_{m=s+1}^{N}(1-\bar{\eta}_{n_m}\tau)\,, \tag{4.183}$$

where $n_{m_1}\neq n_{m_2}$ if $m_1\neq m_2$. Such polynomials generate the solutions to (4.149) with the same $|f(\xi)|$.

4.5.2
Branching of solutions

Let the kernel $K(\xi,\xi')$ depend on a certain real parameter c. Then the solution to equation (4.149) can branch at some values of c. As before, in order to obtain the equation for the branching points, we apply the perturbation method to equation (4.149). Let at $c=c_0$ equation (4.149) have the solution $f_0(\xi)$ in the form (4.151); further we call this solution 'initial'. The perturbation $c=c_0+\varepsilon$ leads to the perturbations of the operator B and the solution f, which, to first approximations, are $B=B_0+\varepsilon B_1$ and $f=f_0+\varepsilon f_1$, respectively.

Keeping the terms of the first order with respect to ε, we obtain

$$\exp(i\arg f)=\exp(i\arg f_0)+i\varepsilon\exp(i\arg f_0)\,\mathrm{Im}\,\frac{f_1}{f_0}\,. \tag{4.184}$$

Substituting the expressions for B, f, and $\exp(i\arg f)$ into (4.149) and equating the terms at ε, we obtain the following nonhomogeneous equation with respect to f_1:

$$\alpha f_1+B_0\left[f_1\right]-iB_0\left[F\exp(i\arg f_0)\mathrm{Im}\,\frac{f_1}{f_0}\right]=B_1\left[F\exp(i\arg f_0)\right]. \tag{4.185}$$

Branching is possible only in the case when the function $f_1(\xi)$ is determined nonuniquelly, that is, if the homogeneous equation

$$\alpha f_1+B_0\left[f_1\right]=iB_0\left[F\exp(i\arg f_0)\mathrm{Im}\,\frac{f_1}{f_0}\right] \tag{4.186}$$

holds. The values c_0, for which this is true, are the possible branching points. After introducing the notations

$$D=(\alpha E+B_0)^{-1}B_0\,, \tag{4.187}$$

$$u(\xi) + iv(\xi) = f_1(\xi)/f_0(\xi)\,. \tag{4.188}$$

Equation (4.186) becomes

$$f_0 \cdot (u + iv) = i\,D_0\left[F \exp(i \arg f_0) v\right]\,. \tag{4.189}$$

Substituting (4.154) and the equality

$$f_0 = |f_0| \frac{P_N}{|P_N|} \tag{4.190}$$

into (4.189) and multiplying both its sides by $\bar{P}_N$, we obtain

$$|f_0|\,|P_N| \cdot (u + iv) = i\,\bar{P}_N\,D_0\left[F \frac{P_N}{|P_N|} v\right]\,. \tag{4.191}$$

Equating the imaginary parts of both side of equation (4.191) leads to the real equation

$$|f_0|\,|P_N| v = \mathrm{Re}\left(\bar{P}_N\,D_0\left[F \frac{P_N}{|P_N|} v\right]\right) \tag{4.192}$$

with respect to the real function v. When this function is found, u is immediately calculated from (4.191).

Equation (4.192) can be considered as an eigenvalue problem, nonlinear with respect to the spectral parameter c. It can be rewritten in the form

$$\lambda_n |f|\,|P_N| v_n = \mathrm{Re}\left(\bar{P}_N\,D\left[F \frac{P_N}{|P_N|} v_n\right]\right) \tag{4.193}$$

as a problem of finding the values of c, at which an eigenvalue λ_n of this linear problem equals one.

According to (4.149) and (4.187), $D\left[F P_N/|P_N|\right] = f$. Substituting this expression into the right-hand side of (4.192) at $v \equiv 1$ we obtain the identity. In this way we have established that the homogeneous equation (4.193) has eigenvalue $\lambda_0 = 1$ with the eigenfunction $v_0 \equiv 1$ for any c. This fact is expected, it describes the freedom of the constant β in (4.151). This means that at the branching point the multiplicity of the eigenvalue $\lambda_n = 1$ in (4.193) must be two or more.

As before, the transcendental equations for determination of the branching points and parameters of the branched solutions to equation (4.150) can be constructed. In the case when the solution of the type (4.151) branches without changing the degree N of the polynomial P_N, the theory of implicit functions of several variables is applied for this purpose. If the polynomial degree is changed in new solutions, then the transcendental equation system is obtained from the equations of the type (4.156) written for the different values of N simultaneously. Of course, integral equation (4.155) should be satisfied as well.

In the case when N does not change at the branching point, the transcendental equation has the same form as (4.53) in Section 4.1.3 with

$$\mathcal{F} = \left\{\vec{\Phi}, \vec{\Psi}, \Upsilon\right\}, \tag{4.194}$$

where the vectors $\vec{\Phi}$ and $\vec{\Psi}$ are given by (4.157),

$$\begin{aligned}
&\Upsilon(|f(\xi)|, \eta_{N1}, \eta_{N2, \dots} \eta_{NN}) \equiv \alpha |f(\xi)| + \frac{\beta}{|P_N(\tau)|} \\
&\times \int_a^b |f(\xi')| K(\xi, \xi') \frac{\operatorname{Re}\left[\bar{P}_N(\tau) P_N(\tau')\right]}{|P_N(\tau')|} \mathrm{d}\xi' \\
&- \frac{1}{|P_N(\tau)|} \int_a^b F(\xi') K(\xi, \xi') \frac{\operatorname{Re}\left[\bar{P}_N(\tau) P_N(\tau')\right]}{|P_N(\tau')|} \mathrm{d}\xi' = 0,
\end{aligned} \tag{4.195}$$

$$\mathcal{F}' = \begin{pmatrix}
\left\{\dfrac{\partial \Phi_{Nj}}{\partial \eta'_{Nk}}\right\}_{j,k=1}^{N} & \left\{\dfrac{\partial \Phi_{Nj}}{\partial \eta''_{Nk}}\right\}_{j,k=1}^{N} & \left\{\dfrac{\partial \Phi_{Nj}}{\partial |f|}\right\}_{j}^{N} \\
\left\{\dfrac{\partial \Psi_{Nj}}{\partial \eta'_{Nk}}\right\}_{j,k=1}^{N} & \left\{\dfrac{\partial \Psi_{Nj}}{\partial \eta''_{Nk}}\right\}_{j,k=1}^{N} & \left\{\dfrac{\partial \Psi_{Nj}}{\partial |f|}\right\}_{j}^{N} \\
\left\{\dfrac{\partial \Upsilon}{\partial \eta'_{Nk}}\right\}_{k=1}^{N} & \left\{\dfrac{\partial \Upsilon}{\partial \eta''_{Nk}}\right\}_{k=1}^{N} & \dfrac{\partial \Upsilon}{\partial |f|}
\end{pmatrix} \tag{4.196}$$

is the Jacobi matrix of these functions. The last block-collumn in the matrix $\mathcal{F}'$ implies the discretized first variations of Φ_{Nj}, Ψ_{Nj}, and Υ with respect to $|f|$.

At the branching points where N increases by one, the parameters η_{Nk} of the initial polynomial $P_N(\tau)$ and parameters $\eta_{N+1,k}$ of the off-branched one $P_{N+1}(\tau)$ are connected by the equalities

$$\eta_{Nk} = \eta_{N+1,k}, \qquad k = 1, \dots, N, \tag{4.197}$$

with $\operatorname{Im} \eta_{N+1,N+1} = 0$. Besides equation (4.155) and system (4.156), the equations

$$\begin{aligned}
&\alpha |f(\xi)| + \frac{1}{|P_{N+1}(\tau)|} \int_a^b |f(\xi')| K(\xi, \xi') \frac{\operatorname{Re}\left[\bar{P}_{N+1}(\tau) P_{N+1}(\tau')\right]}{|P_{N+1}(\tau')|} \mathrm{d}\xi' \\
&= \frac{1}{|P_{N+1}(\tau)|} \int_a^b F(\xi') K(\xi, \xi') \frac{\operatorname{Re}\left[\bar{P}_{N+1}(\tau) P_{N+1}(\tau')\right]}{|P_{N+1}(\tau')|} \mathrm{d}\xi',
\end{aligned} \tag{4.198}$$

and

$$\int_a^b \tau^{n-1} s(\xi) \frac{F(\xi) - |f(\xi)|}{|P_{N+1}(\tau)|} \mathrm{d}\xi, \quad n = 1, 2, \dots, N+1, \tag{4.199a}$$

$$\int_a^b \tau^{n-1} q(\xi) \frac{F(\xi) - |f(\xi)|}{|P_{N+1}(\tau)|} \mathrm{d}\xi, \quad n = 1, 2, \dots, N+1 \tag{4.199b}$$

should hold.

Since the new parameter $\eta_{N+1,N+1}$ is real, the integral equation (4.198) coincides with (4.155). As before, the nth equation of system (4.199a) $n = 1, 2, \ldots, N$, is a linear combination of the corresponding equation of system (4.156) and $(n+1)$th equation of (4.199a), and only two additional equations

$$\int_a^b \frac{\tau^N s(\xi)\,(F(\xi) - |f(\xi)|)}{|P_N(\tau)|(1 - \eta_{N+1,N+1}\tau)} \mathrm{d}\xi\,, \tag{4.200a}$$

$$\int_a^b \frac{\tau^N q(\xi)\,(F(\xi) - |f(\xi)|)}{|P_N(\tau)|(1 - \eta_{N+1,N+1}\tau)} \mathrm{d}\xi \tag{4.200b}$$

should hold. On the whole, we have one real integral equation and $2N + 2$ transcendental ones for determining the real function $|f(\xi)|$, N complex parameters η_{Nk}, $k = 1, 2, \ldots, N$ and real $\eta_{N+1,N+1}$ and c_j.

At the branching points where the polynomial degree increases by two, the equalities

$$\eta_{Nk} = \eta_{N+2,k}\,, \quad k = 1, \ldots, N\,, \tag{4.201}$$

are valid. Besides (4.155) and (4.156), the four additional equations

$$\int_a^b \frac{\tau^{n-1} s(\xi)\,(F(\xi) - |f(\xi)|)}{|P_N(\tau)|(1 - \eta_{N+2,N+1}\tau)(1 - \eta_{N+2,N+2}\tau)} \mathrm{d}\xi = 0\,, \quad n = N+1, N+2\,; \tag{4.202a}$$

$$\int_a^b \frac{\tau^{n-1} q(\xi)\,(F(\xi) - |f(\xi)|)}{|P_N(\tau)|(1 - \eta_{N+2,N+1}\tau)(1 - \eta_{N+2,N+2}\tau)} \mathrm{d}\xi = 0\,, \quad n = N+1, N+2\,, \tag{4.202b}$$

should be fulfilled with $\eta_{N+2,N+1}$, $\eta_{N+2,N+2}$ satisfying the conditions

$$\eta_{N+2,N+1} = \bar{\eta}_{N+2,N+2} \tag{4.203}$$

or

$$\operatorname{Im} \eta_{N+2,N+1} = \operatorname{Im} \eta_{N+2,N+2} = 0\,. \tag{4.204}$$

Hence, we have one real integral equation and $2N + 4$ transcendental ones for determining the real function $|f(\xi)|$, N complex parameters η_{Nk}, $k = 1, 2, \ldots, N$, real c_j, and one complex $\eta_{N+2,N+1}$ or two real $\eta_{N+2,N+1}$, $\eta_{N+2,N+2}$. In whole we have one real unknown function and $2N + 3$ real numbers. As it was mentioned in the preceding subsection, the existence of solutions to such a system is low-probable in general case. However, they may exist in the case when $a = -b$, one of the functions $s(\xi)$ and $q(\xi)$ is even and the second one is odd, and

$$F(\xi) = F(-\xi)\,. \tag{4.205}$$

Then the solutions are possible, which are generated by the polynomials with the even modulus. Such a property should be inhered to $P_{N+2}(\tau)$, that is,

$$|P_{N+2}(\tau)| = |P_{N+2}(-\tau)| \,. \tag{4.206}$$

This equality decreases the number of unknowns twice: the parameters $\eta_{N+2,k}$ become imaginary or appear by pears with opposite signs, whereas, according to (4.100), $\eta_{N+2,k}$, $k = N+1, N+2$ are always imaginary with opposite signs:

$$\mathrm{Re}\,\eta_{N+2,N+1} = \mathrm{Re}\,\eta_{N+2,N+2} = 0\,, \tag{4.207a}$$

$$\eta_{N+2,N+1} = \bar{\eta}_{N+2,N+2}\,. \tag{4.207b}$$

On the other hand, conditions (4.206) decrease the number of equations twice, as well: N equations of system (4.156) and two additional equations (4.202a) become identities, because they have odd integrands on the left-hand side.

Finally, at fulfilling (4.205), (4.206) the solution branching is possible with increasing the polynomial degree by two if the following transcendental equation system holds:

$$\int_a^b \tau^{2n-1} s(\xi) \frac{F(\xi) - |f(\xi)|}{|P_N(\tau)|} \mathrm{d}\xi = 0\,, \quad n = 1, 2, \ldots [N/2]\,, \tag{4.208a}$$

$$\int_a^b \tau^{2n-2} q(\xi) \frac{F(\xi) - |f(\xi)|}{|P_N(\tau)|} \mathrm{d}\xi = 0\,, \quad n = 1, 2, \ldots [(N+1)/2]\,, \tag{4.208b}$$

$$\int_a^b \frac{\tau^{2[(N+2)/2]-1} s(\xi)(F(\xi) - |f(\xi)|)}{|P_N(\tau)|(1 - \eta_{N+2,N+1}\tau)(1 - \eta_{N+2,N+2}\tau)} \mathrm{d}\xi = 0\,, \tag{4.208c}$$

$$\int_a^b \frac{\tau^{2[(N+1)/2]} q(\xi)(F(\xi) - |f(\xi)|)}{|P_N(\tau)|(1 - \eta_{N+2,N+1}\tau)(1 - \eta_{N+2,N+2}\tau)} \mathrm{d}\xi = 0\,, \tag{4.208d}$$

where η_{Nk}, $k = 1, \ldots, N$, are either imaginary or appear by pairs with opposite signs, and $\eta_{N+2,k}$, $k = N+1, N+2$ are subject to conditions (4.207a). As result, we have one real integral equation, $N+2$ transcendental ones for determining the real function $|f(\xi)|$, and $N+2$ real numbers.

4.5.3
Numerical results

The simplest form of the general integral equation (4.150) is the equation arising in the unconditional minimization problem for functional (3.100) with the isometric operator describing the one-dimensional Fourier transform of the finite function (see (3.68)). Since the first addend in (3.100) decreases when the parameter c grows, whereas the second addend should remain almost the same, the parameter t must depend on c, for instance, as $t = \alpha/c$ with constant α. Then the functional to be minimized is written as $\sigma_\alpha(u) = \sigma_0(u) + \sigma_1(u)$, where

$$\sigma_0(u) = \int_{-1}^{1} (|f(\xi)| - F(\xi))^2 \,\mathrm{d}\xi\,, \tag{4.209a}$$

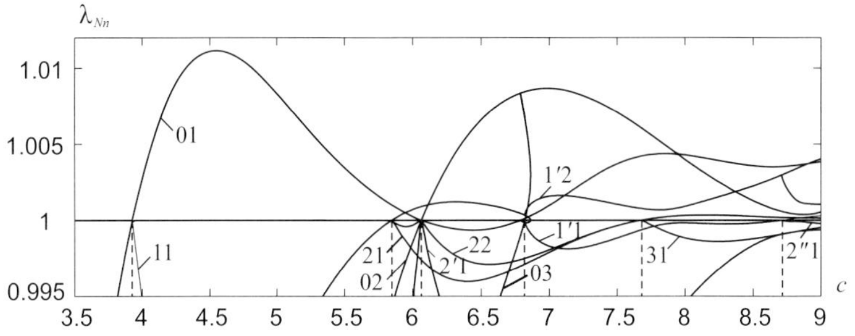

Figure 4.22 Eigenvalues of the homogeneous integral equation (4.192); $F(\xi) = \cos(x\pi/2)$, $\alpha = 0.05$.

$$\sigma_1(u) = \frac{\alpha}{c}\int_{-1}^{1} |u(x)|^2 \mathrm{d}x \,. \tag{4.209b}$$

Using expressions (3.71) and (3.72) for the operators participating in (3.104), we rewrite this equation in the form

$$\alpha f(\xi) + \int_{-1}^{1} \frac{\sin c(\xi - \xi')}{(\xi - \xi')} f(\xi')\mathrm{d}\xi' = \int_{-1}^{1} \frac{\sin c(\xi - \xi')}{(\xi - \xi')} F(\xi') \exp(i \arg f(\xi'))\mathrm{d}\xi' \,. \tag{4.210}$$

This equation is a particular case of (4.150) with $s(\xi) = \sin(c\xi)$, $q(\xi) = \cos(c\xi)$, $\tau = \xi$, $a = -1$ and $b = 1$.

Below we give the numerical results for the case $F(\xi) = \cos(x\pi/2)$, $\alpha = 0.05$. Figure 4.22 shows the eigenvalues of the homogeneous equation (4.192). As before, the curves are labeled by two numbers Nn, with N as the polynomial degree of the solution from which the branching is found, and n is the ordinary number of the solution considered in the solution set corresponding to this N. As noted, there is an eigenvalue $\lambda_0 = 1$ for any c. The points where $\lambda_n = 1$ with $n \neq 0$ are the branching points of solutions to (4.210). In the considered range of the values c, different types of branching were observed: without changing N, changing N by one, and changing N by two. The maximum value of polynomial degree N in this range is three. Recall that we are considering an example of the odd function $F(\xi)$.

The parameters η_{Nn} of the polynomial $P_N(\tau)$ are determined from the integral equation (4.155) and the transcendental equation system (4.156). As before, the system was solved by the Newton method. The results are presented in Figure 4.23.

Figure 4.24 shows the values of both addends of (4.209) in the functional σ. It can be seen that the dependencies of both addends are similar. However, as noted, the value σ_1 varies over a smaller range than σ_0. In contrast to the previous cases, the polynomial of degree N in the optimal solution is not obligatory maximal due to the different nature of the addends in the functional σ.

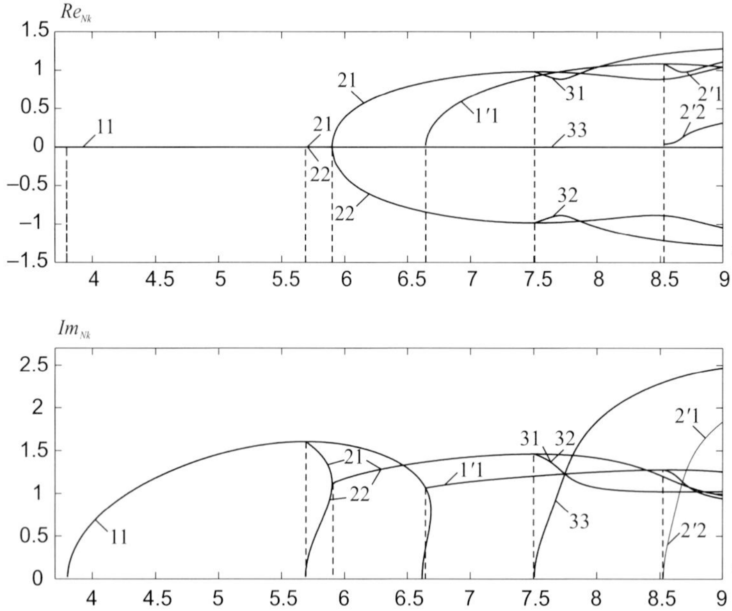

Figure 4.23 Real and imaginary parts of parametrs η_{Nk}; $F(\xi) = \cos(x\pi/2)$, $\alpha = 0.05$.

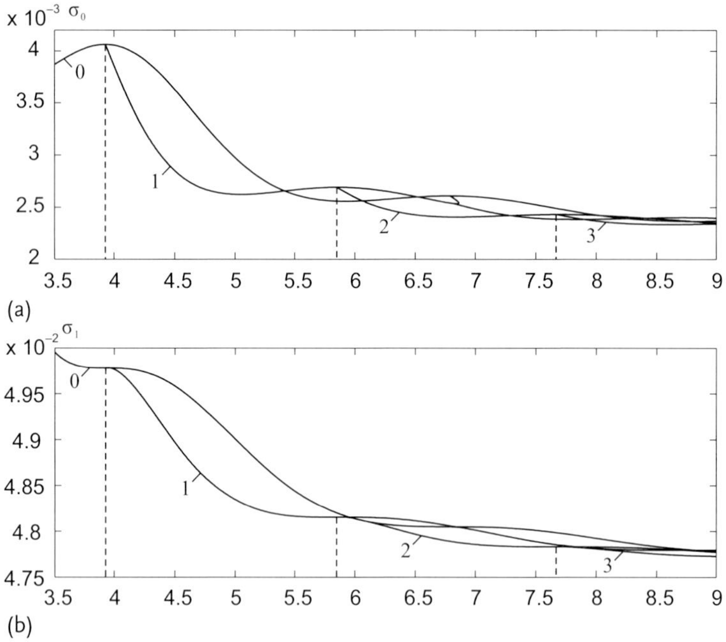

Figure 4.24 Comparison of values of the functional σ for different solutions to (3.100); $F(\xi) = \cos(x\pi/2)$.

The theory concerning the nonlinear integral equations considered in this chapter is not yet complete. The unanswered questions relate, in particular, to: the existence of solutions of different types to (4.151); the hypothesis that the optimal solutions for equations of type (4.150) correspond to the polynomials of maximal degree; to the limitation of the polynomial degree for this equation, etc. In all cases, only solutions having no zeros are considered. The theory of solutions which vanish at some points of their definition domain has not yet been developed. There are only a few papers (e.g. [64] and [77]) on this topic.

5
Numerical Methods, Algorithms and Results

The analytical technique described in the previous chapter can certainly be applied only to a limited part of the problems considered in the book. In this chapter some numerical methods are analyzed and applied to the formulated problems. Section 5.1 describes some theoretical aspects of the iterative methods for direct minimization of the functionals and for the solution of nonlinear integral equations and their systems which were obtained earlier. A modification of the Newton method is described and applied to the concrete problems in Section 5.2. As a rule, this method is most effective when used together with the above iterative ones. A specific numerical technique known as the 'opposite directions' method is utilized in Section 5.3 for the optimization of linearly extended systems. Almost all the methods are demonstrated on two-dimensional problems described by one-dimensional equations. An exception is the method of generalized separation of variables described in Section 5.4 using examples of the optimization problem for two-dimensional antenna arrays. The problems concerning eigenfunctions of certain homogeneous integral equations which describe the eigenwaves and eigenoscillations of physical systems are considered in Section 5.5.

5.1
Theoretical results

In Chapter 3 same simple iteration methods were proposed for the nonlinear integral equations formulated there. Owing to their relaxivity, these methods can be considered as versions of the steepest descent method for the appropriate functionals. The descent step (factor at the gradient) is fixed for all iterations of the method. In this section, the properties of methods with variable descent step are investigated (Section 5.1.1) and the convergence rate of the iterative methods themselves is established for particular cases of the operator (Section 5.1.2).

Phase Optimization Problems. O. O. Bulatsyk, B. Z. Katsenelenbaum, Y. P. Topolyuk, and N. N. Voitovich

ISBN: 978-3-527-40799-6

5.1.1
Direct minimization of functionals

5.1.1.1 The case of an isometric operator

The functionals considered in the book are nonconvex and they can have many extremal points (usually continual sets of them). Therefore a specific technique is required when investigating the convergence of the methods for their minimization.

Consider the functional

$$\sigma(u) = \| |Au| - F \|_2^2 \tag{5.1}$$

introduced in Section 3.1 (see 3.6). Let A be a linear *isometric* operator acting from a complex functional Hilbert space H_1 of L_2-type into another H_2 of the same type; $F \in H_2$ is a given non-negative function. Denote by u_* the function providing a minimum to $\sigma(u)$. The existence of such a function is established in [26].

The main notion of the steepest descent method considered here is the *gradient* of the functional $\sigma(u)$. We denote the gradient of $\sigma(u)$ on the function $u = u_0$ by $g = \operatorname{grad} \sigma(u_0)$ and understand it as the function $g \in H_1$ such that

$$\lim_{t\to 0} \frac{\sigma(u_0 + tg) - \sigma(u_0)}{t\,\|g\|_1} = \max_{v\in H_1} \lim_{t\to 0} \frac{\sigma(u_0 + tv) - \sigma(u_0)}{t\,\|v\|_1} \tag{5.2}$$

for real $t > 0$ and any $v \in H_1$. Writing expression (3.16) for the first variation of $\sigma(u)$ at $\delta u = tv/\|v\|_1$ and using property (3.43) of the isometric operator, we obtain

$$\lim_{t\to 0} \frac{\sigma(u_0 + tv) - \sigma(u_0)}{t\,\|v\|_1} = 2\,\mathrm{Re}\left(\frac{v}{\|v\|_1}, u_0 - A^* F e^{i \arg Au_0}\right)_1 . \tag{5.3}$$

In order to maximize the right-hand side of (5.3), we must take the first multiplier in the scalar product to be proportional to the second one. Finally, we obtain

$$\operatorname{grad} \sigma(u_0) = u_0 - A^*\left[F \exp\left(i \arg Au_0\right)\right] . \tag{5.4}$$

The simplest of the gradient methods for functional minimization is the *method of steepest descent*. In this method the next approximation to the minimum point is calculated as

$$u_{p+1} = u_p - \lambda_p \operatorname{grad} \sigma\left(u_p\right) \tag{5.5}$$

with the parameter $\lambda_p > 0$ depending, in general, on the step number p. We analyze this method in application to the functional $\sigma(u)$.

Theorem 5.1

In a neighborhood of the minimum point u_* of functional $\sigma(u)$, the gradient method based on (5.5) is relaxational for $0 < \lambda_p \le 1$, and $\{u_p\}$ is the minimizing sequence for the functional $\|\operatorname{grad} \sigma(u)\|$, that is,

$$\lim_{p\to\infty} \left\|\operatorname{grad} \sigma\left(u_p\right)\right\| = 0 . \tag{5.6}$$

Proof: When proving this theorem, we will use several facts established in [43] and [26]. It is easy to see that functional (5.1) is increasing, that is,

$$\lim_{\|u\|_1 \to \infty} \sigma(u) = \infty \,. \tag{5.7}$$

This fact follows from the inequality

$$\sigma(u) = \|F\|^2 - 2(F, |Au|)_2 + \|u\|^2 \ge \|F\|^2 - 2\|F\|\,\|u\| + \|u\|^2 \,. \tag{5.8}$$

Owing to this property, the function set $U_0 = \{u : \sigma(u) \le \sigma(u_0)\}$ is bounded; here u_0 is the initial approximation to u.

The functional $(F, |Au|)$ is convex in U_0. Indeed,

$$\begin{aligned}
&(F, |A(\lambda u_1 + (1-\lambda)\, u_2)|)_2 \\
&= (F, |\lambda A u_1 + (1-\lambda)\, A u_2|)_2 \\
&\le (F, |\lambda A u_1|) + (F, |(1-\lambda)\, A u_2|)_2 \\
&= \lambda\, (F, |Au_1|)_2 + (1-\lambda)\, (F, |Au_2|)_2 \,,
\end{aligned} \tag{5.9}$$

where $0 < \lambda < 1$. It is known that the gradient of any convex functional is monotonous, that is,

$$\operatorname{Re}\left(A^* F e^{i \arg A(u+v)} - A^* F e^{i \arg Au}, v\right)_1 \ge 0 \tag{5.10}$$

for any $u + v \in U_0$, $u \in U_0$.

We establish the condition for the sequence (5.5) to be relaxational, that is $\sigma(u_{p+1}) < \sigma(u_p)$.

Using the Lagrange mean value theorem from [43], we write

$$\sigma(u_p) - \sigma(u_{p+1}) = \operatorname{Re}\left(\operatorname{grad} \sigma\left(u_{p+1} + \tau(u_p - u_{p+1})\right), u_p - u_{p+1}\right) , \tag{5.11}$$

where $0 < \tau < 1$. After simple derivations using (5.5) and (5.10), we obtain

$$\begin{aligned}
&\sigma(u_p) - \sigma(u_{p+1}) \\
&= \operatorname{Re}\left(\operatorname{grad} \sigma(u_p), u_p - u_{p+1}\right) \\
&\quad + \operatorname{Re}\left(\operatorname{grad} \sigma\left(u_{p+1} + \tau(u_p - u_{p+1})\right) - \operatorname{grad} \sigma(u_p), u_p - u_{p+1}\right) \\
&= \operatorname{Re}\left(\operatorname{grad} \sigma(u_p), u_p - u_{p+1}\right) \\
&\quad + \operatorname{Re}\left(u_{p+1} - u_p + \tau(u_p - u_{p+1}), u_p - u_{p+1}\right) \\
&\quad - \operatorname{Re}\left(B\left[u_{p+1} + \tau(u_p - u_{p+1})\right] - B\left[u_p\right], u_p - u_{p+1}\right) \\
&= \operatorname{Re}\left(\operatorname{grad} \sigma(u_p), u_p - u_{p+1}\right) \\
&\quad + \operatorname{Re}\left(u_{p+1} - u_p + \tau(u_p - u_{p+1}), u_p - u_{p+1}\right) \\
&\quad + \frac{1}{1-\tau} \operatorname{Re}\left(B\left[u_p + (1-\tau)(u_{p+1} - u_p)\right]\right. \\
&\qquad \left. - B\left[u_p\right], (1-\tau)(u_{p+1} - u_p)\right)
\end{aligned}$$

$$\geq \lambda_p \left\| \operatorname{grad} \sigma\left(u_p\right)\right\|^2 + \left(-\lambda_p^2 + \tau\lambda_p^2\right) \left\| \operatorname{grad} \sigma\left(u_p\right)\right\|^2$$

$$= \lambda_p \left\| \operatorname{grad} \sigma\left(u_p\right)\right\|^2 (1 - \lambda_p + \tau\lambda_p) > 0\,, \tag{5.12}$$

where $B[u_p] = A^*[F \exp(i \arg Au_p)]$.

It can be seen that the final inequality in (5.12) holds for any $\tau < 1$ if

$$\lambda_p \leq 1\,. \tag{5.13}$$

The first part of the theorem is proved.

The sequence $\{\sigma(u_p)\}$ is bounded and monotonic, and hence it is convergent. From (5.12) and (5.13) we have

$$\lim_{p\to\infty} \left\| \operatorname{grad} \sigma\left(u_p\right)\right\|^2 \leq \lim_{p\to\infty} \frac{1}{\lambda_p(1 - \lambda_p + \tau\lambda_p)} \lim_{p\to\infty} \left[\sigma\left(u_p\right) - \sigma\left(u_{p+1}\right)\right]\,. \tag{5.14}$$

As for any convergent sequence, the second multiplier on the right-hand side of (5.14) is zero. Since the first multiplier is bounded, (5.6) is true, and the proof is completed. □

We need the following lemma (see [43], lemma 10.1).

Lemma 5.1

Let a functional $g(x)$, given in a normed space E, be growing, bounded from below, and $d = \inf_{x\in E} g(x)$. Then any minimizing sequence $\{x_n\}$: $\lim_{n\to\infty} g(x_n) = d$ is bounded.

Theorem 5.2

If the sequence $\{u_p\}$ is minimizing, then the iterative process (5.5) generates a weakly convergent sequence, and its weak limit belongs to the set of solutions to the equation

$$u = A^*\left[F \exp\left(i \arg Au\right)\right]. \tag{5.15}$$

Proof: As a bounded set in the Hilbert space H_1, the set of solutions to (5.15) is a weak compact. The sequence $\{u_p\}$ satisfies the condition of the above lemma, and hence it is bounded. Therefore, a subsequence, weakly convergent to a certain function $\tilde{u}$, can be separated from $\{u_p\}$. Then, according to [66], (page 45, Theorem 2), the entire sequence also weakly converges to $\tilde{u}$.

Let $\sigma_* = \inf \sigma(u) = \sigma(u_*)$. Since the sequence $\{u_p\}$ is minimizing and the functional $\sigma(u)$ is weakly lower semicontinuous (see [26], page 25), then

$$\sigma_* = \lim_{p\to\infty} \sigma(u_p) \geq \sigma(\tilde{u}) \geq \sigma(u_*) = \inf \sigma = \sigma_*\,. \tag{5.16}$$

From this, $\sigma(\tilde{u}) = \sigma_*$, that is, the function $\tilde{u}$ belongs to the set of functions on which the infimum of functional (5.1) is reached, and hence $\tilde{u}$ is a solution to (5.15). □

We have obtained the following two results: the sequence (5.5) is minimizing for the norm of the gradient (5.4); if this sequence is minimizing for the functional (5.12), then the gradient method generates a weakly convergent sequence.

Below, the strong convergence of the gradient method is investigated. It is affected by the fact that, in general, the gradient of the functional to be minimized does not satisfy the Lipschitz condition.

Consider the set of functions

$$G = \left\{ u : \sigma(u) \leq \lim_{p \to \infty} \inf_{u_p \in H_1} \sigma\left(u_p\right) \right\} . \tag{5.17}$$

Since functional (5.1) is increasing, the set G is bounded [43].

Definition 5.1 [68] A sequence $\{u_p\}$ is *quasi-Fejer convergent* to a set G if for any $u \in G$ there exists a numerical sequence $\{\varepsilon_p\}$ such that $\varepsilon_p \geq 0$, $\sum_{p=1}^{\infty} \varepsilon_p < \infty$, and $\|u_{p+1} - u\|^2 \leq \|u_p - u\|^2 + \varepsilon_p$.

Theorem 5.3

Let the set G defined by (5.17) contain a global minimum u_* of the functional $\sigma(u)$, the sequence $\{u_p\}$ of the iterative scheme (5.5) with $\lambda_p = \lambda = const$, $0 < \lambda < 1$ be minimizing for $\sigma(u)$, and $\sigma(u)$ be convex in a neighborhood of u_*. Then $\{u_p\}$ converges to one of the global minimum points and

$$\left\| u_p - u_* \right\|^2 \leq 4 d_G\left(u_p\right) + 2 \sum_{k \geq p} \|u_{k+1} - u_k\|^2 . \tag{5.18}$$

Proof: Consider the equality

$$\left\| u_{p+1} - u \right\|^2 - \left\| u_p - u \right\|^2 - \left\| u_{p+1} - u_p \right\|^2 = -2 \operatorname{Re}\left(u - u_p, u_{p+1} - u_p\right) \tag{5.19}$$

for $\{u_p\}$ obtained by (5.5), $u \in G$. Using the convexity of σ and (5.19), we can continue (5.19) as follows:

$$-2 \operatorname{Re}\left(u - u_p, u_{p+1} - u_p\right) = 2\lambda \operatorname{Re}\left(u - u_p, \operatorname{grad} \sigma\left(u_p\right)\right) \leq 2\lambda[\sigma(u) - \sigma\left(u_p\right)] \leq 0 . \tag{5.20}$$

Then

$$\left\| u_{p+1} - u \right\|^2 \leq \left\| u_p - u \right\|^2 + \left\| u_{p+1} - u_p \right\|^2 . \tag{5.21}$$

From (5.12) we get the inequality

$$\sigma\left(u_{p+1}\right) - \sigma\left(u_p\right) \leq -\gamma \left\| u_p - u_{p+1} \right\|^2 \tag{5.22}$$

for the gradient method with $0 < \lambda < 1$ and $\gamma > 0$.

Summing inequalities (5.22) for all p from 0 to a certain m, we obtain

$$\sum_{p=0}^{m} \left\| u_p - u_{p+1} \right\|^2 \leq \frac{1}{\gamma} \left(\sigma\left(u_0\right) - \sigma\left(u_m\right)\right) \leq \frac{1}{\delta} \left(\sigma\left(u_0\right) - \sigma\left(u_*\right)\right) , \tag{5.23}$$

where u_0 is the initial approximation to u, $\delta > 0$. As $m \to \infty$, we get

$$\sum_{p=0}^{\infty} \left\| u_p - u_{p+1} \right\|^2 \leq \infty .$$

Hence, $\{u_p\}$ is a quasi-Fejer sequence convergent to the set G. This fact extends the result of Theorem 5.2. In particular (see [67]), the sequence $\{\| u_p - u\|^2\}$ converges for all $u \in G$. If the sequence $\{u_p\}$ is minimizing for the functional (5.1), then $\{u_p\}$ is weakly convergent and has a unique accumulation point.

Functional (5.1) is weakly lower semi-continuous and increasing [26]. This fact guarantees the weak closeness of the set G [43] and its weak compactness. If the sequence $\{u_p\}$ is minimizing for functional (5.1), then its weak accumulation point belongs to G.

According to [68], Proposition 3.3, (iii), if the gradient method is quasi-Fejer convergent to a set G, then the sequence $\{\| u_p - u\|\}$ converges for any $u \in G$. If the sequence of the gradient method is minimizing, then it has a strong accumulation point.

Since the weak accumulation point belongs to G and the strong accumulation point exists, then the sequence $\{u_p\}$ strongly converges to a point in G (see [68], Theorem 3.11).

Obviously, G is closed. According to [68], Theorem 3.11, (iv), the strong convergence to a point in G is equivalent to $\underline{\lim}_{p\to\infty} d_G(u_p) = 0$, where $d_G(u_p)$ is the distance between G and the sequence $\{u_p\}$. Then, according to Theorem 3.13 from [68], estimation (5.18) holds. □

5.1.1.2 **The case of a compact operator**

If the linear operator A acting from a complex functional Hilbert space H_1 into H_2 is compact, then the functional

$$\sigma_t(u) = ||F - |Au|||^2 + t\|u\|^2 \tag{5.24}$$

should be used instead of (5.1), where $t > 0$ is a real parameter and, as before, $F \in H_2$ is a given non-negative function, (see 3.100). The function u_* providing a global minimum to $\sigma_t(u)$ exists ([26], Theorem 1.3.2). In a similar way to the above case, the expression for the gradient of functional (5.24) can be obtained:

$$\operatorname{grad} \sigma_t\left(u\right) = tu + A^* Au - A^*\left[F \exp\left(i \arg Au\right)\right] . \tag{5.25}$$

We considered the gradient method based on the recurrent scheme

$$u_{p+1} = u_p - \lambda_p \operatorname{grad} \sigma_t \left(u_p\right) \tag{5.26}$$

with $\lambda_p > 0$ depending, in general, on the step number p.

Theorem 5.4

In a neighborhood of a global minimum u_* of the functional (5.24), the gradient method (5.26) is relaxational if $0 < \lambda_p < 1/(t + \|A\|^2)$; moreover, the sequence $\{u_p\}$ is minimizing for the functional $\|\operatorname{grad} \sigma_t(u)\|$, that is

$$\lim_{p\to\infty} \left\|\operatorname{grad} \sigma_t \left(u_p\right)\right\| = 0\,. \tag{5.27}$$

Proof: In a similar way to (5.8), the functional (5.24) can be shown to be increasing. Therefore, the set $U_0 = \{u : \sigma_t(u) \le \sigma_t(u_0)\}$ is bounded, where u_0 is the initial approximation to u.

Owing the convexity of the functional $(F, |Au|)$ in U_0, inequality (5.10) is valid, that is,

$$\operatorname{Re}\left(A^* F \exp\left(i \arg Au\right) - A^* F \exp\left(i \arg Av\right), u - v\right) \ge 0 \tag{5.28}$$

for any $u, v \in U_0$.

We find the condition for the sequence (5.26) to be relaxational, that is, $\sigma_t(u_{p+1}) < \sigma_t(u_p)$. Using the Lagrange mean value theorem, we can write

$$\sigma_t\left(u_p\right) - \sigma_t\left(u_{p+1}\right) = \operatorname{Re}\left(\operatorname{grad} \sigma_t\left(u_{p+1} + \tau\left(u_p - u_{p+1}\right)\right), u_p - u_{p+1}\right), \tag{5.29}$$

where $0 < \tau < 1$.

After performing derivations similar to (5.12), and using (5.10) and (5.26) and the obvious inequality

$$\left\|A\left(\operatorname{grad} \sigma_t\left(u_p\right)\right)\right\|_2^2 \le \|A\|^2 \left\|\operatorname{grad} \sigma_t\left(u_p\right)\right\|_1^2, \tag{5.30}$$

we have

$$\sigma_t\left(u_p\right) - \sigma_t\left(u_{p+1}\right) \ge \left((\tau - 1)\,\|A\|^2 \lambda_p^2 + t(-\lambda_p^2 + \tau\lambda_p^2) + \lambda_p\right) \times \left\|\operatorname{grad} \sigma_t\left(u_p\right)\right\|^2. \tag{5.31}$$

The right-hand side of (5.31) is larger than zero if

$$\lambda_p < \frac{1}{t + \|A\|^2}\,. \tag{5.32}$$

This completes the proof of the first part of the theorem.

The sequence $\{\sigma_t(u_p)\}$ is bounded and monotonic; hence it converges. Since the left-hand side of (5.31) tends to zero as $p \to \infty$ and the first multiplier on the right-hand side is limited, the second multiplier tends to zero as well, that is, (5.27) is true. This completes the proof of the theorem. □

Theorem 5.5

If the sequence $\{u_p\}$ is minimizing for the functional $\sigma_t(u_p)$, then the iterative process (5.26) generates a weakly convergent sequence, and its weak limit belongs to the set of solutions to the equation

$$tu + A^* Au = A^*\left[F \exp\left(i \arg Au\right)\right] . \tag{5.33}$$

Proof: First, we prove that the sequence $\{u_p\}$ is a weak compact. To this end, we show that the functional σ_t is weakly lower semi-continuous. It is known (see [43], Theorem 8.1) that any functional $g(x)$ in a normed space is weakly lower semi-continuous if and only if the set $E_c = \{x : g(x) \leq c\}$ is weakly closed for any constant c.

The functional $\sigma_t(u)$ can be written as

$$\sigma_t(u) = (F, F)_2 - 2(F, |Au|)_2 + (Au, Au)_2 + t\|u\|^2 . \tag{5.34}$$

The weak continuity of the functional $(Au, Au)_2 = (|Au|, |Au|)_2$ is obvious. Let $u_n \to u_0$ weakly. From the compactness of the operator A, it follows that $Au_n = f_n \to f_0 = Au_0$ strongly. The weak continuity of the functional $(F, |Au|)$ follows from the inequality

$$|(F, |Au|) - (F, |Au_0|)| = |(F, |f_n| - |f_0|)| \leq \|F\| \|f_n - f_0\| . \tag{5.35}$$

The weakly lower semi-continuity of the functional (u, u) is known. The functional $\sigma_t(u)$ is weakly lower semi-continuous as the sum of the weakly lower semi-continuous and weakly semi-continuous functionals.

The functional $\sigma_t(u)$ is increasing. This fact follows from the inequality

$$\sigma_t(u) \geq (F, F) - 2\|F\| \|A\| u + (Au, Au) + t\|u\|^2 . \tag{5.36}$$

We introduce the set $U_0 = \{u : \sigma_t(u) \leq \sigma_t(u_0)\}$, where u_0 is the initial approximation for the iterative process (5.26). From the above mentioned Theorem 8.1 from [43], the weak closure of the set U_0 follows. Since the functional $\sigma_t(u)$ is growing, the set U_0 is bounded [43]. It is known [73] that any bounded weakly closed set of a reflexive Banach space is a weak compact. Since H_1, as any Hilbert space, is reflexive and the set U_0 is a bounded weakly closed set, then U_0 is a weak compact. Consequently, the minimizing sequence $\{u_p\}$ belongs to a weak compact, and a weakly convergent subsequence can by separated from it, such that its limiting function belongs to this compact, too.

Further proof of the theorem repeats that from Theorem 5.2 for the isometric operator A. □

If (5.24) is convex in a neighborhood of its global minimum, then the minimizing sequence $\{u_p\}$ converges in norm to this point (see [67] and [68]).

5.1.2
The simplest iteration method

5.1.2.1 The case of an isometric operator

Consider the functional (3.6) for the case of the isometric operator A. The Lagrange–Euler equation for this functional can be written in the form (3.45) or (3.46). The latter equation has an obvious property: when f_* solves (3.46), then $f_* \exp(i\alpha)$ with any real constant α also solves (3.46).

In Chapter 3, the iterative method (3.60) for solving this equation was proposed and its relaxivity was established. Here we investigate its convergence in detail.

The iteration scheme of the method is written in the two following forms (see (3.61b) and (3.60)):

$$u_{p+1} = A^*\left[F \exp\left(i \arg Au_p\right)\right], \tag{5.37}$$

and

$$f_{p+1} = AA^*\left[F \exp\left(i \arg f_p\right)\right], \tag{5.38}$$

where $f_p = Au_p$.

Lemma 5.2

(See [69]). Let $A : H_1 \rightarrow H_2$ be a linear isometric operator and $F \geq 0$ be a given real function from H_2. Then the sequence $\{u_p\}$, $p = 1, 2, \ldots$, $u_1 \neq \{0\}$, obtained by (5.37), has the two equivalent properties

$$\lim_{p \to \infty} \left\| u_{p+1} - u_p \right\| = 0\,, \tag{5.39}$$

$$\lim_{p \to \infty} \left\| \operatorname{grad} \sigma\left(u_p\right) \right\| = 0\,. \tag{5.40}$$

It is known that any isometric operator A is continuous. Hence (5.39) gives

$$\lim_{p \to \infty} \left\| f_{p+1} - f_p \right\| = 0\,. \tag{5.41}$$

Let both f_* and

$$f_{*p} = f_* e^{i\alpha_p} \tag{5.42}$$

with a real constant α_p be the solutions to (5.15). The constant α_p will be calculated at each step of iterations (5.38).

We write $f_* = f_*' + i f_*''$ and introduce the operators

$$B'v = \operatorname{Re}\left(iAA^*\left[F \exp(i \arg f_*)\right]v\right) \tag{5.43a}$$

$$B''v = \operatorname{Im}\left(iAA^*\left[F \exp(i \arg f_*)\right]v\right), \tag{5.43b}$$

$$Bv = B'v + iB''v \tag{5.43c}$$

for any real v.

Theorem 5.6

Let the conditions of Lemma 5.2 be satisfied. If the equation

$$f'_* B'' \varphi_m - f''_* B' \varphi_m = \lambda_m \varphi_m |f_*|^2 \tag{5.44}$$

has the eigenvalue $\lambda_0 = 1$ of multiplicity one. Then the sequence $\{f_p\}$ converges to a set of solutions to (5.15) in the norm of the space H_2. In the first approximation, $h_p = f_p - f_{*p}$ satisfies the inequality

$$\|h_p\| \leq \left(1 + \frac{\|B\| \|F\|}{|1 - \lambda_k|}\right) \|\eta_p\| , \tag{5.45}$$

where λ_k is the eigenvalue of (5.44), closest to unity,

$$\eta_p = AA^*\left[F \exp\left(i \arg f_p\right)\right] - f_p . \tag{5.46}$$

Proof: According to Lemma 5.2 and notations (5.46) and (5.38), we have $\lim_{p\to\infty} \|\eta_p\| = 0$. Substituting $f_p = f_{*p} - h_p$ into the right-hand side of (5.46) and keeping the linear part of h_p, we get

$$h_p - iAA^*\left[F \exp\left(i \arg f_{*p}\right) \operatorname{Im}\left(h_p / f_{*p}\right)\right] = \eta_p . \tag{5.47}$$

We show that there exists a constant C such that $\|h_p\| \leq C\|\eta_p\|$. First, we reduce the complex equation (5.47) for the complex function f_{*p} to a real equation for the real function $\operatorname{Im} z_p$, where

$$z_p = \frac{h_p}{f_{*p}} . \tag{5.48}$$

Introduce the operator B_p, as follows

$$B_p[\operatorname{Im}(z)] = iAA^*\left[\left(F \exp\left(i \arg f_{*p}\right)\right) \operatorname{Im}(z)\right] . \tag{5.49}$$

Then (5.47) becomes of the form

$$z_p f_{*p} - B_p z''_p = \eta_p . \tag{5.50}$$

Separating the real and imaginary parts from (5.50), we reduce it to the equation system

$$z'_p f'_{*p} - z''_p f''_{*p} - B'_p z''_p = \eta'_p , \tag{5.51a}$$

$$z'_p f''_{*p} + z''_p f'_{*p} - B''_p z''_p = \eta''_p , \tag{5.51b}$$

where $z_p = z'_p + i z''_p$,

$$B'_p v = \operatorname{Re}(iAA^*[F \exp(i \arg f_{*p}) v]) , \tag{5.52a}$$

$$B''_p v = \operatorname{Im}(iAA^*[F \exp(i \arg f_{*p}) v]) . \tag{5.52b}$$

Eliminating the function z'_p from (5.51), we obtain

$$|f_{*p}|^2 z''_p + f''_{*p} B'_p z''_p - f'_{*p} B''_p z''_p = \eta''_p f'_{*p} - f''_{*p} \eta'_p = \mathrm{Im}\left(\bar{f}_{*p} \eta_p\right). \quad (5.53)$$

It is easy to show that the function $\varphi_0 = const$ is an eigenfunction corresponding to the eigenvalue $\lambda_0 = 1$ of the homogeneous equation

$$\lambda_m \varphi_m |f_{*p}|^2 + f''_{*p} B'_p \varphi_m - f'_{*p} B''_p \varphi_m = 0 . \quad (5.54)$$

Equations (5.54) and (5.44) are equivalent, that is, their eigenvalues and eigenfunctions are the same.

It is known, that the restriction of the linear operator D_p,

$$D_p u = |f_{*p}|^2 u - f''_{*p} B'_p u - f'_{*p} B''_p u \quad (5.55)$$

onto $H_2 \backslash \{\varphi_0\}$ is continuously invertible when the right-hand side of (5.53) is orthogonal to the corresponding eigenfunction of the adjoint operator D_p^*. According to the definition of the adjoint operator,

$$\left(D_p u, v\right)_2 = \left(u, D_p^* v\right)_1 . \quad (5.56)$$

From (5.56) we obtain

$$D_p^* v = |f_{*p}|^2 v + B_p'^*\left[f''_{*p} v\right] - B_p''^*\left[f'_{*p} v\right], \quad (5.57)$$

where

$$B_p'^*\left[f''_{*p} v\right] = -\frac{F f''_{*p}}{|f_{*p}|} AA^*\left[f''_{*p} v\right], \quad (5.58)$$

$$B_p''^*\left[f'_{*p} v\right] = \frac{F f'_{*p}}{|f_{*p}|} AA^*\left[f'_{*p} v\right]. \quad (5.59)$$

The right-hand side of (5.57) vanishes on function $v_0 = F/|f_{*p}|$, which means that $v_0 \in \mathrm{Ker}\, D_p^*$.

The condition for the continuous invertibility of the linear operator from (5.53) has the form

$$\left(\mathrm{Im}\left(\bar{f}_{*p} \eta_p\right), \frac{F}{|f_{*p}|}\right)_2 = 0 . \quad (5.60)$$

This condition holds if

$$\alpha_p = \arg\left(\eta_p \exp\left(-i \arg f_*\right), F\right)_2 . \quad (5.61)$$

In this case, the restriction of the operator D_p^{-1} is bounded on the closed subspace $H_2/\{v_0\}$, where $\{v_0\}$ is the subspace spanned on v_0. From (5.53), we have

$$\left\|z''_p\right\| \le \left\|D_p^{-1}\right\| \left\|\eta_p\right\| \left\|f_{*p}\right\| . \quad (5.62)$$

Using the spectral norm of the operator, we get

$$\|z_p''\| \leq \frac{1}{|1-\lambda_k|} \|\eta_p\| \|f_{*p}\|, \tag{5.63}$$

where λ_k is the eigenvalue of (5.44), closest to unity.

Using the equivalence of (5.54) and (5.44), from (5.48) and (5.50), we obtain

$$\|h_p\| = \|f_{*p} z_p\| = \|\eta_p + B_p z_p''\|. \tag{5.64}$$

Substituting (5.63) into (5.64) and using the norm equality $\|B_p\| = \|B\|$ which follows from (5.49), (5.42), (5.43c), we have

$$\|h_p\| \leq \left(1 + \frac{\|B\| \|f_{*p}\|}{|1-\lambda_k|}\right) \|\eta_p\|. \tag{5.65}$$

It can be seen that, if f_{*p} is a solution to (5.15), then the inequality

$$\sigma(u_*) = \|F\|^2 - \|f_{*p}\|^2 \geq 0 \tag{5.66}$$

holds. This means that $\|f_{*p}\|$ in (5.65) can be substituted by $\|F\|$, which gives (5.45). □

Since, at the branching points $\lambda_k = 1$, then estimation (5.45) means that the convergence rate rapidly decreases in the neighborhood of these points.

5.1.2.2 The case of a compact operator

Consider functional (5.24) for the case of the compact operator A. The Lagrange–Euler equation for this functional can be written in the form (see 3.102)

$$tu + A^* A u = A^* \left[F \exp\left(i \arg Au\right)\right], \tag{5.67}$$

where A^* is the operator adjoint to A.

To solve (5.67), we use the iterative method

$$t u_{p+1} + A^* A u_{p+1} = A^* \left[F \exp\left(i \arg A u_p\right)\right], \quad (p = 0, 1, 2, \ldots), \quad u_0 \neq 0. \tag{5.68}$$

For the iterative process (5.68) the following lemma holds (cf. [26], p. 228).

Lemma 5.3

Let A be a linear compact operator acting from H_1 into H_2, $F \in H_2$ be a given real non-negative function, and the initial approximation u_0 in (5.68) satisfy the condition $Au_0 \neq 0$. Then the sequence $\{u_p\}$, $p = 1, 2, \ldots$, calculated from (5.68) is minimizing for the functional $\|\text{grad}\sigma_t(u)\|$, that is

$$\lim_{p\to\infty} \|\text{grad}\,\sigma_t(u_p)\| = \lim_{p\to\infty} \left\|t u_p + A^* A u_p - A^* \left[F \exp\left(i \arg A u_p\right)\right]\right\| = 0. \tag{5.69}$$

Proof: Denote $f_p = Au_p$, $\Psi_p = \arg f_p$. Similar to (3.65), it can be shown (see [26], p. 230) that

$$\sigma_t(u_p) - \sigma_t(u_{p+1}) \geq 2\left(F\left[1 - \cos\left(\Psi_p - \Psi_{p+1}\right)\right], \left|Au_{p+1}\right|\right) \geq 0\,. \qquad (5.70)$$

This gives

$$\lim_{p\to\infty} \left[\sigma_t(u_p) - \sigma_t(u_{p+1})\right] = 0\,. \qquad (5.71)$$

From (5.70), we obtain

$$\lim_{p\to\infty} \left(F\left[1 - \cos\left(\Psi_p - \Psi_{p+1}\right)\right], \left|Au_{p+1}\right|\right)_2 = 0\,. \qquad (5.72)$$

Owing to the non-negativity of both multipliers in the scalar product, either

$$\lim_{p\to\infty} \left\|Au_{p+1}\right\| = 0 \qquad (5.73)$$

or

$$\lim_{p\to\infty} \left\|F\left(1 - \cos\left(\Psi_p - \Psi_{p+1}\right)\right)\right\| = 0 \qquad (5.74)$$

is true. Let equality (5.73) hold. We take any $v_0 \neq 0$ such that $|Av_0| \neq 0$ identically, and consider the function $u_0 = Cv_0$, where the constant C is chosen from the condition that $\sigma_t(u_0)$ as the function of C is minimal. The optimal value of C is

$$C = \frac{(F, |Av_0|)_2}{t\,(|Av_0|, |Av_0|)_2}\,. \qquad (5.75)$$

With the aid of (5.75), we obtain

$$\|F\|^2 - \sigma_t(u_0) > 0\,. \qquad (5.76)$$

Due to the monotonicity of $\{\sigma_t(u_p)\}$, the following inequality takes place:

$$\|F\|^2 - \sigma_t(u_0) < \left(2F, \left|Au_p\right|\right) + t\left(Au_p, Au_p\right)\,. \qquad (5.77)$$

Applying the Cauchy–Bunyakovski inequality to $(2F, |Au_p|)$ and (Au_p, Au_p), we obtain the quadratic inequality with respect to $\|Au_p\|$:

$$t\|Au_p\|^2 + 2\|F\|\|Au_p\| - \|F\|^2 + \sigma_t(u_0) > 0\,. \qquad (5.78)$$

The solution to (5.78) is

$$\|Au_p\| > -\frac{\|F\|}{t} + \sqrt{\frac{\|F\|^2}{t^2} + \frac{1}{t}\left(\|F\|^2 - \sigma_t(u_0)\right)}\,. \qquad (5.79)$$

From this, if $\lim_{p\to\infty} \|Au_p\|$ exists, then it is larger than zero. According to (5.68), the function u_p does not depend on the constant C (5.75). Hence, (5.79) means that (5.73) cannot be true, and (5.74) holds. Choosing, for the uniqueness, $-\pi \le \Psi_p < \pi,\;\; p = 1, 2, \ldots,$ we obtain (5.74).

The iteration process (5.68) can be written in the form:

$$u_{p+1} = \left(tE + A^*A\right)^{-1} A^*\left[F \exp\left(i \arg Au_p\right)\right]. \tag{5.80}$$

Then

$$\begin{aligned} &u_{p+1} - u_p \\ &= \left(tE + A^*A\right)^{-1} A^*\left[F\left(\exp\left(i \arg Au_p\right) - \exp\left(i \arg Au_{p-1}\right)\right)\right] \\ &= 2i\left(tE + A^*A\right)^{-1} A^*\left[F \sin\left(\left(\Psi_p - \Psi_{p-1}\right)/2\right) \exp\left(i\left(\Psi_p - \Psi_{p-1}\right)/2\right)\right], \end{aligned} \tag{5.81}$$

and

$$\left\|u_{p+1} - u_p\right\| \le 2\left\|\left(tE + A^*A\right)^{-1}\right\|\left\|A^*\right\|\left\|F\right\|\left\|\sin\left(\left(\Psi_p - \Psi_{p-1}\right)/2\right)\right\|. \tag{5.82}$$

Together with (5.74), this inequality gives

$$\lim_{p\to\infty}\left\|u_{p+1} - u_p\right\| = 0. \tag{5.83}$$

On the other hand, from (5.25) and (5.80) it follows that

$$\operatorname{grad}\sigma_t\left(u_p\right) = \left(tE + A^*A\right)\left(u_{p+1} - u_p\right). \tag{5.84}$$

Hence, (5.69) is true, which completes the proof of the lemma. □

Consider the other form of (5.67) (see 3.103)

$$tf + AA^*f = AA^*\left[F \exp\left(i \arg(f)\right)\right], \tag{5.85}$$

and write the iterative process

$$tf_{p+1} + AA^*f_{p+1} = AA^*\left[F \exp\left(i \arg\left(f_p\right)\right)\right], \tag{5.86}$$

identical, in fact, to (5.68).

Before formulating the theorem that establishes the convergence conditions for this process, we introduce the operator D acting from H_2 into H_2, as follows:

$$D\varphi = |f_*|^2\varphi + f_*'' K^{-1} B'\varphi - f_*' K^{-1} B''\varphi, \tag{5.87}$$

where

$$K = tE + AA^* \,, \tag{5.88a}$$

$$Bu = B'u + iB''u \,, \tag{5.88b}$$

$$B'u = \mathrm{Re}\left(iAA^*\left[F\exp\left(i\arg f_*\right)u\right]\right)\,, \tag{5.88c}$$

$$B''u = \mathrm{Im}\left(iAA^*\left[F\exp\left(i\arg f_*\right)u\right]\right)\,, \tag{5.88d}$$

$$f_* = f'_* + if''_*\,, \quad f_* = Au_*\,. \tag{5.88e}$$

Theorem 5.7

If the kernel of the operator D contains only the function $\varphi_0 = const$, then there exists a bounded set of functions containing all elements of the sequence $\{f_p\}$ convergent to the set of solutions to (5.85) in the norm of the space H_2 with the rate of the geometrical progression with the denominator $q < 1$, and the following estimations take place:

$$\|f_p - f_{*p}\| \le \left(1 + \|D_r^{-1}\|\|B\|\|K^{-1}\|\|f_*\|\right)\|K^{-1}\|\|f_{p+1} - f_p\|\,; \tag{5.89}$$

$$\|f_{p+2} - f_{p+1}\| \le q\|f_{p+1} - f_p\|\,; \tag{5.90}$$

where f_*, f_{*p} are certain solutions to (5.85),

$$q = 1 - \frac{1}{1 + \|D_r^{-1}\|\|B\|\|K^{-1}\|\|f_*\|}\,, \tag{5.91}$$

D_r is the restriction of the operator D onto the bounded subspace of the space H_2 without linear shell, spanned on the function φ_0.

Proof: The proof is similar to that for Theorem 5.6. Let a function f_* solve equation (5.85). Together with f_* we consider the functions

$$f_{*p} = f_*\exp\left(i\alpha_p\right)\,, \tag{5.92}$$

with constants α_p chosen at each step of iterations (5.86); these functions also solve (5.85).

It can be seen that

$$f_{p+1} - f_p = tf_p + AA^*f_p - AA^*\left[F\exp\left(i\arg\left(f_p\right)\right)\right]\,. \tag{5.93}$$

Denote

$$h_p = f_p - f_{*p}\,, \tag{5.94}$$

$$\eta_p = f_{p+1} - f_p\,. \tag{5.95}$$

From the above lemma and the boundedness of the operator A, it follows that

$$\lim_{p\to\infty} \|\eta_p\| = \lim_{p\to\infty} \left\| t f_p + AA^* f_p - AA^*\left[F \exp\left(i \arg\left(f_p\right)\right)\right]\right\| = 0 \,. \tag{5.96}$$

After substituting $f_p = f_{*p} - h_p$ into the right-hand side of (5.93), extracting the main part with respect to h_p in $\exp\left(i \arg\left(f_p\right)\right)$, and using (5.95), we obtain

$$t h_p + AA^* h_p - i AA^* \left[F \exp\left(i \arg f_{*p}\right) \operatorname{Im}\left(h_p / f_{*p}\right)\right] = \eta_p \,. \tag{5.97}$$

We introduce the new function

$$z_p = \frac{h_p}{f_{*p}} \tag{5.98}$$

and denote

$$B_p v = i AA^* \left[F \exp\left(i \arg f_{*p}\right) v\right] . \tag{5.99}$$

Then (5.97) attains the form

$$\left(t E + AA^*\right) z_p f_{*p} - B_p \left(z_p''\right) = \eta_p \,. \tag{5.100}$$

Here the notation $z_p = z_p' + i z_p''$ is used.

Separating the real and imaginary parts in (5.100) leads to the following equation system:

$$\left(t E + AA^*\right)\left(z_p' f_{*p}' - z_p'' f_{*p}''\right) - B_p' z_p'' = \eta_p' \,, \tag{5.101a}$$

$$\left(t E + AA^*\right)\left(z_p' f_{*p}'' + z_p'' f_{*p}'\right) - B_p'' z_p'' = \eta_p'' \,. \tag{5.101b}$$

After eliminating the unknown function z_p' from (5.101), we obtain

$$\left|f_{*p}\right|^2 z_p'' + f_{*p}'' K^{-1} B_p' z_p'' - f_{*p}' K^{-1} B_p'' z_p'' = \operatorname{Im}\left(\bar{f}_{*p} K^{-1} \eta_p\right) , \tag{5.102}$$

where $K = t E + AA^*$.

The function $z_p'' = const$ makes the left-hand side of (5.102) zero identically. The restriction of the operator from the left-hand side of (5.102) onto the closed subspace of H_2 with excluding the linear shell spanned onto the function $z_p'' = const$ is continuously invertible if the right-hand side of (5.102) is orthogonal to the eigensubspace of the respective adjoint operator and the dimension of his kernel equals the unity. We find the kernel of this operator. By the definition

$$\left(D_p u, v\right) = \left(u, D_p^* v\right) , \tag{5.103}$$

where

$$D_p u = \left|f_{*p}\right|^2 u + f_{*p}'' K^{-1} B_p' u - f_{*p}' K^{-1} B_p'' u \,. \tag{5.104}$$

Operators D and D_p are equivalent, that is, their eigenvalues and eigenfunctions are the same.

From (5.103), we immediately obtain

$$D_p^* v = \left| f_{*p} \right|^2 v + \left(K^{-1} B_p' \right)^* \left[f_{*p}'' v \right] - \left(K^{-1} B_p'' \right)^* \left[f_{*p}' v \right], \tag{5.105}$$

where

$$\left(K^{-1} B_p' \right)^* \left[f_{*p}'' v \right] = - \frac{F f_{*p}''}{\left| f_{*p} \right|} A A^* \left(K^{-1} \right)^* \left[f_{*p}'' v \right], \tag{5.106a}$$

$$\left(K^{-1} B_p' \right)^* \left[f_{*p}' v \right] = - \frac{F f_{*p}'}{\left| f_{*p} \right|} A A^* \left(K^{-1} \right)^* \left[f_{*p}' v \right]. \tag{5.106b}$$

The function $v_0 = F / \left| f_{*p} \right|$ makes the right-hand side of (5.105) zero identically, hence v_0 belongs to the kernel of D_p^*. The condition for the operator in the left-hand side to be continuously invertible is

$$\left(\mathrm{Im} \left(\bar{f}_{*p} K^{-1} \eta_p \right), \frac{F}{\left| f_{*p} \right|} \right) = 0 . \tag{5.107}$$

This condition is fulfilled if the constant α_p is chosen as

$$\alpha_p = \arg \left(K^{-1} \eta_p \exp \left(-i \arg f_* \right), F \right) . \tag{5.108}$$

From (5.102), it follows that

$$\left\| z_p'' \right\| \le \left\| D_r^{-1} \right\| \left\| K^{-1} \right\| \left\| \eta_p \right\| \left\| f_{*p} \right\| . \tag{5.109}$$

Similarly as in the case on isometric operator, here we use the spectral norm of the operator D_r^{-1}. Equations (5.98), (5.100) give

$$\left\| h_p \right\| = \left\| f_{*p} z_p \right\| = \left\| K^{-1} B_p z_p'' + K^{-1} \eta_p \right\| . \tag{5.110}$$

From this, using (5.109) the triangular inequality and the norm equality $\| B_p \| = \| B \|$, we obtain

$$\left\| h_p \right\| \le \left(1 + \left\| D_r^{-1} \right\| \left\| B \right\| \left\| K^{-1} \right\| \left\| f_* \right\| \right) \left\| K^{-1} \right\| \left\| \eta_p \right\| . \tag{5.111}$$

After taking into account (5.94) and (5.95), this inequality coincides with (5.89). In order to prove inequality (5.90), it is sufficient to write, from (5.89):

$$\left\| f_* - f_{p+1} \right\| - \left\| f_* - f_p \right\| \le L \left(\left\| f_{p+2} - f_{p+1} \right\| - \left\| f_{p+1} - f_p \right\| \right) , \tag{5.112}$$

where $L = (1 + \| D_r^{-1} \| \| B \| \| K^{-1} \| \| f_* \|)$. Then (5.90) holds with $q = 1 - L^{-1}$. □

In a similarly way to the previous case, since the norm of the operator D_r^{-1} tends to infinity when approaching the branching points, estimation (5.111) means that the convergence rate rapidly decreases in the neighborhood of these points.

5.2
Methods of Newton type

The Newton method with its modifications is a powerful technique for solving nonlinear equations of a different nature. Its main advantages are simplicity, universality, and, except in the vicinity of specific points, fast divergence. The disadvantages are the need to deal with matrices of large dimensions in multi-dimensional problems, the necessity for initial approximations from the 'area of influence' of the solution to be found in multi-solution problem, and the previously mentioned, slow divergence near specific points, such as branching points. The last disadvantage is essential for the problems considered here. There are various approaches to weaken this effect (see, e.g. [16]). The use of the singular value decomposition seems to be the most useful for this purpose [74]. It allows us not only to exclude the influence of the small singular values near such points and pass through them, but also to detect these points and to determine the directions in which new solutions branch off.

In this section this modification of the method is briefly described and applied for the solution of several concrete problems considered in the book. As a rule, it is applied together with the simple iteration methods which possess the relaxation property and permit us to find the 'optimal solutions', that is, the extreme points of the functionals for which the considered equation is the Lagrange–Euler equation. As shown above in Theorems 5.6 and 5.7, the convergence of these methods rapidly decreases near the branching points.

5.2.1
General scheme of the method

Let the nonlinear equation to be solved have the form

$$\Phi(f, c) = 0\,. \tag{5.113}$$

Here Φ is a nonlinear operator-function, f is an unknown complex function, c is a real parameter. If at certain c a function f exists which satisfies (5.113), then we call the pair (f, c) a solution to this equation. Assume that f depends on c continuously. Then any solution (f, c) of (5.113) belongs to a branch of solutions (is a point of this branch). Our purpose is to construct a method of the Newton type to determine such branches and to recognize the branching points on them, if they exist.

Following the general idea of the Newton method, we assume that (c_p, f_p) is the current (pth) approximation to a point on the branch and find the next approximation from the requirement that (5.113) be satisfied in the first order of the increments $\varepsilon = c_{p+1} - c_p$, $\delta = f_{p+1} - f_p$. This leads to the equation

$$\Phi_c \cdot \varepsilon + \Phi_f(f_p, c_p, \delta) = -\Phi_p\,, \tag{5.114}$$

where

$$\Phi_c = \left.\frac{\partial \Phi}{\partial c}\right|_{c=c_p}\,, \tag{5.115}$$

$\Phi_f(f_0, c_0, \delta)$ is defined from the relation

$$\Phi(f, c) = \Phi_p + \Phi_c \cdot \varepsilon + \Phi_f(f_p, c_p, \delta) + o(\varepsilon, \delta), \qquad (5.116)$$

$\Phi_p = \Phi(f_p, c_p)$. Equation (5.114) is linear with respect to ε and δ. After it has been solved, the next approximation is calculated as

$$c_{p+1} = c_p + \varepsilon, \quad f_{p+1} = f_p + \delta. \qquad (5.117)$$

Of course, (5.113) is underdetermined. Therefore, if the iterative procedure converges, then its limiting point can be, in general, any point on the branch. Equation (5.114) is underdetermined as well, and it has a family of solutions. If there are no additional conditions, then the solution with minimal norm should be chosen from them, that is, the next approximation should be the closest to the previous one. Such a choice allows us to avoid the situation where the new point on the branch is either very far from or very close to the previous known point. It is possible to keep the constant step between the points on the branch during the calculations, defining this step, for instance, by the relation

$$|\Delta c|^2 + \|\Delta f\|^2 = h^2, \qquad (5.118)$$

where $\Delta c = c_m - c_{m-1}$, $\Delta f = f_m - f_{m-1}$ and the indices represent the values for the previous already calculated point on the branch $(m-1)$, or the new one, to be calculated (m). This equation supplements (5.113). In this case the linear equation (5.114) is supplemented by the equation

$$\Delta c_p \cdot \varepsilon + (\Delta f_p, \delta) = 0, \qquad (5.119)$$

where $\Delta c_p = c_p - c_{m-1}$, $\Delta f_p = f_p - f_{m-1}$. Then the linear equation system (5.113), (5.119) is determined.

Equation (5.114) (itself or together with (5.119)) can best be solved by the singular value decomposition technique. For definiteness, we assume that the operator-function Φ, as well as the function f should be discretized and expressed as n-dimensional vectors. Then (5.114) is the linear algebraic system with an $n \times (n+1)$-dimensional matrix B and the $(n+1)$-dimensional vector of the unknown $x = \{\varepsilon; \delta\}$, as follows

$$Bx = -\Phi_p. \qquad (5.120)$$

As any rectangular complex matrix, B can be expressed in the form

$$B = USV^*, \qquad (5.121)$$

where U and V are $n \times n$ and $(n+1) \times (n+1)$ orthogonal matrices, respectively, S is the $(n+1) \times n$ diagonal matrix with non-negative elements (the last row of S is zero); the asterisk denotes the conjugate matrix. In other words, there are two systems of orthogonal vectors, u_k, $k = 1, 2, \dots, n$, and v_k, $k = 1, 2, \dots, n+1$, and n numbers $s_k \geq 0$, such that

$$B v_k = s_k u_k \,, \tag{5.122a}$$

$$B^* u_k = s_k v_k \,. \tag{5.122b}$$

The solutions to (5.120) are easily expressed by the vectors v_k, as follows

$$x = -\sum_{k=1}^{n} \frac{(u_k, \Phi_p)}{s_k} v_k + C v_{n+1} \,, \tag{5.123}$$

where C is an arbitrary constant. This arbitrariness is caused by the underdetermination of equation (5.120). If the solution of the minimal norm is to be found, then $C = 0$.

The form of solution (5.123) allows us to follow its behavior in the case when a singular number s_k tends to zero. Such a situation occurs, in particular, near the branching points. The singular vector v_k corresponding to the small s_k (zero at the branching point) shows the direction in which the branched off solution moves away from the parent one. This is a key idea in finding the branching points and their analysis. The problem is complicated by the need to separate this singular value from other small ones (and even zero as in the case when continual families of the solutions exist). As a rule, such difficulties can be successively overcome in concrete situations in appropriate ways. In any case, the vectors v_k corresponding to very small s_k should be excluded from the expression (5.123) in order to avoid instability of the algorithm.

In the case when the initial equation (5.113) is supplied by (5.118), then the linear equation system (5.114) for one step of the Newton method is supplied by (5.119). In this case the matrix B is square $(n+1) \times (n+1)$-dimensional, the vector Φ_p is supplied by the zero element, and the sum in (5.123) is extended to $k = n+1$, whereas the last addend, with an arbitrary factor C, is absent.

5.2.2
The particular case of a compact operator[1]

The amplitude-phase optimization problem for the compact operator A from the direct problem was formulated in Section 3.1.8. It is based on the functional

$$\sigma_t(u) = \big\| |Au| - F \big\|_2^2 + t \| u \|_1^2 \tag{5.124}$$

(see 3.100). Here t is a fixed positive real parameter. The Lagrange–Euler equation formulated for the function $f = Au$ has the form (5.113) with

$$\Phi(f, c) = t f + B\big[f - F \exp(i \arg f)\big] \,, \tag{5.125}$$

1) This section was written by M. I. Andriychuk and contains his results.

$B = AA^*$. For the $(p+1)$th step of the Newton method, the variations of Φ caused by perturbations $\varepsilon = c_{p+1} - c_p$, $\delta = f_{p+1} - f_p$ of the pth Newton approximation (c_p, f_p) to the point on the solution branch should be calculated. These variations are

$$\Phi_c = \left.\frac{\partial B}{\partial c}\right|_{c=c_p} \left[f_p - F\exp(i\arg f_p)\right], \tag{5.126a}$$

$$\Phi_f(f_p, c_p, \delta) = t\delta + B_p\left[\delta - iF\exp(i\arg f_p)\,\mathrm{Im}\,\frac{\delta}{f_p}\right]. \tag{5.126b}$$

Since the unknown complex function δ is involved under the operator B_p only in the form (δ/f_p), it is convenient to introduce two real functions y and z, as follows: $\delta = f_p \cdot (y + iz)$. Then (5.114) for ε, δ has the form

$$\Phi_c \cdot \varepsilon + (tI + B_p)[f_p \cdot y] + it(f_p \cdot z) + iB_p[(f_p - F\exp(i\arg f_p))\,z] = -\Phi(f_p, c_p), \tag{5.127}$$

where I is the unit operator. Acting on (5.127) with the operator $D = (tI + B_p)^{-1}$ yields

$$D[\Phi_c] \cdot \varepsilon + f_p \cdot y + if_p \cdot z - iDB_p[F\exp(i\arg f_p) \cdot z] = -D[\Phi_0(f_p, c_p)]. \tag{5.128}$$

After elimination of the function y, we get

$$\mathrm{Im}\{\bar{f}_p \cdot D[\Phi_c]\} \cdot \varepsilon + |f_p|^2 \cdot z - \mathrm{Re}\{\bar{f}_p \cdot DB_p[F\exp(i\arg f_p) \cdot z]\} = -\mathrm{Im}\{\bar{f}_p \cdot D[(f_p, c_p)]\}. \tag{5.129}$$

This is the linear equation for ε, z. After it has been solved, the unknown function δ is calculated from (5.128) and the next approximation to the point on the branch is found.

As a particular case of the problem with compact operator we consider the amplitude-phase optimization problem for the two-dimensional circular antenna of the radius a (see Section 3.1.8.5). In this case the radiation pattern $f(\varphi)$ is expressed (to within a nonessential constant factor) by the current $u(\theta)$ on the antenna, as follows

$$f(\varphi) = Au \equiv \int_0^{2\pi} u(\theta)\exp(ika\cos(\varphi - \theta))\mathrm{d}\theta\,, \tag{5.130}$$

where θ and φ are the angular coordinates on the antenna and in the far-field zone, respectively. The concrete forms of functions (5.125) and (5.126) are

$$\Phi(f, c) = tf(\varphi) + \int_0^{2\pi} J_0(c\sin((\varphi - \varphi')/2)) \cdot [f(\varphi) - F(\varphi)\exp(i\arg f(\varphi))]\,\mathrm{d}\varphi'\,, \tag{5.131a}$$

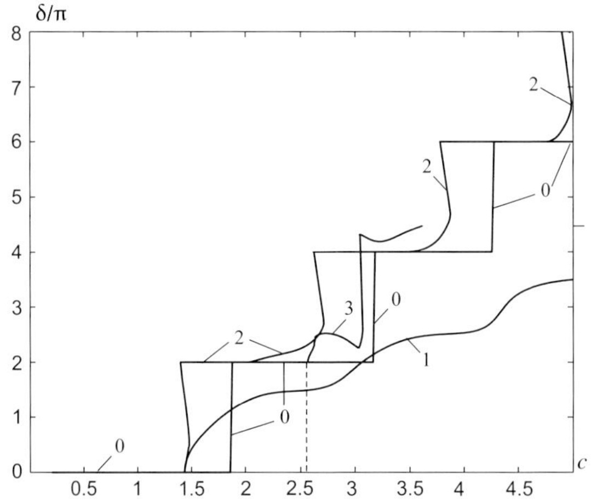

Figure 5.1 Phase increment of different solutions. $F = \sin^2(\varphi/2) + 0.1$; $t = 1$.

$$\Phi_c = -\int_0^{2\pi} \sin((\varphi - \varphi')/2)\cdot J_1\,(c\sin((\varphi - \varphi')/2)) \cdot \left[f_p(\varphi) - F(\varphi)\exp(i\arg f_p(\varphi))\right]\mathrm{d}\varphi'\,, \qquad (5.131b)$$

$$\Phi_f(f_p, c_p, \delta) = t\delta(\varphi) + \int_0^{2\pi} J_0(c\sin((\varphi - \varphi')/2)) \cdot \left[\delta(\varphi') - iF(\varphi')\exp(i\arg f_p(\varphi'))\,\mathrm{Im}\,\frac{\delta(\varphi')}{f_p(\varphi')}\right]\mathrm{d}\varphi'\,, \qquad (5.131c)$$

where J_0 is the Bessel function of the first kind, $c = 2ka$ is the variable parameter of the problem and, as before, a nonessential factor is omitted in (5.131).

The calculations were made for the problem with $F(\varphi) = \sin^2(\varphi/2) + 0.1$, $t = 1$. In Figure 5.1 the branching process is shown schematically. For identification of a solution its phase increment

$$\Delta = \max_{\varphi_1,\varphi_2 \in (0,2\pi)} \left|\arg(f(\varphi_1)) - \arg(f(\varphi_2))\right| \qquad (5.132)$$

over the circle is chosen. The curves describing different solutions f_j, $j = 0, 1, 2, 3$ are marked by their indices. A real solution f_0 exists at all c. As c increases the sign alternations appear in this solution; we assume that its phase increment grows jump-wise. There can exist several solutions of this type, with different points at which the sign alters. Here we consider only one of them. At $c = c_1 = 1.43$ two new solutions f_1, f_2 branch off from f_0, with different phase evenness: arg f_1 is even, arg f_2 is odd. In the first case the increment is a smooth function of φ, in the second one it has unexpected jumps. At $c = c_2 = 2.56$ the solution f_3 having already two sign alternations branches off from f_0.

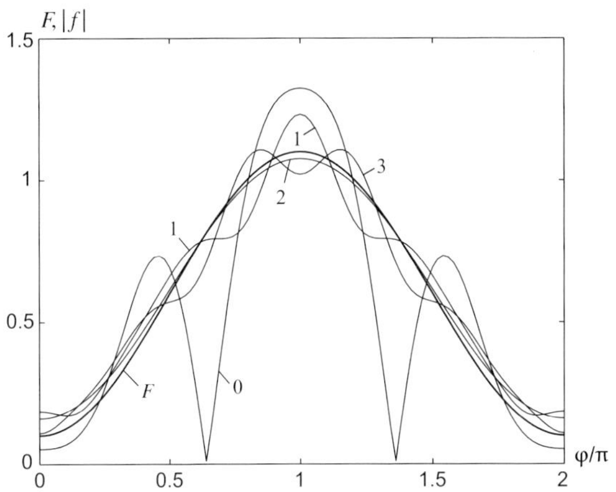

Figure 5.2 Amplitude distributions of different solutions at $c = 3.0$.

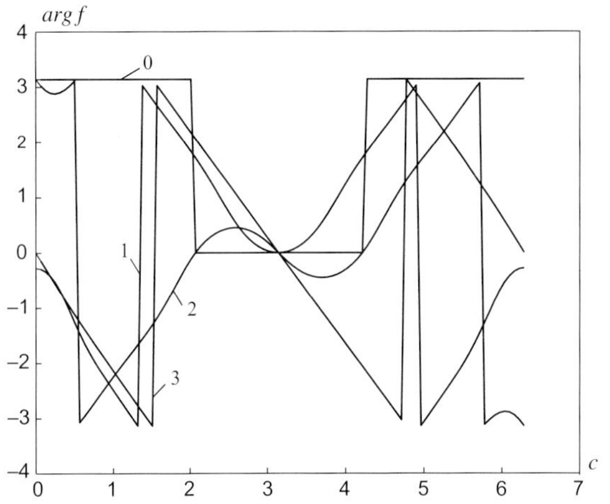

Figure 5.3 Phase distributions of different solutions at $c = 3.0$.

For comparison, the amplitude and phase distributions of different solutions are shown in Figure 5.2. It is significant that the amplitude distribution of the real solution approaches the desired function $F(\varphi)$ noticeably less than the complex ones. On the other hand, the phase distribution of the solution f_2 closest to $F(\varphi)$ in the modulus, is smooth but has a large increment, about 3π. As it was expected, this solution branches off from f_0 at the first branching point and has an odd phase distribution.

The phase distributions of different solutions at $c = 3.0$ are shown in Figure 5.3. The phase distribution of the solution varies essentially when increasing the value of c. This variations is clearly seen in Figure 5.4 in which the phase portrait (dependence of the imaginary part on the real one) of $f_1(\varphi)$ is shown for c varying from 0.5 to 5.0 with the step 0.1.

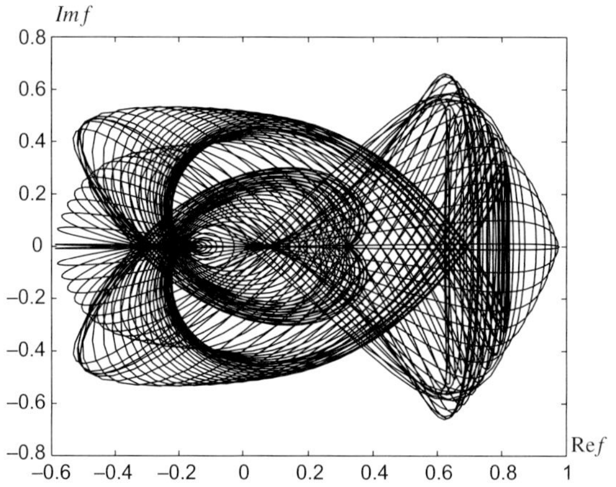

Figure 5.4 Typical phase portrait of solution $f_1(\varphi)$; $t = 1$; $c = 0.05$ to 5.0 with step 0.1.

5.2.3 Phase optimization problems

As the next example for the application of the modified Newton method we consider the phase optimization problem formulated in Section 3.2.2 and based on functional (3.163):

$$\chi(\psi) = \left(|A[U \exp(i\psi)]|, F\right)_2 , \tag{5.133}$$

where U and F are given real positive functions. For our purpose we take the Lagrange–Euler equation for this functional in the form (3.173) of a system of two nonlinear equations

$$\Phi(w, f, c) = 0 , \tag{5.134a}$$

$$\Psi(w, f, c) = 0 , \tag{5.134b}$$

where

$$\Phi(w, f, c) = w - A^* \left[F \exp(i \arg f)\right] , \tag{5.135a}$$

$$\Psi(w, f, c) = f - A\left[U \exp(i \arg w)\right] . \tag{5.135b}$$

The functions w, f are unknown in these equations. After they are found, the required real function ψ is calculated as $\psi = \arg w$. As before, we assume that the operator A depends on a real parameter c.

Following Section 5.2.1, we denote the corrections to w, f, c at the pth step of the Newton method by $\delta_w = w_{p+1} - w_p$, $\delta_f = f_{p+1} - f_p$, $\varepsilon = c_{p+1} - c_p$, respectively, and write the equations for δ_w, δ_f, ε as

$$\Phi_c \cdot \varepsilon + \Phi_w(w_p, f_p, c_p, \delta_w) + \Phi_f(w_p, f_p, c_p, \delta_f) = -\Phi(w_p, f_p, c_p) , \tag{5.136a}$$

$$\Psi_c \cdot \varepsilon + \Psi_w(w_p, f_p, c_p, \delta_w) + \Psi_f(w_p, f_p, c_p, \delta_f) \\ = -\Psi(w_p, f_p, c_p)\,, \quad (5.136\text{b})$$

where the functions Φ_c, Φ_w, Φ_w, Ψ_c, Ψ_ψ, Ψ_φ are given in Section 3.2.2 by expressions (3.183) and (3.182) with the notations $\delta_w = \varepsilon w_1$ and $\delta_f = \varepsilon f_1$.

If the step between the neighboring points on the branch is fixed during the calculations,

$$|\Delta c|^2 + \|\Delta w\|^2 + \|\Delta f\|^2 = h^2\,, \qquad (5.137)$$

where $\Delta c = c_m - c_{m-1}$, $\Delta w = w_m - w_{m-1}$, $\Delta f = f_m - f_{m-1}$, m is the point number on the branch, then (5.137) supplements the equation system (5.134), and the linear equation system (5.136) should be supplemented by the equation

$$\Delta c_p \cdot \varepsilon + (\Delta w_p, \delta_w) + (\Delta f_p, \delta_f) = 0\,. \qquad (5.138)$$

In the matrix form the linear equation system can be written as

$$\left\{ \begin{matrix} \Phi_c & \Phi_w & \Phi_f \\ \Psi_c & \Psi_w & \Psi_f \\ \Delta c & \Delta w & \Delta f \end{matrix} \right\} \times \left\{ \begin{matrix} \varepsilon \\ \delta_w \\ \delta_f \end{matrix} \right\} = \left\{ \begin{matrix} -\Phi \\ -\Psi \\ 0 \end{matrix} \right\}. \qquad (5.139)$$

Here the matrix blocks are denoted by the same letters as the functions generating them.

The unknown functions δ_w, δ_f are complex in this system. Following the considerations made in Section 3.2.2 when analyzing homogeneous equations (3.185), the two upper equation blocks of this system can be reduced to equations for the real constant ε and functions δ_1, δ_2 only. We explain this possibility for the case of the fixed parameter c (i.e. $\varepsilon = 0$), when (5.137) and the last equation in (5.139) generated by it are absent. The reduced system has the form

$$\left\{ \begin{matrix} \hat{\Phi}_w & \hat{\Phi}_f \\ \hat{\Psi}_w & \hat{\Psi}_f \end{matrix} \right\} \times \left\{ \begin{matrix} \delta_1 \\ \delta_2 \end{matrix} \right\} = \left\{ \begin{matrix} -\hat{\Phi} \\ -\hat{\Psi} \end{matrix} \right\}, \qquad (5.140)$$

where $\delta_1 = \mathrm{Im}(\delta_w/w)$, $\delta_2 = \mathrm{Im}(\delta_f/f)$, and the elements of the block-matrices are formed by the functions

$$\hat{\Phi}_w = |w|^2\delta_1\,, \qquad (5.141\text{a})$$

$$\hat{\Phi}_f = -\mathrm{Re}\left(\bar{w} \cdot A^*\left[F \exp(i \arg f_0)\delta_2\right]\right), \qquad (5.141\text{b})$$

$$\hat{\Psi}_w = -\mathrm{Re}\left(\bar{f} A\left[U \exp(i \arg w_0)\delta_1\right]\right), \qquad (5.141\text{c})$$

$$\hat{\Psi}_f = |f|^2\delta_2\,, \qquad (5.141\text{d})$$

$$\hat{\Phi} = \mathrm{Im}\left(\bar{w}\Phi\right), \qquad (5.141\text{e})$$

$$\hat{\Psi} = \mathrm{Im}\left(\bar{f}\Psi\right). \qquad (5.141\text{f})$$

If at least one of the functions w and f has no zeros in its definition domain, then system (5.141) can be simplified by using the diagonality of the matrices $\hat{\Phi}_w$, $\hat{\Psi}_f$ (similar to homogeneous system (3.186) in Section 3.2.2.2). In this case one of the functions δ_1 or δ_2 can be eliminated from the system. We will specify this possibility when considering a particular case of the operator A below.

5.2.3.1 Continuous Fourier transform: theoretical results

Consider the case when the operator A describes the Fourier transform of the finite function. In this case

$$Au = \sqrt{\frac{c}{2\pi}} \int_{-1}^{1} u(x) \exp(icx\xi) \mathrm{d}x \,, \tag{5.142a}$$

$$A^* \left[F \exp(i\varphi) \right] = \sqrt{\frac{c}{2\pi}} \int_{-1}^{1} F(\xi) \exp(i\varphi(\xi)) \exp(-icx\xi) \mathrm{d}\xi \,, \tag{5.142b}$$

(see, in somewhat different notation, (3.68), (3.71)). Since the operator $A : L_2(-1, 1) \to L_2(-\infty, \infty)$ is isometric, then, owing to the Parseval equality, the problem of maximization of the functional

$$\chi(\psi) = \int_{-1}^{1} |f(\xi)| \, F(\xi) \mathrm{d}\xi \tag{5.143}$$

is equivalent to the problem of minimization of the functional

$$\sigma(\psi) = \int_{-\infty}^{\infty} \left[|f(\xi)| - F(\xi) \right]^2 \mathrm{d}\xi \,, \tag{5.144}$$

where

$$f(\xi) = \sqrt{\frac{c}{2\pi}} \int_{-1}^{1} U(x) \exp(i\psi(x)) \exp(icx\xi) \mathrm{d}x \,. \tag{5.145}$$

Here we assume that the given real positive functions $U(x)$ and $F(\xi)$ are normalized by the conditions $\int_{-1}^{1} U^2(x)\mathrm{d}x = 1$, $\int_{-1}^{1} F^2(\xi)\mathrm{d}\xi = 1$; in (5.144) we imply that $F(\xi) = 0$ for $|\xi| > 1$.

The problem is related, in particular, to the following physical problems:

a) The two-dimensional variant of Problem **Bp** for the power transmitting line. The functions $U(x)$ and $F(\xi)$ describe the amplitude distributions $|U_0(x)|$ and V_1 of the field on the antenna and rectenna, respectively; x and ξ are the normalized coordinates X_0/a_0 and X_1/a_1 in the domains (intervals) D_0 and D_1; $2a_0$, $2a_1$ are the linear sizes of these domains, $c = ka_0a_1/d$, d is the distance between the antenna and rectenna. The function $f(\xi)$ describes the obtained field distribution on the rectenna as $U_1(X_1) = f(\xi)\exp(-icx^2/2)$. The required function $\psi(x)$ is connected with the phase distribution of the field U_0 on the antenna by the relation $\arg U_0(X_0) = \psi(x) + icx^2/2$. The functionals σ and χ are connected as

$$\sigma = 1 - \chi^2 \,. \tag{5.146}$$

b) The two-dimensional variant of Problem **Fp** for the two-element field transformer. The functions $U(x)$ and $F(\xi)$ are the given and desired amplitude field distributions $U_0(x)$ and $V_1(\xi)$ on the input and output apertures, respectively. Other notations are analogous to the case (a). The phase corrections to be provided by the lenses on the apertures are $\Psi_0(X_0) = \psi(x) + icx^2/2$, $\Psi_1(X_1) = \arg f(x) + ic\xi^2/2$.
c) The two-dimensional variant of the antenna synthesis Problem **Jp** (linear antenna case). The function $U(x)$ is the given amplitude distribution $|U(X)|$ of the current on the antenna, $F(\xi)$ denotes the desired amplitude radiation pattern, $\xi = ka \sin\theta_0$ is the dimensionless coordinate on the antenna, $2a$ is the antenna length, $2\theta_0$ is the angular size of the domain Ω in which the desired amplitude pattern differs from zero. The relation between the functionals σ and χ is given by (5.146).

Note that, due to the duality principle, the results obtained for the above problems can be easily transmitted onto the respective 'dual' problems, in which the given functions $U(x)$ and $F(\xi)$ are interchanged. For instance, in the dual case of the antenna problem the given amplitude current distribution is described by $F(x)$ and the desired amplitude pattern by $U(\xi)$. In the same way the functions $f(\xi)$, $\psi(x)$ are interchanged with $w(x)$, $\varphi(\xi)$, introduced below, respectively. The values of the functionals χ and σ remain the same in the dual problem.

The Lagrange–Euler equations for (5.143) can be written in the form (5.134) with

$$\Phi(w, f, c) = w(x) - \sqrt{\frac{c}{2\pi}} \int_{-1}^{1} F(\xi) \exp(i \arg f(\xi)) \exp(-icx\xi) \mathrm{d}\xi\,, \quad (5.147a)$$

$$\Psi(w, f, c) = f(\xi) - \sqrt{\frac{c}{2\pi}} \int_{-1}^{1} U(x) \exp(i \arg w(x)) \exp(icx\xi) \mathrm{d}x\,. \quad (5.147b)$$

The required functions are $\psi = \arg(w)$, $\varphi = \arg(f)$.

In order to compare the possibilities and properties of different numerical methods, we apply to these equations both the Newton method and the direct iteration method described in Section 3.2.2.

The properties of the required phase functions $\psi(x)$ and $\varphi(\xi)$ essentially depend on the evenness of the given functions U and F. We confine ourselves to the case when they are even functions of their arguments, $|U(-x)| = |U(x)|$ and $F(-\xi) = F(\xi)$.

Similar to the corresponding amplitude-phase optimization problem investigated in detail in Section 4.2, the equation system (5.147) has nonunique solutions which can branch when the parameter c varies. Unfortunately, no analytical representation of the solutions like (4.71) has been found for equations (5.147); all these solutions and their branching should be investigated numerically. For the case of symmetrical data these solutions can be classified into the following three groups.

a) The functions

$$w(x) = \sqrt{\frac{c}{2\pi}} \int_{-1}^{1} F(\xi) \exp(i\varphi(\xi)) \exp(-icx\xi) d\xi \tag{5.148}$$

and $f(\xi)$ given by (5.145) are real at $x \in (-1, 1)$ and $\xi \in (-1, 1)$, respectively.

b) One of these function, say, $w(x)$, is real, and the other, $f(\xi)$, has an odd (to within a constant addend) phase $\varphi(-\xi) = -\varphi(\xi)$.

c) The functions $f(\xi)$ and $w(x)$ have even phases: $\psi(-x) = \psi(x), \varphi(-\xi) = \varphi(\xi)$. Solutions of other forms were not found for such problems.

Below we discuss the methods for determination of these solutions and compare their properties using particular examples. It can be seen that, if the real functions

$$w_0(x) = \int_{-1}^{1} F(\xi) \cos(cx\xi) d\xi \,, \quad -1 < x < 1 \,, \tag{5.149a}$$

$$f_0(\xi) = \int_{-1}^{1} U(x) \cos(cx\xi) dx \,, \quad -1 < \xi < 1 \,, \tag{5.149b}$$

have no zeros, then the evident (*'trivial'*) solution to (5.147) is $\psi(x) \equiv 0$, $\varphi(\xi) \equiv 0$ (the nonessential constant factor is omitted here). This solution exists for any given $U(x)$ and $F(\xi)$ and sufficiently small c. Other solutions of the group (a) can exist (i.e., be its continuation, branch off from it or exist independently). They are defined by the zero points x_n, $n = \pm 1 \ldots \pm 2N$, and ξ_m, $m = \pm 1 \ldots \pm 2M$, of the functions $w(x)$ and $f(\xi)$, respectively. The double signs of their indices imply that these points exist as symmetrical pairs. The numbers N and M of these pairs is a priori unknown, they should be determined during the calculations. In particular, the values $N = 0$ or $M = 0$ are possible.

According to (5.147), the required functions $\psi(x)$, $\varphi(\xi)$ are defined by the signs of the functions $w(x)$, $f(\xi)$, as follows:

$$\psi(x) = \begin{cases} 0 \,, & w(x) \geq 0 \,, \\ \pi \,, & w(x) < 0 \,; \end{cases} \tag{5.150a}$$

$$\varphi(x) = \begin{cases} 0 \,, & f(x) \geq 0 \,, \\ \pi \,, & f(x) < 0 \,. \end{cases} \tag{5.150b}$$

Therefore, $\exp(i\psi(x)) = \mathrm{sgn}(w(x))$, $\exp(i\varphi(\xi)) = \mathrm{sgn}(f(\xi))$. In order to avoid discontinuities under the integrals, the equations for $\{x_n\}$, $\{\xi_m\}$ following from (5.143) can conveniently be written in the form

$$w(c, \xi_1, \ldots, \xi_M; x_n) \equiv \sum_{m=1}^{M+1} (-1)^{m-1} \int_{\xi_{m-1}}^{\xi_m} F(\xi) \cos(cx_n\xi) d\xi = 0 \,, \quad n = 1, 2, \ldots, N \,, \tag{5.151a}$$

$$f(c, x_1, \ldots, x_N; \xi_m) \equiv \sum_{n=1}^{N+1} (-1)^{n-1} \int_{x_{n-1}}^{x_n} U(x) \cos(c\xi_m x) \mathrm{d}x = 0 ,$$

$$m = 1, 2, \ldots, M . \quad (5.151b)$$

Here $x_0 = 0$, $x_{N+1} = 1$, $\xi_0 = 0$, $\xi_{M+1} = 1$ are assigned. The additional conditions $0 < x_1 < x_2 < \cdots < x_N < 1$, $0 < \xi_1 < \xi_2 < \cdots < \xi_M < 1$ should hold; they allow, in particular, determination of the numbers N and M. The parameter c is included into the list of arguments of the functions in (5.151) in order to use it in the Newton method to determine the branching points.

First, we describe the direct iterative procedure (3.175) proposed in Section 3.2.2 as an application to the equation system (5.151). Starting from initial values $x_1^{(0)}, \ldots, x_N^{(0)}$ (e.g. with $N^{(0)} = 0$) at fixed c and following this procedure, we calculate the first approximations $\{\xi_m^{(1)}\}$ to the values $\{\xi_m\}$ as the sequence of zeros of the function $f_m(c, x_1, \ldots, x_N; \xi)$ in the segment $\xi \in (0, 1)$, calculated by (5.151a). If $N^{(0)} = 0$ is chosen, then only the integral over the entire segment $(0, 1)$ is present on the right-hand side of (5.151a). The function $w(c, \xi_1, \ldots, \xi_M; x)$ is calculated by (5.151b) at $\xi_m = \xi_m^{(1)}$, $m = 1, \ldots, M$. Its zeros $x = x_n^{(1)}$, $n = 1, 2, \ldots, N$, are the next approximations to the numbers $\{x_n\}$. These two operations are repeated step by step until the necessary accuracy is reached. As was shown in Section 3.2.2, this procedure increases the value of the functional χ at each step and therefore produces the convergent sequence $\{\chi_p\}$. Numerical results show that the sequences $\{\xi_m^{(p)}\}$ and $\{x_n^{(p)}\}$ are convergent as well. The convergence is fast enough. However, it noticeably slows near the values of c which are the branching points at which new solutions with complex $w(x)$ and/or $f(\varphi)$ appear.

The Newton method can be applied to (5.151) as well. For this we write these equations in the form analogous to (5.134), using the notations $\Phi_n(c, \xi_1, \ldots, \xi_M; x_n) = w(c, \xi_1, \ldots, \xi_M; x_n)$, $\Psi_m(c, x_1, \ldots, x_N; \xi_m) = f(c, x_1, \ldots, x_N; \xi_m)$, and denote the corrections to the unknown values in the pth step as $\varepsilon = c_{p+1} - c_p$, $\delta_x = \{x_n^{(p+1)}\} - \{x_n^{(p)}\}$, $\delta_\xi = \{\xi_m^{(p+1)}\} - \{\xi_m^{(p)}\}$; here ε is a real number, δ_x and δ_ξ are vectors of the dimensionality N and M, respectively. The elements of the Jacobi matrix are calculated by the expressions

$$\frac{\partial \Phi_n}{\partial c} = x_n \sum_{m=1}^{M+1} (-1)^m \int_{\xi_{m-1}}^{\xi_m} \xi F(\xi) \sin(c x_n \xi) \mathrm{d}\xi = 0 , \quad (5.152a)$$

$$\frac{\partial \Phi_n}{\partial x_n} = c \sum_{n=1}^{M+1} (-1)^m \int_{\xi_{m-1}}^{\xi_m} \xi F(\xi) \sin(c x_n \xi) \mathrm{d}\xi , \quad (5.152b)$$

$$\frac{\partial \Phi_n}{\partial \xi_m} = 2 \cdot (-1)^{m-1} F(\xi_m) \cos(c x_n \xi_m) , \quad (5.152c)$$

$$\frac{\partial \Psi_m}{\partial c} = \xi_m \sum_{n=1}^{N+1} (-1)^n \int_{x_{n-1}}^{x_n} x U(x) \sin(c \xi_m x) \mathrm{d}x = 0 , \quad (5.152d)$$

$$\frac{\partial \Psi_m}{\partial x_n} = 2 \cdot (-1)^{n-1} U(x_n) \cos(c \xi_m x_n) , \quad (5.152e)$$

$$\frac{\partial \Psi_m}{\partial \xi_m} = c \sum_{n=1}^{N+1} (-1)^n \int_{x_{n-1}}^{x_n} x\, U(x) \sin(c\xi_m x) \mathrm{d}x \,. \tag{5.152f}$$

The right-hand side of the linear equation system for the pth step of the method is formed from the vectors $\{-\Phi_m(c, x_1, \ldots, x_N; \xi_m)\}$, $\{-\Psi_n(c, \xi_1, \ldots, \xi_M; x_n)\}$. For simplicity, here we omit the index (p) denoting the iteration number. Finally, the linear algebraic system can be written in the following symbolic form:

$$\begin{Bmatrix} \Phi_c & \Phi_x & \Phi_\xi \\ \Psi_c & \Psi_x & \Psi_\xi \\ \Delta c & \Delta x & \Delta \xi \end{Bmatrix} \times \begin{Bmatrix} \varepsilon \\ \delta_x \\ \delta_\xi \end{Bmatrix} = \begin{Bmatrix} -\Phi \\ -\Psi \\ 0 \end{Bmatrix} . \tag{5.153}$$

The notation of the matrix blocks in (5.153) is analogous to that in (5.139). In particular, the values Δc, Δx, $\Delta \xi$ are determined similarly to (5.137) for the case when the step between the neighbor points on the branch are fixed during calculations. If the parameter c is fixed, then the first column and last row in the matrix should be removed together with the unknown ε and the last element on the right-hand side.

Numerical results show that the solutions of the type considered are, as a rule, less effective. They may be used only in the case when the local half-wave phase correctors realizing the correction (5.150) are only permissible on both apertures of the two-element field transformer. The solutions of the second type are preferable if the continuously variable correction can be created on one aperture. This applies also to the antenna synthesis problem, in which only one phase function, namely the current phase distribution ψ has to be physically created, whereas the second function, the phase pattern φ can be arbitrary. In such solutions one of the functions f or w is real, that is, the required phase φ or ψ has the form (5.150), whereas the second function is asymmetrically complex, that is, its phase is an odd function. Solutions of this type do not exist at small values of c, they arise (in particular, branch off from the solutions of the first type) at certain values of $c = c_j$. Recall that we consider the case when $U(-x) = U(x)$, $F(-\xi) = F(\xi)$.

Let us seek for solutions to equation system (5.143), in which

$$\varphi(-\xi) = -\varphi(\xi) , \tag{5.154}$$

and $\psi(x)$ satisfy condition (5.150a) providing, in this way, the reality of the function $w(x)$. Owing to property (5.148), the function $w(x)$ defined by (5.148) ceases to be even, its zeros x_n are located nonsymmetrically in the interval $(-1, 1)$. Taking into account this fact together with (5.148), we obtain from (5.143) the following system of equations for $\{x_n\}$ and $\varphi(\xi)$:

$$\int_0^1 F(\xi) \cos(\varphi(\xi) - c x_n \xi) \mathrm{d}\xi = 0 , \quad n = 1, 2, \ldots, N . \tag{5.155a}$$

$$\varphi(\xi) = \arg \sum_{n=1}^{N+1} (-1)^{n-1} \int_{x_{n-1}}^{x_n} U(x) \exp(ic\xi_m x) \mathrm{d}x \,. \tag{5.155b}$$

The direct iterative method can be used to solve this system. This method consists in determining from (5.155a) by turns first the points x_n and their number N in $(-1, 1)$ with $\varphi(\xi)$ taken from the previous iteration, and then the next approximation to $\varphi(\xi)$ from (5.155b) with N and $\{x_n\}$ as obtained before. An appropriate nonconstant initial approximation $\varphi^{(0)}(\xi)$ to the function $\varphi(\xi)$ should be chosen in this procedure. If $\varphi^{(0)}(\xi) = const$ (or 0, which is equivalent) is chosen, then the left-hand side of (5.155a) (being proportional to the function $w(x)$ given by (5.148)) becomes even, which in the next half-step gives a symmetrical function $f(\xi)$ and the procedure leads to the symmetrical real solution.

For all other initial approximations the iterative process converges to a solution with property (5.154), except for the case when the value of the functional χ_s on the limiting function is larger than that on a solution of the first group. In this case we can obtain the limiting function $\varphi(\xi)$ of type (5.150b) with symmetrical location of the points ξ_m, which is equivalent to an odd function of type (5.154) if in condition (5.150b) for $\xi < 0$ the value π is replaced by $-\pi$ (which does not change the function $\exp(i\varphi)$).

For applying the Newton method, it is convenient to substitute equation (5.155b) by its weaker form following from (5.147b), and write the equation system as

$$\Phi_n(c, \varphi, x_n) \equiv \int_0^1 F(\xi)\cos(\varphi(\xi) - c x_n \xi)\mathrm{d}\xi = 0\,, \qquad n = 1, 2, \ldots, N\,. \quad (5.156a)$$

$$\Psi(c, \varphi, x_n) \equiv \mathrm{Im}\left[\exp(-i\varphi(\xi))\sum_{n=1}^{N+1}(-1)^{n-1}\int_{x_{n-1}}^{x_n} U(x)\exp(i c \xi_m x)\mathrm{d}x\right] = 0. \quad (5.156b)$$

Then the derivatives and variations in the elements of the Jacobi matrix at each step of the method have the forms

$$\frac{\partial \Phi_n}{\partial c} = x_n \int_0^1 \xi F(\xi)\sin(\varphi(\xi) - c x_n \xi)\mathrm{d}\xi\,, \qquad n = 1, 2, \ldots, N\,, \quad (5.157a)$$

$$\frac{\partial \Phi_n}{\partial x_n} = c\int_0^1 \xi F(\xi)\sin(\varphi(\xi) - c x_n \xi)\mathrm{d}\xi\,, \qquad n = 1, 2, \ldots, N\,, \quad (5.157b)$$

$$\left[\Phi_n\right]_\varphi = -\int_0^1 F(\xi)\sin(\varphi(\xi) - c x_n \xi)\cdot\delta_\varphi(\xi)\mathrm{d}\xi\,, \qquad n = 1, 2, \ldots, N\,, \quad (5.157c)$$

$$\frac{\partial \Psi}{\partial c} = \mathrm{Re}\left[\xi_m \exp(-i\varphi(\xi)) \sum_{n=1}^{N+1} (-1)^{n-1} \int_{x_{n-1}}^{x_n} x\, U(x) \exp(ic\xi_m x) \mathrm{d}x\right], \tag{5.157d}$$

$$\frac{\partial \Psi}{\partial x_n} = 2\,\mathrm{Im}\left[\exp(-i\varphi(\xi))(-1)^{n-1} U(x_n) \exp(ic\xi_m x_n)\right], \quad n = 1, 2, \ldots, N, \tag{5.157e}$$

$$\Psi_\varphi = -\mathrm{Re}\left[\delta_\varphi(\xi) \cdot \exp(-i\varphi(\xi)) \sum_{n=1}^{N+1} (-1)^{n-1} \int_{x_{n-1}}^{x_n} U(x) \exp(ic\xi_m x) \mathrm{d}x\right]. \tag{5.157f}$$

Here, as before, the upper index (p) in the numbers $x_n^{(p)}$ and function $\varphi^{(p)}(\xi)$ denoting the iteration number, is dropped.

The linear algebraic system for $\delta_1 = \psi_{p+1} - \psi_p$, $\delta_2 = \varphi_{p+1} - \varphi_p$, $\varepsilon = c_{p+1} - c_p$ has the form

$$\left\{ \begin{matrix} \Phi_c & \Phi_x & \Phi_\varphi \\ \Psi_c & \Psi_x & \Psi_\varphi \\ \Delta c & \Delta x & \Delta\varphi \end{matrix} \right\} \times \left\{ \begin{matrix} \varepsilon \\ \delta_x \\ \delta_\varphi \end{matrix} \right\} = \left\{ \begin{matrix} -\Phi \\ -\Psi \\ 0 \end{matrix} \right\}, \tag{5.158}$$

where Δc, Δx, $\Delta\varphi$ have the meaning analogous to that given in (5.137).

Of course, besides the solutions described above, there can exist solutions of the same group (b), with the function $\psi(x)$ satisfying the condition $\psi(-\xi) = -\psi(\xi)$, and real, nonsymmetrical function $w(x)$ having zeros in the interval $(-1, 1)$. All the above formulas can be rewritten for such solutions in an obvious way.

The solutions of type (c), in which the functions $f(\varphi)$ and $v(\xi)$ are complex, exist for both symmetrical and nonsymmetrical given functions $U(x)$, $F(\xi)$. Moreover, the global extremum points of the functional χ_s are contained among these solutions (except for the case when they are described by functions (5.149) in the case of symmetrical data). For these reasons we consider here the problem in the general form, without any assumptions about the data symmetry.

The Newton method is convenient to be applied to the weak form (5.134), (5.147) of the Lagrange–Euler equations. The derivatives and variations forming the elements of the Jacobi matrix in (5.139) are

$$\Phi_c(\psi_p, \varphi_p, c_p, \delta_1) = -\mathrm{Re}\Big\{x \exp(-i\psi_p(x)) \times \int_{-1}^{1} F(\xi) \exp(i\varphi(\xi)) \exp(-ic_p x\xi)\xi \mathrm{d}\xi\Big\}, \tag{5.159a}$$

$$\Phi_\psi(\psi_p, \varphi_p, c_p, \delta_1) = -\mathrm{Re}\Big\{\exp(-i\psi_p(x)) \times \int_{-1}^{1} F(\xi) \exp(i\varphi_p(\xi)) \exp(-ic_p x\xi) \mathrm{d}\xi\Big\} \cdot \delta_1(x), \tag{5.159b}$$

$$\Phi_\varphi(\psi_p, \varphi_p, c_p, \delta_1) = \mathrm{Re}\Big\{\exp(-i\psi_p(x)) \times \int_{-1}^{1} F(\xi)\exp(i\varphi_p(\xi))\exp(-ic_p x\xi)\delta_2(\xi)\mathrm{d}\xi\Big\}, \quad (5.159c)$$

$$\Psi_c(\psi_p, \varphi_p, c_p, \delta_1) = \mathrm{Re}\Big\{\xi\exp(-i\varphi_p(\xi)) \times \int_{-1}^{1} |U(x)|\exp(i\psi_p)\exp(ic_p x\xi)x\mathrm{d}x\Big\}, \quad (5.159d)$$

$$\Psi_\psi(\psi_p, \varphi_p, c_p, \delta_1) = \mathrm{Re}\Big\{\int_{-1}^{1} |U(x)|\exp(i\psi_p(x))\exp(icx\xi)\delta_1(x)\mathrm{d}x \times \exp(-i\varphi_p(\xi))\Big\}, \quad (5.159e)$$

$$\Psi_\varphi(\psi_p, \varphi_p, c_p, \delta_2) = -\mathrm{Re}\Big\{\int_{-1}^{1} |U(x)|\exp(i\psi_p(x))\exp(icx\xi)\mathrm{d}x \times \exp(-i\varphi_p(\xi))\Big\} \cdot \delta_\varphi(\xi). \quad (5.159f)$$

As can be seen from (5.159b) and (5.159f), the matrices Φ_ψ and Ψ_φ in (5.139) are diagonal. This fact allows us to simplify this linear system by reducing its dimension. If, for instance, the function $\mathrm{Re}\{f_p(\xi)\exp(-i\varphi_p(\xi))\}$ with

$$f_p(\xi) = \int_{-1}^{1} |U(x)|\exp(i\psi_p(x))\exp(icx\xi)\mathrm{d}x\,, \quad (5.160)$$

forming, according to (5.159f), the diagonal of the matrix Ψ_φ, has no zeros at $\xi \in [-1, 1]$, then δ_φ can be explicitly expressed as

$$\delta_\varphi = \left[\Psi_\varphi\right]^{-1}\left[-\Psi - \Psi_c \cdot \varepsilon - \Psi_\psi \cdot \delta_\psi\right], \quad (5.161)$$

which, after substituting into (5.139), allows us to eliminate this unknown function from the system and rewrite it in the form

$$\begin{Bmatrix} \tilde{\Phi}_c & \tilde{\Phi}_\psi \\ \tilde{\Delta}c & \tilde{\Delta}\psi \end{Bmatrix} \times \begin{Bmatrix} \varepsilon \\ \delta_\psi \end{Bmatrix} = \begin{Bmatrix} -\tilde{\Phi} \\ -\tilde{\Psi} \end{Bmatrix}, \quad (5.162)$$

where

$$\tilde{\Phi}_c = \Phi_c - \Phi_\varphi\left[\Psi_\varphi\right]^{-1}\Psi_c\,, \quad (5.163a)$$

$$\tilde{\Phi}_\psi = \Phi_\psi - \Phi_\varphi\left[\Psi_\varphi\right]^{-1}\Psi_\psi\,, \quad (5.163b)$$

$$\tilde{\Phi} = \Phi + \Phi_\varphi\left[\Psi_\varphi\right]^{-1}\Phi^{(2)} \quad (5.163c)$$

$$\tilde{\Delta}c = \Delta c - \Delta\varphi\left[\Psi_\varphi\right]^{-1}\Psi_c\,, \quad (5.163d)$$

$$\tilde{\Delta}\psi = \Delta\psi - \Delta\varphi\left[\Psi_\varphi\right]^{-1}\Psi_\psi\,, \quad (5.163e)$$

$$\tilde{\Psi} = \Delta\varphi\left[\Psi_\varphi\right]^{-1}\Phi^{(2)}\,. \quad (5.163f)$$

This simplification decreases the size of the total matrix by almost twice what is significant for its singular decomposition and analysis.

If at least one of the functions $\mathrm{Re}\{w_p \exp(-i\psi_p)\}$ and $\mathrm{Re}\{f_p \exp(-i\varphi_p)\}$ has zeros in its definition domain, then the homogeneous problem associated with (5.139) has obvious nontrivial solutions. If, for instance, $\mathrm{Re}\{w_p(x_n)\exp(-i\psi_p(x_n))\} = 0$, where x_n is a point of the discretization in $[-1, 1]$, then such a solution is

$$\varepsilon = 0\,, \quad \delta_1(x_m) = \delta_{nm}\,, \quad \delta_2 \equiv 0\,. \tag{5.164}$$

One of the singular values of the matrix equals zero in this case, and the corresponding vector has the δ-shaped form. Therefore, the value $\delta_1(x_m)$ in the solution to (5.139) cannot be uniquely determined in this case. This fact can be taken into account during practical calculations. The simplest action in this case is to take $\delta_1(x_m)$ as the average value between $\delta_1(x_{m-1})$ and $\delta_1(x_{m+1})$, or use a higher-order spline approximation.

5.2.3.2 **Continuous Fourier transform: numerical results**

As a concrete example, we show numerical results for the case $U(x) \equiv const$, $F(\xi) = \cos(\pi\xi/2)$. The value $\sigma = 1 - \chi^2$ versus the parameter c for various solutions $\{f_n, w_n\}$ to equation system (5.134) is shown in Figure 5.5. The curves are labeled with the indices n of the solutions. Below we mark the functions $\psi_n = \arg w_n$, $\varphi_n = \arg f_n$ with the same indices.

For small values of the parameter c (i.e. small sizes of the rectenna and antenna apertures or large distance between them in the problem of field transformation), the only trivial solution to (5.134) with $\psi_0 \equiv 0$, $\varphi_0 \equiv 0$ exists. This means that no transformation of the fields can be made; in this case only the maximum energy may be directed from one aperture into the other one in order to minimize the total

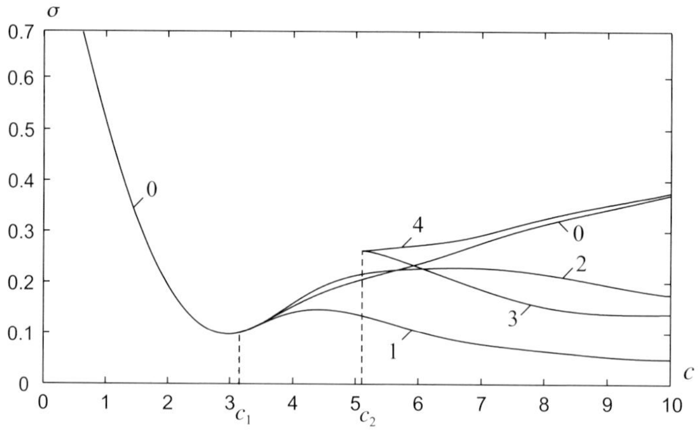

Figure 5.5 Value of functional $\sigma = 1 - |\chi|^2$ for various solutions; $U(x) = const$; $F(\xi) = \cos(\pi\xi/2)$.

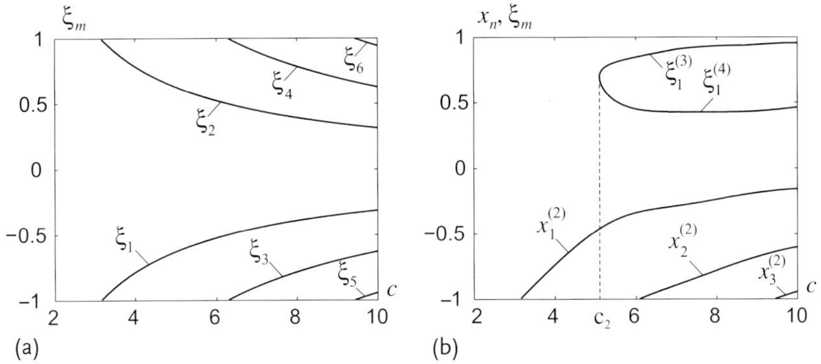

Figure 5.6 Zero points of real solution $f_0(\xi)$ (a) and complex asymetrical ones $f_2(\xi)$, $f_3(\xi)$, $f_4(\xi)$ (b); $U(x) = const$; $F(\xi) = \cos(\pi\xi/2)$.

losses. This situation takes place up to $c = \pi$, while both functions

$$f_0 = AU\,, \tag{5.165a}$$

$$w_0 = A^* F \tag{5.165b}$$

(being the real functions for even U and F) do not change their signs at $\xi \in (-1, 1)$ and $x \in (-1, 1)$, respectively.

At $c = \pi$ (we denote this value by c_1) the function $f_0(\xi)$ becomes zero at $|\xi| = 1$. The value $c = c_1$ is a branching point of the trivial solution to (5.134): in addition to $\{f_0, w_0\}$, two new solutions with indices 1 and 2 arise at this point. The trivial solution also changes its form at $c > c_1$, the function f_0 alternates its sign, so that

$$\psi_0(x) \equiv 0\,, \tag{5.166a}$$

$$\varphi_0(\xi) = \begin{cases} 0\,, & f_0(\xi) \geq 0\,; \\ \pi, & f_0(\xi) < 0\,. \end{cases} \tag{5.166b}$$

Both fields $f_0(\xi)$ and $w_0(x)$ remain real, the value of χ increases only by in-phasing of the functions F and f. However, this solution is no longer optimal. It remains optimal only in the class of functions with real $f(\xi)$, $w_n(x)$. The zeros ξ_m of the function $f_0(\xi)$ versus c are shown in Figure 5.6a. The number of zeros increases with increasing c.

If this solution has to be realized for one of the concrete problems (a), (b) or (c) described at the beginning of Section 5.2.3.1, then the results of Figure 5.6 may be used in the problem (b) concerning the two-element phase field transformer. On the segments $\xi_2 < |\xi| < \xi_4$ and $\xi_6 < |\xi| < 1$ of the output aperture, the half-wave supplements to the confocal phase correctors should be made (here the symmetry property $\xi_1 = -\xi_2$, $\xi_3 = -\xi_4$, $\xi_5 = -\xi_6$ is taken into account). In the dual problems in which the functions $U(x)$ and $F(\xi)$ are interchanged, these corrections should be made on the antenna (problems (a), (c)) or on the input aperture (problem (b)).

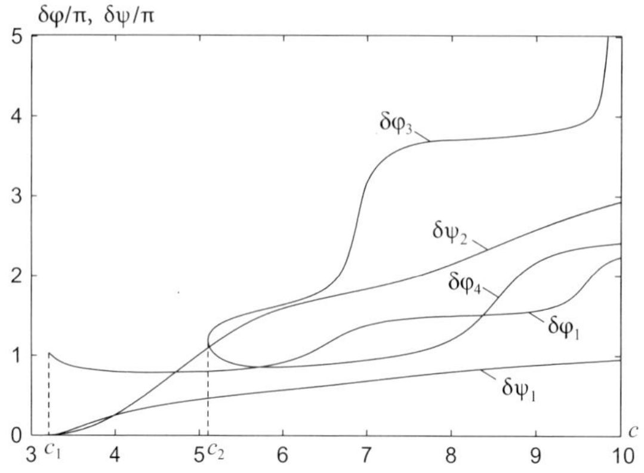

Figure 5.7 Phases variations for various solutions; $U(x) = const$; $F(\xi) = \cos(\pi\xi/2)$.

For $c > c_1$ the optimal solution is indexed by the number '1'. In this solution the functions $\psi_1(x)$ and $\varphi_1(\xi)$ are variable and all fields are complex. In a similar way to Section 5.2.2, for identification of the functions ψ_1, φ_1 in the figures we use their maximum variations $\Delta\psi_1$, $\Delta\varphi_1$ in the segment $(-1, 1)$. In Figure 5.7 these variations are shown together with other curves which will be described further. Note that $\Delta\varphi_1$ starts at $c = c_1$ from the value $\Delta\varphi_1 = \pi$. This is connected with the fact that, according to (5.166), the function $\varphi_0(\xi)$ in the real solutions gets the jump π at this point.

Another solution indexed as '2' branches off from the trivial one at $c = c_1$. This solution is of type (b) described in the previous section (see the text below (5.148)). In this solution, the function $w_2(x)$ is complex, with the even modulus $|w_2(-x)| = |w_2(-x)|$ and odd phase $\psi_2(-x) = -\psi_2(x)$, whereas $f_2(\xi)$ is real but nonsymmetrical. As can be seen from Figure 5.5, this solution is not optimal, in some range of values c it is even less effective than the real one. However, this solution together with the two other ones described below has a certain mathematical interest.

The variation of the function $\psi_2(x)$ is given in Figure 5.7. In the considered range of values c, the function $f_2(\xi)$ has one to three zeros $\xi_m^{(2)}$. These zeros are shown in Figure 5.6b.

In the case considered, the equation system (5.134) has two more solutions of type (b); we denote them by the indices '3' and '4'. In these solutions $|f_{3,4}(-\xi)| = |f_{3,4}(\xi)|$, $\varphi_{3,4}(-\xi) = -\varphi_{3,4}(\xi)$. The phase variations $\delta\varphi_3$, $\delta\varphi_4$ for these solutions are shown in Figure 5.7. The functions $w_3(x)$, $w_4(x)$ are real, with zeros $x_n^{(3)}$ and $x_n^{(4)}$(see Figure 5.6b). These solutions are specific. In contrast to others, they do not branch off from any solution existing before the point $c = c_2$, at which they arise (this fact is obtained numerically and needs theoretical justification). As can be seen from Figures 5.7 and 5.6b, they belong to the same branch of the solutions. In a graph form similar to Figure 4.5 they would be presented as an isolated ('hanging') branch not connected with the root. This type of complex solution is not likely to have been obtained for the nonlinear equations of the type considered. The real solutions of this type were described in [77].

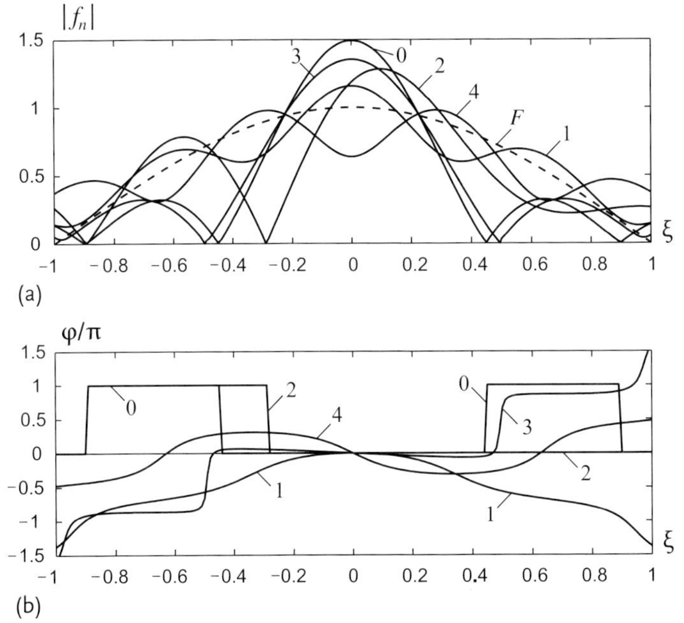

Figure 5.8 Amplitude (a) and phase (b) distributions of $f_n(\xi)$ for various solutions; $c = 7.0$, $U(x) = const$; $F(\xi) = \cos(\pi\xi/2)$.

In Figures 5.8 and 5.9 we show the amplitudes and phases of various solutions obtained at $c = 7$.

Somewhat different behavior of the solutions takes place in the case when both the functions $U(x)$ and $F(\xi)$ are the same. In the next figures the results are given for $U(x) = const$ and $F(\xi) = const$. Figure 5.10 shows the value of σ for various solutions to equation system (5.134). As before, the trivial solution (5.165) exists up to the point $c = \pi$ (now we denote this point by c_2), until these functions are positive. However, this solution ceases to be optimal earlier, at $c = c_1 < c_2$. This value (approximately equal to 2.80) is the first branching point of the real solution. We will return to this point later, after a complete description of the real solutions.

At $c = c_2$ the solution $\{f_0, w_0\}$ branches into four. Three of them are real. The first real solution (index 0) remains symmetrical, with the functions $f_0(\xi)$ and $w_0(x)$ coinciding in shape. The functions $\varphi_0(\xi)$ and $\psi_0(x)$ are also the same and of the form (5.166b). Their zeros are shown in Figure 5.11a as $x_{1,2}^{(0)}$, $\xi_{1,2}^{(0)}$. Two other real solutions (index 0′) have a different $f_{0'}(\xi)$ and $w_{0'}(x)$. They are mirror-symmetrical, that is, these functions are interchanged in them (the duality principle). One of these functions, say, $f_{0'}(\xi)$ remains as alternating. It coincides with f_0 from the previous example (for $U(x) = const$, $F(\xi) = \cos(\pi\xi/2)$), and its zeros are given in Figure 5.6a as ξ_m, $m = 1, 2, \dots, 6$. The second function ($f_{0'}(\xi)$) keeps a constant sign (at least in the considered range of the parameter c). These three real solutions correspond to the three local maxima of the functional χ (minima of σ) in the

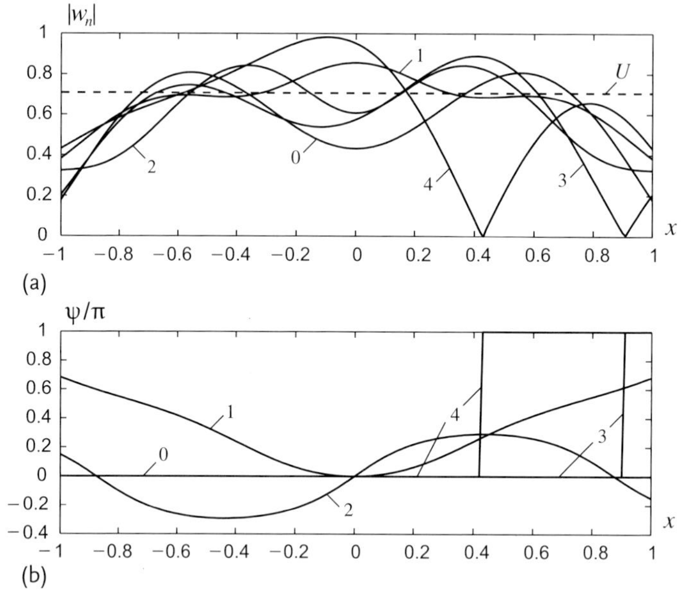

Figure 5.9 Amplitude (a) and phase (b) distributions of $w_n(x)$ for various solutions; $c = 7.0$, $U(x) = const$; $F(\xi) = \cos(\pi\xi/2)$.

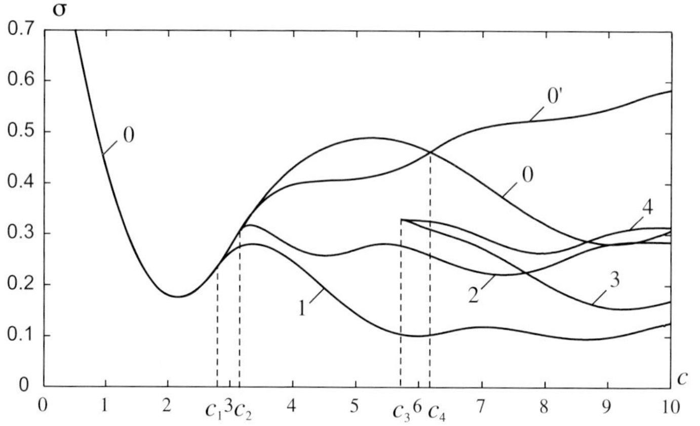

Figure 5.10 Value of functional $\sigma = 1 - |\chi|^2$ for various solutions; $U(x) = const$; $F(\xi) = const$.

class of of real functions. The global maximum in this class is reached on different solutions in different ranges of values c (see Figure 5.10). First, immediately after $c = c_2$, the solution $\{f_0, w_0\}$ is the best. Later, after $c = c_4 \approx 6.175$, the solution $\{f_{0'}, w_{0'}\}$ becomes better.

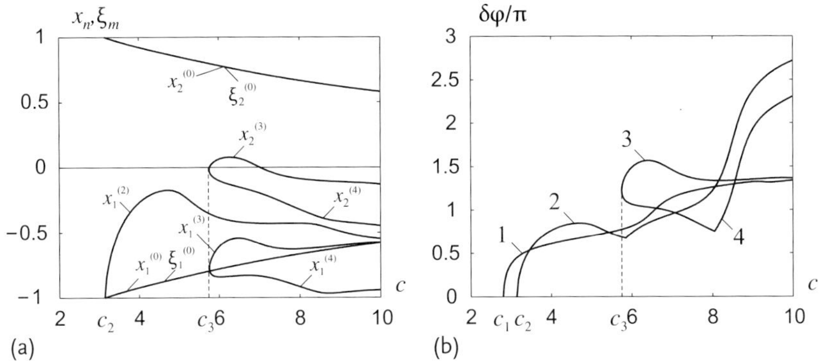

Figure 5.11 Zero lines of functions $w(x)$, $f(\xi)$ for symmetric real and asymmetric complex solutions (a) and phase variations of $f(\xi)$ for asymmetric solutions (b); $U(x) = const$; $F(\xi) = const$.

The optimal, complex solution denoted by the index '1' arises at $c = c_1 \approx 2.80$. It is symmetrical, that is, the functions $f_1(\xi)$ and $w_1(x)$ (and, respectively, $\varphi_1(\xi)$ and $\psi_1(x)$) coincide in shape. Their phase variations $\Delta\varphi_1$, $\Delta\psi_1$ are shown in Figure 5.11b as curve 1.

Similar to the previous example, besides the real solution (type (a)) and the complex one (type (c)), the asymmetrical solutions (type (b)) exist in the considered case. In such solutions one of the unknown functions, say, $w(x)$ is real and asymmetrical, whereas the second one, $f(\xi)$ has even amplitude and odd phase. Three such mirror-symmetrical pairs of solutions were found. We analyze the representatives of these pairs denoted by the indices '2', '3' and '4', respectively. The solution $f_2(\xi)$, $w_2(x)$ arises at $c = c_2$.The function $w_2(x)$ is real, its unique zero is shown in Figure 5.11a as $x_1^{(2)}$. The function $f_2(\xi)$ has even modulus, $|f_2(-\xi)| = |f_2(\xi)|$, and odd phase, $\varphi_2(-\xi) = -\varphi_2(\xi)$. The phase variation $\Delta\varphi_2$ is shown in Figure 5.11b as curve 2.

As in the previous case, two other asymmetrical solutions, index '3' and '4', have not branched off from any existing solutions. They arise at $c = c_3 \approx 5.75$ and belong to the same isolated branch (see curves 3 and 4 in Figure 5.10). The functions $w_3(\xi)$ and $w_4(\xi)$ are real. Each of them has two zeros x_1, x_2 described by the corresponding curves in Figure 5.11a with the superscripts (3) and (4), respectively. The variations $\Delta\varphi_3(\xi)$, $\Delta\varphi_4(\xi)$ of their odds phases are shown in Figure 5.11b as curves 3 and 4. Note that, for c larger than $c_5 \approx 7.685$, the solution $f_3(\xi)$, $w_3(x)$ is the best of all the nonoptimal ones. At $c = 10$ it yields losses of only about 30 percent in comparison with the optimal solution.

5.2.3.3 Discrete Fourier transform[2)]

Consider the case when the operator A describes the discrete Fourier transform of a finit-dimensional vector. The functions Φ and Ψ in (5.134) are the following

2) This subsection is written by M. I. Andriychuk and contains his results.

function and vector, respectively:

$$\Phi_n(w, f, c) = w_n - \int_{-1}^{1} F(\xi)\exp(i \arg f(\xi) - icn\xi)\mathrm{d}\xi \,,$$

$$n = -M, \ldots, M\,; \quad (5.167\text{a})$$

$$\Psi(w, f, c) = f(\xi) - \sum_{n=-M}^{M} U_n \exp\left(i \arg w_n - cn\xi\right)\,, \quad (5.167\text{b})$$

where $w = \{w_n\}$. Here the factor $\sqrt{c/(2\pi)}$ is omitted at the integrals; it should be taken into account when calculating the value of the functional χ below. The given vector $U = \{U_n\}$, $n = -M, \ldots, M$, and function $F(\xi)$ are real and positive; we norm them as $\|U\| = \sum_{n=-M}^{M} |U_n|^2 = 1$, $\|F\| = \int_{-1}^{1} |F(\xi)|^2 \mathrm{d}\xi = 1$. The functions $f(\xi)$, $F(\xi)$, and $\varphi(\xi)$ are defined for $\xi \in [-\pi/c, \pi/c]$, $c < \pi$; it is assumed that $F(\xi) \equiv 0$ for $|\xi| > 1$; c is a given physical parameter.

The problem is to find the real vector $\psi = \{\psi_n\}$, $\psi_n = \arg w_n$, $n = -M, \ldots, M$, which provide a maximum for the functional

$$\chi(\psi) = \int_{-1}^{1} |f(\xi)|\, F(\xi)\mathrm{d}\xi \,. \quad (5.168)$$

This problem is equivalent to to the problem of minimization of the functional

$$\sigma(\psi) = \int_{-\pi/c}^{\pi/c} \left[|f(\xi)| - F(\xi)\right]^2 \mathrm{d}\xi = 1 - \chi^2(\psi)\,. \quad (5.169)$$

The problem is related, in particular, to Problem **O′p** for the linear antenna array with $T = 2M + 1$ equidistantly positioned elements. In this case, the vector $\{U_n \exp(i\psi_n)\}$ describes the currents on the elements, $F(\xi)$ and $f(\xi)$ are the desired amplitude pattern and the complex one obtained in the period $\xi \in [-\pi/c, \pi/c]$, $\xi = kd \sin\theta$ is generalized angular coordinate, d is the distance between the elements. The parameter c is defined as $c = kd \sin\theta_0$, where $2\theta_0$ is the angle, outside which $F(\xi) \equiv 0$.

Similar to the case of the continuous Fourier transform, for the symmetrical data, $U_{-n} = U_n$, $n = 1, \ldots, M$, $F(-\xi) - F(\xi)$, the solutions to (5.134) can be classified into the three groups:

a) the vector w and function $f(\xi)$ are real;
b) the function $f(\xi)$ and vector $\{w\}$ have even phases: $\psi_{-n} = \psi_n$, $n = 1, \ldots, M$; $\varphi(-\xi) = \varphi(\xi)$;
c) one of them is real but nonsymmetrical, with alternating sign; the other has an odd (to within a constant addend) phase.

As the numerical results show, the solutions providing a maximum for the functional χ belong, as a rule, to group (b). They can be calculated by the iterative procedure (3.179) having in this case the form

$$f^{(p)}(\xi) = \sum_{n=-M}^{M} U_n \exp\left(i\left(\arg w_n^{(p)} + cn\xi\right)\right), \tag{5.170a}$$

$$w_n^{(p+1)} = \int_{-1}^{1} F(\xi)\exp\left(i \arg f^{(p)}(\xi) - icn\xi\right) \mathrm{d}\xi\,, \qquad n = -M, \ldots, M\,. \tag{5.170b}$$

Group (c) contains solutions of two different types. In the first of these the function $f(\xi)$ is complex with the odd phase $\varphi(-\xi) = -\varphi(\xi)$, $\varphi(\xi) = \arg f(\xi)$, and the vector $w = \{w_n\}$ is real asymmetrical, with alternating sign. For the not too large M, the simplest way of calculating these solutions is by direct combinatorial exhaustion. Solutions of the second type have a complex vector w with asymmetrical phase distribution $\psi_{-n} = -\psi_n$, $\psi_n = \arg w_n$. For solutions of this group, (5.134a) is rewritten in the form

$$w_n = 2\int_0^1 F(\xi)\cos(\arg(f(\xi) - cn\xi)\mathrm{d}\xi\,, \quad n = -M, \ldots, M\,. \tag{5.171}$$

The function Ψ in (5.134b) remains of the form (5.167b). The solutions providing a maximum for χ in this class can be calculated by an iterative procedure analogous to (5.170) using (5.171) to calculate $w_n^{(p+1)}$.

In all cases, except the first type of solution from group (b), the problem can be solved by the Newton method. The Jacobi matrix for it can be written in a similar way to (5.159) for the case of the continuous Fourier transform.

The behavior of solutions and their branching in the discrete case somewhat differs from the continuous one considered above. However, the main properties

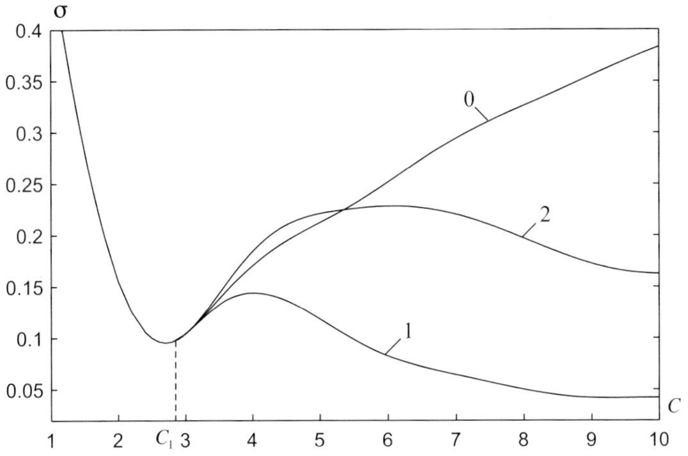

Figure 5.12 Value of functional $\sigma = 1 - |\chi|^2$ for various solutions; $M = 5$; $U_n = 1$; $F(\xi) = \cos(\pi\xi/2)$.

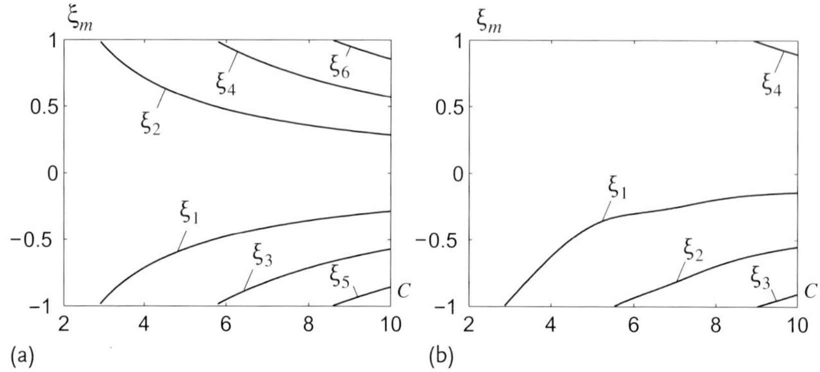

Figure 5.13 Zeros of functions $f_0(\xi)$ (a) and $f_2(\xi)$ (b) for real and asymmetrical solutions, respectively; $M = 5$; $U_n = 1$; $F(\xi) = \cos(\pi\xi/2)$.

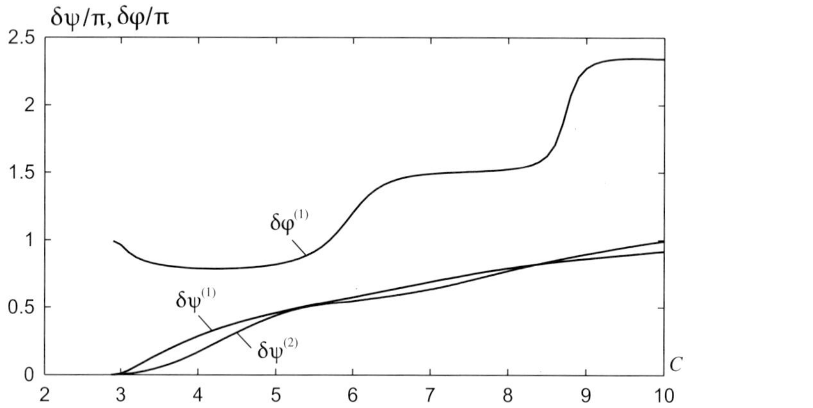

Figure 5.14 Phase increment for various solutions; $M = 5$; $U_n = 1$; $F(\xi) = \cos(\pi\xi/2)$.

of physically interesting solutions are kept. Here we give only three solutions from their sets, one of each mentioned group, for the case $M = 5$; $U_n = 1$; $F(\xi) = \cos(\pi\xi/2)$. Figure 5.12 shows the dependencies on the parameter $C = Mc$ of the values of $\sigma = 1 - \chi^2$ for a real solution (curve 0), a symmetrical complex (curve 1), and an asymmetrical complex one f_2 with 'odd' phase vector $w^{(2)}$, $w_{-n}^{(2)} = -w_n^{(2)}$ and real $f_2(\xi)$ having zeros at certain $\xi = \xi_m$ (curve 2). As before, the only solution for $C < C_1$ is the real one $f_0(\xi)$, having no zeros for $-1 \le \xi \le 1$. At $C = C_1$ two solutions f_1 and f_2 of different types branch off from $f_0(\xi)$, and f_0 have symmetrical zeros. Zeros of the functions $f_0(\xi)$ and $f_2(\xi)$ are shown in Figure 5.13a,b, respectively. The increments of the phase vectors $\psi^{(1)} = \{\psi_n^{(1)}\}$ and $\psi^{(2)} = \{\psi_n^{(2)}\}$ for both complex solutions, as well as the decrement of the function $\varphi_2(\xi)$ of the asymmetrical solution are shown in Figure 5.14.

The amplitude of the functions f_n obtained at $C = 7$ in all solutions are presented in Figure 5.15. For the chosen input data the required phase vector $\psi^{(0)} = \{\psi_n^{()}\}$

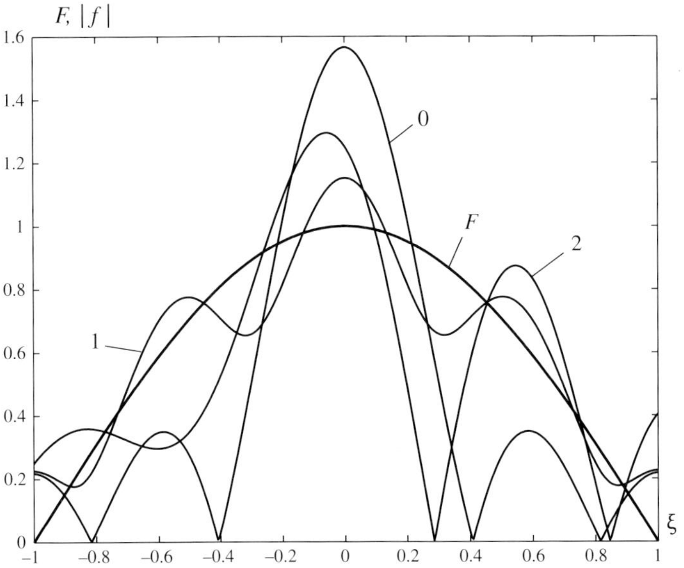

Figure 5.15 Amplitude distributions $|f_n(\xi)|$ for various solutions; $M = 5$; $U_n = 1$; $F(\xi) = \cos(\pi\xi/2)$; $C = 7.0$.

in the real solution f_0 is identical to zero at $C = 7$. The corresponding vectors for the optimal complex solution f_1 and nonsymmetrical one f_2 are

$$\psi^{(1)}/\pi = \{-2.1601, -1.7409, -1.2927, -0.7249, -0.1942,\\ 0, -0.1942, -0.7249, -1.2927, -1.7409, -2.1601\}\ ;$$
$$\psi_n^{(2)}/\pi = \{0.2125, -0.5341, -0.9616, -0.9991, -0.7267,\\ 0, 0.7267, 0.9991, 0.9616, 0.5341, -0.2125\}\ .$$

5.3
The method of opposite directions

Under the *opposite directions method* we imply an approach to optimization of linearly extended systems (waveguides, quasi-optical transmitting lines, etc.) based on the local variations of the optimization parameters. It was successively applied to the optimization and reconstruction of various systems (both discrete and continuous), such as the lens beam lines [75], quasi-optical radiating systems [31], flat resonant antennas with semitransparent apertures [20], and irregular waveguides [76].

The main idea of the approach consists in reducing the functional to the form in which its dependence on the local varied parameters (segment d_n in Figure 5.16) is presented in an explicit form. The information concerning fixed parameters is compactly concentrated in the two (left and right) groups. The position of varied parameters is moved shuttle-wise from left to right and backwards. When moving from left to right, the left group of the information is gradually enriched, whereas

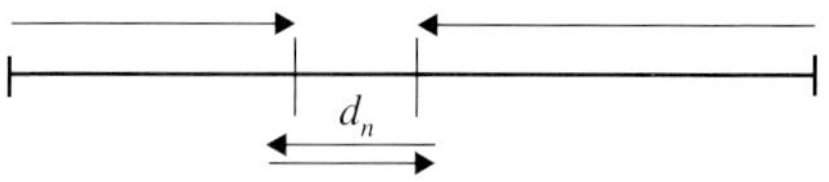

Figure 5.16 Illustration of the opposite direction method.

the right group is depleted. Conversely, when moving from right to left, a small amount of information is added to the right group, whereas another part is subtracted from the left one.

In this section the optimization problems concerning two multi-element systems is numerically solved. In Section 5.3.1 the multi-element phase field transformer is considered (Problems **Gp** and **Hp**). This system is described by a multi-operator phase optimization problem with operators of the same type (Section 3.2.2.3) and a the similar one with operators of two different types (Section 3.2.2.4). In both cases the parameters of one element of the system (lens or semitransparent mirror) are varied at one step and the information about other elements is concentrated in two (for lenses) or three (for mirrors) complex functions. As before, we confine ourselves by two-dimensional problems.

5.3.1
Optimization of multi-element phase field transformers

As in other sections of the book, we are equally interested in problems of two types. The first of these is more practical; it consists in solving the phase optimization problem formulated as the variational one with the objective function in the form of a functional to be minimized, and then in analyzing the numerical results obtained. The second problem is more theoretical; it consists in investigating a nonlinear integral equation system generated by this functional and establishing the number, types and properties of its solutions. Both problems are closely connected, because the considered functional may have several local extrema which correspond to different solutions of the nonlinear equation system. Separation of the global maximum (or maxima) among them needs a priori information concerning the solutions. In this subsection we analyze both problems using the example of a concrete physical problem.

5.3.1.1 Problem formulation. Property of solutions
Consider the physical Problem **Gp** formulated in Section 2.1.2. It is a particular case of the multi-operator phase optimization problem with single-type operators. This problem is mathematically formulated in Section 3.2.2.3 and reduces to the nonlinear equation system (3.196) as the Lagrange–Euler equation for the functional (3.190). A simple iteration procedure (3.206) is proposed for solving this system. Each step of this procedure increases (or more exactly, does not decrease) the value of the minimized functional. The procedure can be successfully realized in the scheme of the opposite directions method. We demonstrate it on the two-dimensional version of Problem **Gp**.

Consider the multi-element phase field transformer consisting of $M+1$ lenses (see Figure 2.3). For simplicity, we assume that the lenses are of the same size $2a$ and that they are equidistantly located with a distance d between them. Then the unique physical parameter describing the system is $c = ka^2/d$. In the problem considered here, the operators A_m, $m = 0, 1, \dots, M-1$, in (3.189) describe the Fourier transform of finite functions.

Introduce the local dimensionless transverse coordinates $x_m = X_m/a$, $m = 0, 1, \dots, M$, on the apertures. The phase corrections provided by the lenses are described by the functions $\Psi_m(x_m)$, $m = 0, 1, \dots, M$. The two functions (in general, complex) are given in the problem: the generating field $U_0(x_0)$ on the input aperture D_0 and the desired field $V_M(x_M)$ on the output one, D_M. We assume that they are normed by the conditions

$$\int_{-1}^{1} |U_0(x_0)|^2 \, dx_0 = 1 \,, \tag{5.172a}$$

$$\int_{-1}^{1} |V_M(x_M)|^2 \, dx_M = 1 \,. \tag{5.172b}$$

The functional to be minimized has the form

$$\chi(\psi_1, \psi_2, \dots, \psi_M) = \left| \int_{-1}^{1} U_M(x_M) \bar{V}(x_M) dx_M \right| \,, \tag{5.173}$$

where $U_M(x_M)$ is the field on the output of the transformer. According to (2.23), the optimization functions in this functional are connected with the required phase corrections as follows

$$\psi_0(x_0) = \Psi_0(x_0) - \frac{c}{2}|x_0|^2 \,, \tag{5.174a}$$

$$\psi_m(x_m) = \Psi_m(x_m) - c|x_m|^2 \,, \quad m = 1, 2, \dots, M-1 \,, \tag{5.174b}$$

$$\psi_M(x_M) = \Psi_M(x_M) - \frac{c}{2}|x_M|^2 \,. \tag{5.174c}$$

Following Section 3.2.2.3, we introduce the operators A_m, A_m^*, $m = 0, 1, \dots, M-1$, and the auxiliary functions $u_m(x_m)$, $v_m(x_m)$, $m = 0, 1, \dots, M$, by the recursive relations

$$u_{m+1}(x_{m+1}) = A_m[u_m \exp(i\psi_m)] = \sqrt{\frac{c}{2\pi}} \int_{-1}^{1} \exp(icx_m x_m) \times u_m(x_m) \exp(i\psi_m(x_m)) dx_m \,, \quad m = 0, 1, \dots, M-1 \,, \tag{5.175a}$$

$$v_{m-1}(x_{m-1}) = A_{m-1}^*[v_m \exp(-i\psi_m)] = \sqrt{\frac{c}{2\pi}} \int_{-1}^{1} \exp(-icx_m x_{m-1}) \times v_m(x_m) \exp(-i\psi_m(x_m)) dx_m \,, \quad m = M, M-1, \dots, 1 \,. \tag{5.175b}$$

The functions

$$u_0(x_0) = U_0(x_0)\,, \tag{5.176a}$$

$$v_M(x_M) = V_M(x_M) \tag{5.176b}$$

are fixed. The functions $U_M(x_M)$ and $V_0(x_0)$ are calculated by $u_M(x_M)$ and $v_0(x_0)$, as follows:

$$U_M(x_M) = u_M(x_M)\exp(i\psi_M)\,, \tag{5.177a}$$

$$V_0(x_0) = v_0(x_0)\exp(-i\psi_0)\,. \tag{5.177b}$$

According to Section 3.2.2.3, the functions u_m, $m = 1, 2, \ldots, M$, v_m, $m = 0, 1, \ldots, M-1$ are subject to the nonlinear integral equations system

$$v_{m-1}(x_{m-1}) = \sqrt{\frac{c}{2\pi}}\int_{-1}^{1}\exp(-icx_m x_{m-1}) \times |v_m(x_{m-1})|\exp(-i\arg(u_m(x_m)))\mathrm{d}x_m\,, \quad m = 1, 2, \ldots, M\,, \tag{5.178a}$$

$$u_{m+1}(x_{m+1}) = \sqrt{\frac{c}{2\pi}}\int_{-1}^{1}\exp(-icx_m x_{m+1}) \times |u_m(x_{m+1})|\exp(i\arg(v_m(x_m)))\mathrm{d}x_m\,, \quad m = 0, 1, \ldots, M-1 \tag{5.178b}$$

(see (3.198)). The functions $\psi_m(x_m)$, maximizing functional (5.173), are calculated term-wise by the appropriate functions u_m, v_m:

$$\psi_m(x_m) = \arg(v_m(x_m)) - \arg(u_m(x_m))\,. \tag{5.179}$$

It can be seen that the function u_m does not depend on ψ_n, $n = m, m+1, \ldots, M$, that is, it contains the information about all the correctors located on the left from the mth one ('left group' of the information). In its turn, the function v_m does not depend on ψ_n, $n = 0, 1, \ldots, m$, it contains the information about all the correctors located to the right from the mth one ('right group' of the information).

The properties of solutions to equation system (5.178) are different depending on whether the given functions $U_0(x_0)$ and $V_M(x_M)$ are even or not. First, we consider the case $U_0(-x_0) = U_0(x_0)$, $V_M(-x_M) = V_M(x_M)$. In this case, the 'trivial' solution $\psi_m(x_m) \equiv 0$, $m = 0, 1, \ldots, M$ to system (5.178) exists if all the functions $u_m(x_m)$, $v_m(x_m)$, $m = 0, 1, \ldots, M$, have no zeros in the interval $[-1, 1]$. If at least one of these functions has zeros, then solutions (possibly, nonunique) may exist with real u_m, v_m. In these solutions, the functions $\psi_m(x_m)$ have the values 0 and π in different segments of x_m. We call such solutions real; they exist for all c. For their practical realization, it is sufficient to put the half-wave 'patches' on the confocal correctors in the segments where $\psi_m(x_m) = \pi$. These solutions are optimal in the class of real functions $u_m(x_m)$, $v_m(x_m)$; however, they do not provide the global maximum to the functional χ.

In addition to real solutions, at least two types of complex solution may also exist. In solutions of the first type, all the functions u_m, v_m are even: $u_m(-x_m) = u_m(x_m)$, $m = 1, 2, \ldots, M$; $v_m(-x_m) = v_m(x_m)$, $m = 0, 1, \ldots, M - 1$. According to (5.179), the functions $\psi_m(x_m)$, $m = 0, 1, \ldots, M - 1$, are also even. The operators A_m, A_m^* are real on such functions, and equation system (5.178) takes the form

$$v_{m-1}(x_{m-1}) = \sqrt{\frac{2c}{\pi}} \int_0^1 \cos(cx_m x_{m-1}) |v_m(x_m)|$$
$$\times \exp(-i \arg(u_m(x_m))) dx_m, \quad m = 1, 2, \ldots, M, \quad (5.180a)$$

$$u_{m+1}(x_{m+1}) = \sqrt{\frac{2c}{\pi}} \int_{-1}^1 \cos(cx_m x_{m+1}) |u_m(x_m)|$$
$$\times \exp(i \arg(v_m(x_m))) dx_m, \quad m = 0, 1, \ldots, M - 1. \quad (5.180b)$$

Theorem 3.3 is true for these equations, so that any solution generates an equivalent group of solutions, providing the same value for the functional.

Solutions of the second type are not even. There are two sets of solutions of this type. In the first of these the moduli of the functions $u_m(x_m)$, $v_m(x_m)$ with even m are even, and their arguments are odd:

$$|u_{2n}(-x_{2n})| = |u_{2n}(x_{2n})|, \quad \arg(u_{2n}(-x_{2n})) = -\arg(u_{2n}(x_{2n})), \quad (5.181a)$$

$$|v_{2n}(-x_{2n})| = |v_{2n}(x_{2n})|, \quad \arg(v_{2n}(-x_{2n})) = -\arg(v_{2n}(x_{2n})), \quad (5.181b)$$

for all $n = 0, 1, \ldots, [M/2]$. Then, according to (5.179),

$$\psi_{2n}(-x_{2n}) = -\psi_{2n}(x_{2n}), \quad n = 0, 1, \ldots, [M/2], \quad (5.182)$$

and (5.178) take the form

$$v_{2n-1}(x_{2n-1}) = \sqrt{\frac{2c}{\pi}} \int_0^1 |v_{2n}(x_{2n})| \cos(cx_{2n} x_{2n-1}$$
$$- \arg(u_{2n}(x_{2n}))) dx_{2n}, \quad n = 1, 2, \ldots [M/2], \quad (5.183a)$$

$$u_{2n+1}(x_{2n+1}) = \sqrt{\frac{2c}{\pi}} \int_0^1 |u_{2n}(x_{2n})| \cos(cx_{2n} x_{2n+1})$$
$$+ \arg(v_{2n}(x_{2n})) dx_{2n}, \quad n = 0, 1, \ldots [(M - 1)/2]. \quad (5.183b)$$

It can be seen that the functions $u_m(x_m)$, $v_m(x_m)$ with odd $m = 2n + 1$, $n = 0, 1, \ldots, [M - 1]$, are real, but nonsymmetrical in both amplitude and phase. At some points they may become zero, so that the functions $\arg(u_{2n}(x_{2n+1}))$, $\arg(u_{2n}(x_{2n+1}))$, $n = 0, 1, \ldots, [(M + 1)/2]$ take the values 0 or π only. The other equations of (5.178) are

$$v_{2n}(x_{2n}) = \sqrt{\frac{c}{2\pi}} \int_{-1}^1 \exp(-icx_{2n} x_{2n+1}) |v_{2n+1}(x_{2n+1})|$$
$$\times \text{sgn}(u_{2n+1}(x_{2n+1})) dx_{2n+1}, \quad n = 0, 1, \ldots, [(M - 1)/2], \quad (5.184a)$$

$$u_{2n+2}(x_{2n+2}) = \sqrt{\frac{c}{2\pi}} \int_{-1}^{1} \exp(icx_{2n+1}x_{2n+2}) \left| u_{2n+1}(x_{2n+1}) \right| \\ \times \operatorname{sgn}(v_{2n+1}(x_{2n+1}))dx_{2n+1}, \quad n = 0, 1, \ldots, [(M-1)/2] . \tag{5.184b}$$

It can be seen that these equations are consistent with condition (5.182).

In the second set of noneven solutions, the moduli of the functions $u_m(x_m)$, $v_m(x_m)$ with odd m are even, and their arguments are odd. For this case equations (5.181)–(5.184) are rewritten in an obvious way.

It was established in Section 3.2.2.3 that, if the operators A_0, A_1 satisfy condition (3.207), then the problem considered here for the case $M = 2$, $|U_0(x)| \equiv |V_2(x)|$, can be reduced to the nonlinear equation (3.212), coinciding with (3.46) after simple renotation. The operators given by (5.175) satisfy (3.207). Indeed,

$$A_1^* u = \sqrt{\frac{c}{2\pi}} \int_{-1}^{1} \exp(-icx_2x_1)u(x_2)dx_2 \\ = \sqrt{\frac{c}{2\pi}} \overline{\int_{-1}^{1} \exp(icx_0x_1)\bar{u}(x_0)dx_0} = \overline{A_0\bar{u}} . \tag{5.185}$$

In our case, (3.46) has the form (3.74). It belongs to a general class of nonlinear equations (4.1) which have analytical solutions of type (4.3). These solutions are analyzed in detail in Section 4.2. It remains to establish the correspondence between these results and the functions used here.

According to the analogy mentioned, the required functions ψ_0, ψ_2 are calculated as

$$\psi_0(x_0) = \psi(x_0) - \arg U_0(x_0) , \tag{5.186a}$$

$$\psi_2(x_2) = \psi(x_2) + \arg V_2(x_2) . \tag{5.186b}$$

The function $\psi = \arg f$ is determined from the nonlinear integral equation (3.74) with $F(x) = |U_0(x)| \equiv |V_2(x)|$ (here we introduce a common independent variable x for all considered functions). In our notation, the function $f(x)$ coincides in modulus with the functions $u_2(x)$ and $v_0(x)$ determined by (5.178). According to (5.179), the function $\psi_1(x_1)$ is calculated as

$$\psi_1(x_1) = 2 \arg \int_{-1}^{1} \exp(icx_0x_1)F(x_0)\exp(i\psi(x_0))dx_0 . \tag{5.187}$$

According to Theorem 4.5, if the solution $f(x)$ to (3.74) has no zeros at $x \in [-1, 1]$, then

$$\exp(i\psi(x)) = \frac{P_N(x)}{|P_N(x)|} , \tag{5.188}$$

where $P_N(x) = \Pi_{n=1}^{N}(1 - \eta_n x)$ is the polynomial of finite degree N with complex nonconjugated zeros η_n^{-1}. The inverse complex zeros η_n (further – parameters) of these polynomials are calculated from the transcendental equation system (4.74).

We overview some properties of these solutions. At small c only the solution with $N = 0$ exists which corresponds to $\psi(x) \equiv 0$. The value $c = c_1$, at which solutions with $N = 1$ arise the first time, is calculated from the transcendental equation system (4.107). For the case of even functions $F(x)$ considered here, the pair of equivalent solutions with $\eta_1 = \pm i\eta_1''$ arise at $c = c_1$. Such solutions correspond to the class of asymmetrical solutions described above: the function $\psi(x)$ is odd: $\psi(-x) = -\psi(x)$. The functions $u_1(x_1)$, $v_1(x_1)$ are real and identical, but nonsymmetrical, they may have zeros; $\psi_1(x_1) \equiv 0$. The value c_1 is calculated from the transcendental equation

$$\int_{-1}^{1} F(x)\cos(cx)\mathrm{d}x = 0\,. \tag{5.189}$$

For $c > c_1$, the solutions with polynomials of different N can exist simultaneously. In particular, the solution with $N = 2$ arises at $c = c_2$. The parameters of this polynomial are first imaginary, $\eta_{1,2} = \pm i\eta_{1,2}''$ (the sign '$\pm$' relates to both η_1 and η_2); it describes the two different solutions of the same equivalent group. According to Theorem 4.7, η_2 starts from 0 at $c = c_2$. The values c_2 and $\eta_1''(c_2)$ are determined from the following system of transcendental equations

$$\int_{-1}^{1} \frac{F(x)\cos(cx)}{\sqrt{1 + \left(\eta_1''\right)^2 x^2}}\mathrm{d}x = 0\,, \tag{5.190a}$$

$$\int_{-1}^{1} \frac{F(x)x\sin(cx)}{\sqrt{1 + \left(\eta_1''\right)^2 x^2}}\mathrm{d}x = 0\,. \tag{5.190b}$$

As was mentioned in Section 4.2, the global minimum for the functional $\sigma(u)$ (3.69) (which corresponds to the maximum of χ) is reached at the solutions with maximum N.

Up to $c = c_3$, the only optimal solution is the solution with odd $\psi(x)$. At this point, the imaginary parameters η_1 and η_2 generating one of the equivalent solutions, coincide: $\eta_1(c_3) = \eta_2(c) = i\eta''$. The values c_3, η'' solve the system of transcendental equations

$$\int_{-1}^{1} \frac{F(x)\cos(cx)}{1 + (\eta'')^2 x^2}\mathrm{d}x = 0\,, \tag{5.191a}$$

$$\int_{-1}^{1} \frac{F(x)x\sin(cx)}{1 + (\eta'')^2 x^2}\mathrm{d}x = 0\,. \tag{5.191b}$$

After $c = c_3$ the above parameters become complex, with the same imaginary part and opposite-signed real one: $i\eta_1 = \overline{i\eta_2}$. They generate the solution with the property $\psi(-x) + \pi = -[\psi(x) + \pi]$ which belongs to the same class of solutions with odd $\psi(x)$ (the common constant addend π can be rejected). The two parameters corresponding to the second solution become conjugated at $c = c_3$; the solution becomes real ($\psi(x) \equiv 0$). For $c > c_3$ the phases of this solution are even, $\psi(-x) = \psi(x)$. The solutions with odd and even phases of $f(x)$ exist simultaneously. They provide the same value of the functional χ.

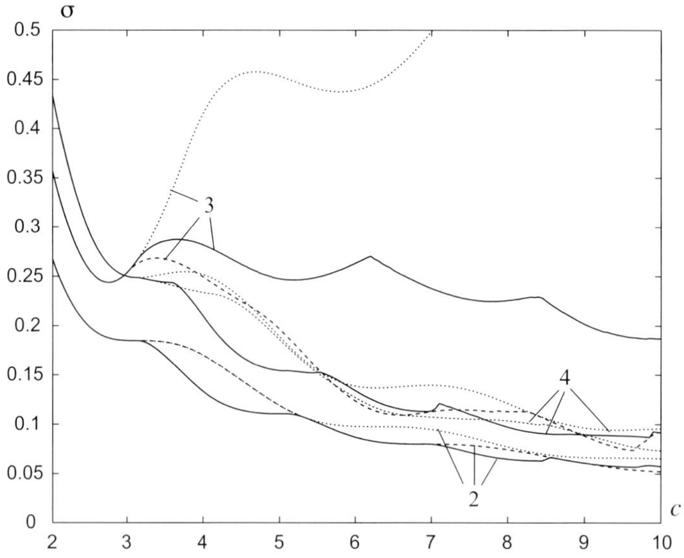

Figure 5.17 Optimal values of σ versus parameter c for different types of solutions; $U_0 = V_M = const.$

Equation (3.74) with $F(x) = const$ was investigated in detail in [56]. The results can be immediately applied to the problem concerning the three-element phase transformer with $|U_0(x_0)| \equiv |V_2(x_2)| = const$. In Figure 5.17 the total (diffraction and transformation) losses

$$\sigma = 1 - \chi^2 \tag{5.192}$$

versus the parameter c are shown for the three types of solution: trivial ($\psi_m \equiv 0$, $m = 0, 1, 2$, – dotted lines), odd optimal ($\psi_m(-x_m) = \psi_m(x_m)$, $m = 0, 2$; $\psi_1 = \{0, \pi\}$, – solid lines), and the best even (not always optimal) ($\psi(-x_m) = \psi(x_m)$, $m = 0, 1, 2$, – dashed lines). The curves related to the considered case $M = 2$ are marked by the label 2. Other results, given in the figure, relate to the larger transformers. They will be analyzed in the next subsection.

Note that the value of σ for the three-element transformer differs from that used in [56] and in Section 4.2, because it describes total losses in the whole transformer but not in one segment only.

For the considered function $F(x)$, the values of c at which the type of optimal solution is changed or the solutions of different types become optimal, are: $c_1 = \pi$, $c_2 = 5.09$, $c_3 = 6.89$, $c_4 = 8.55$. It can be seen that the solutions with odd $\psi_{0,2}$ belong to the optimal ones for $c > c_1$. In the segments where the maximal degree N of the polynomials P_N is even (i.e. for $c_2 < c < c_3$, $c > c_4$), the solutions with even $\psi_{0,1,2}$ are optimal as well. Recall that, according to Theorem 4.7, the set of optimal solutions providing the same maximal value of χ (minimal σ) does not only contain solutions of the above considered types.

The conclusion about the optimality of solutions with even $\psi_m(x_m)$ relates only to the case $M = 2$, $|U_0(x)| \equiv |V_M(x)|$. In other cases the comparative analysis of different solutions of the problem should be made separately. In the next subsection an analysis based on the numerical results obtained by the opposite directions method will be provided for several examples.

To analyze the numerical results related to the multi-element phase transformers, it is convenient to use the theory of periodical beam waveguides: the lines consisting of identical phase correctors (lenses). It is known (see, e.g. Section 3.3.2.2) that the first (main) eigenwave of such a line has the least losses if the phase correctors in the line provide the correction

$$\Psi(x) = cx^2 \tag{5.193}$$

(these correctors are named *confocal*). This means that the maximum coupling coefficient χ (and, correspondingly, the minimal losses σ) would be reached in the case when the given functions U_0 and V_M coincide with the field of this eigenwave.

The field distributions of the eigenwaves over the 'mean section' of the correctors in line with confocal correctors (later, *confocal line*) are described by the eigenfunctions φ_n of the integral equation

$$\lambda_n \varphi_n(x) = \sqrt{\frac{c}{2\pi}} \int_{-1}^{1} \varphi_n(x') \exp(icxx')\mathrm{d}x' \,. \tag{5.194}$$

These functions are the *prolate spheroidal wave functions* [78]. They are real, even for odd numbers $n = 1, 3, \ldots$, and odd for even $n = 2, 4, \ldots$. All together, they make up a complete orthogonal set of functions in $L_2(-1, 1)$. The main eigenwave is described by the function $\varphi_1(x) = s_0(cx)$ which is asymptotically close to the function $\exp(-cx^2/2)$.

An important role in the process of the field transformation belongs to the phase distributions of the eigenvalues of (5.194):

$$\arg \lambda_n = \frac{(n-1)\pi}{2} \,. \tag{5.195}$$

5.3.1.2 Description of the algorithm. Numerical results

In order to apply the opposite directions method to the problem considered, we must slightly modify the iterative procedure described in Section 3.2.2.3. In its simplest form (3.206) the phase corrections ψ_m are improved only at stage (3.206c), when passing from right to left along the transformer, whereas stage (3.206a) (passing from left to right) only recalculates the functions u_{m+1} and does not use the functions v_m, calculated in the previous stage. In the opposite directions method, the phase correctors are improved at each step of both stages, which makes the algorithm symmetrical and more effective. In contrast to the scheme (3.206), besides the initial approximations $\psi_m^{(0)}$, $m = 0, 1, \ldots, M-1$, for the phase correctors, the first approximation $u_m^{(1)}$, $m = 1, 2, \ldots, M$, for the functions u_m, and $\psi_M^{(1)}$ for ψ_M must be calculated by (3.206a) and (3.206b) at $p = 0$ before starting the iterative procedure, that is,

$$u_{m+1}^{(1)} = A_m \left[u_m^{(1)} \exp\left(i\psi_m^{(0)} \right) \right] , \quad m = 0, 1, \ldots, M-1 , \tag{5.196a}$$

$$\psi_M^{(1)} = \arg(v_M) - \arg\left(u_M^{(1)} \right) . \tag{5.196b}$$

Starting from $p = 1$, each exterior step of the method consists of two stages realizing the 'right to left' and 'left to right' passages, respectively. The phase corrections are improved at each stage of the algorithm, as follows:

$$\begin{aligned} v_m^{(2p-1)} &= A_m^* \left[v_{m+1}^{(2p-1)} \exp\left(-i\psi_{m+1}^{(2p-1)} \right) \right] , \\ \psi_m^{(2p-1)} &= \arg\left(v_m^{(2p-1)} \right) - \arg\left(u_m^{(2p-1)} \right) , \quad m = M-1, M-2, \ldots, 0 . \end{aligned} \tag{5.197a}$$

$$\begin{aligned} u_m^{(2p)} &= A_{m-1} \left[u_{m-1}^{(2p)} \exp\left(i\psi_{m-1}^{(2p-1)} \right) \right] , \\ \psi_m^{(2p)} &= \arg\left(v_m^{(2p)} \right) - \arg\left(u_m^{(2p)} \right) , \quad m = 1, 2, \ldots, M . \end{aligned} \tag{5.197b}$$

In the described algorithm the correctors are improved successively, in a shuttle-wise manner. Let all the correctors be fixed except the mth one. Then the function u_m contains the information about all the correctors from the zeroth to the $(m-1)$th ones, whereas v_m contains the information about the correctors from the $(m+1)$th to Mth ones. The improved correction ψ_m is calculated by the explicit formula (5.197a) or (5.197b) using these two functions. At each step of stage (5.197a) describing the passage 'right to left' the information about the right group of correctors is enriched by the function v_m; the function u_m is then no longer needed and can be deleted from the left group. Similarly, in stage (5.197b) describing the passage 'left to right' the information about the left group is enriched by the function u_m; the function v_m can be deleted from the right group.

It can be seen that the modified iterative method described has the same property as the simplest method from Section 3.2.2.3: each step increases the value of the functional $\chi(\psi_1, \psi_2, \ldots, \psi_M)$.

Now we discuss the numerical results obtained for several examples of the convertors considered here. First we return to Figure 5.17 and consider the results related to the case $U_0(x) \equiv V_M(x)$ for $M = 3$ and $M = 4$ (curves labeled by 3 and 4, respectively). It can be seen that the results for the convertor with an even number of elements ($M = 3$) qualitatively differ from those for the odd-element convertors ($M = 2$, $M = 4$). This fact can be simply explained on the example of 'trivial' solutions with $\psi_m \equiv 0$, that is, when the convertor is a segment of a confocal quasi-optical line. The incident field U_0 may be expanded by the eigenfunctions of integral equation (5.194) with certain coefficients C_n. If the function $U_0(x_0)$ is even, then only the coefficients C_n with odd n differ from zero. After transmitting between two correctors, the field of the nth eigenwave obtains the factor $\lambda_n = |\lambda_n| \exp(i \arg \lambda_n)$. According to (5.195), all odd terms are summed up on

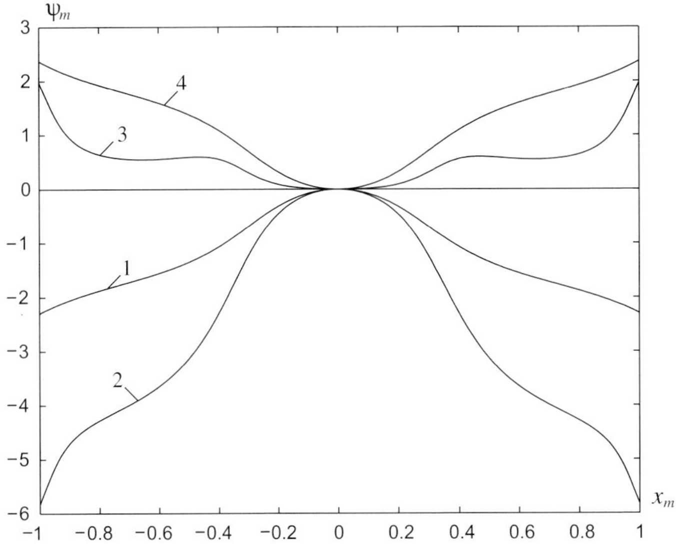

Figure 5.18 Phase corrections for $U_0(x_0) \equiv 1$, $V_M(x_M) \equiv 1$ at $M = 4$, $c = 7.0$.

the next corrector, whereas the even ones alternate their signs. As a result, the 'image' of the field distorts. If the number of periods is even, then all terms arrive at the corrector in-phase and the image restores with accuracy to the addends corresponding to the eigenwaves with small $|\lambda_n|$. The number of λ_n, close to unity in modulus, increases with n increasing, hence at large c the total losses are comparatively small even in the convertors with all $\psi_m \equiv 0$.

The above effect of image distortion also takes place for convertors with an even numbers of elements corresponding to the 'odd' solutions (for the case $M = 3$ considered here, the functions $\psi_{0,2}$ are odd, and $\psi_{1,3} = \{0|\pi\}$ in these solutions). Since the functions $u_m(x_m)$ are no longer even, they contain the eigenfunctions φ_n with odd n, hence the image can be restored only after four (or multiple) periods. For this reason, solutions of this type are not optimal at any c for $M = 3$; for all $c > c_1$, at which nontrivial solutions exist, the solutions with even ψ_m are optimal.

We find that there are no solutions of equation system (5.178) with odd M and $U_0(x) \equiv V_M(x)$, which have the geometrical symmetry $\psi_m(x) = \psi_{M-m}(x)$ (see Figure 5.18). This means that any solution $\{\psi_m\}$, $m = 0, 1, \ldots, M$, has its mirror twin $\{\tilde{\psi}_m\}$, $\tilde{\psi}_m = \psi_{M-m}$, $m = 0, 1, \ldots, M$. Due to the duality principle (see Section 5.2.3.1), these solutions are equivalent: they provide the same value for the functional χ.

In contrast to the considered case $M = 2$, $U_0(x) \equiv V_2(x)$, when asymmetrical solutions were optimal for all $c > c_1$, solutions of this type may not be presented in the set of optimal ones for the longer even-number-element transformers. Numerical results obtained for $M = 4$ show that there are ranges of c, in which the optimal solutions are only either symmetrical or asymmetrical. Moreover, the solutions that are optimal immediately after $c = c_1$, are symmetrical.

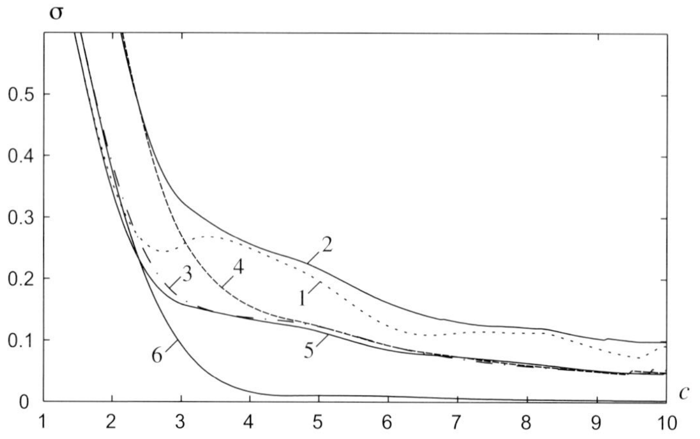

Figure 5.19 Optimal values of σ versus parameter c for different input data.

Below we consider several transformers with various numbers of elements and several functions to be transformed. All numerical results are obtained by the opposite direction method. In all cases the solutions were sought in the class of symmetrical functions $\psi_m(x_m)$, $m = 0, 1, \ldots, M$.

In Figure 5.19 we show the value of the functional $\sigma = 1 - \chi^2$ (energy losses) for optimal solutions of the equation system (5.180) with different variants of the input data. The curves are labeled according to the following table:

U_0	V_M	M	Label
$1/\sqrt{2}$	$1/\sqrt{2}$	3	1
$1/\sqrt{2}$	$1/\sqrt{2}$	8	2
$1/\sqrt{2}$	$\cos(\pi x_M/2)$	3	3
$1/\sqrt{2}$	$\cos(\pi x_M/2)$	8	4
$1/\sqrt{2}$	$\exp(-cx^2/2)$	3	5
$\cos(\pi x_M/2)$	$\exp(-cx^2/2)$	3	6

Owing to the duality principle, the results for the cases with nonidentical input and output fields can be easily recalculated for the problems with interchanged $U_0(x_0)$ and $V_M(x_M)$: $U_0(x_0) = \cos(\pi x_0/2)$, $V_M(x_M) \equiv 2^{-1/2}$.

As was mentioned before, the function $\exp(-cx^2/2)$ is close to the field of the main (with minimal losses) eigenwave of the confocal quasi-optical lines. Optimal transformation of a field into this wave is equivalent to the best excitation of the wave by the chosen field. However, this transformation (and the inverse one) is performed when transforming fields one into other by long transformers of the type considered.

When the fields are transformed one into another, the two types of loss are presented in the total ones: the transformation and diffraction losses. The transformation losses, by necessity, change the configuration of the field when transmitting it along the transformer, or keep it if the given and desired fields are the same. This process requires maximum degrees of freedom, that is, as many as possible phase correctors. The diffraction losses are caused by the energy bypassing the correctors because of their limited sizes. These losses increase with an increasing number of correctors.

For small c (i.e. a small size of the correctors or a large distance between them), the diffraction losses are large, and the short transformers are more efficient than the long ones (cf. curves 3 and 4). As c increases, when the diffraction losses become negligibly small, the efficiencies of the short and long transformers approach each other. In this case the efficiency is mainly influenced by the shapes of the functions to be transformed (cf. curves 2 and 4). In each concrete case, the optimal values of M and c may be determined numerically.

Similarly as in the case of the two- and three-element transformers, for sufficiently small values of $c < c_1$ only 'trivial' solution $\psi_m(x_m) \equiv 0$, $m = 0, 1, \ldots, M$ exists for any U_0 and V_M. The value c_1 is the first branching point of this solution. In contrast to the case $M = 2$, $U_0(x) \equiv V_M(x)$ considered above, in which exact transcendental equations for determining such points exist (see (5.189)–(5.191)). In the general case, they should be determined by solving the appropriate homogeneous integral equations. For multielement transformers, such equations have a very awkward form and are not given in this book. The value c_1 can be approximately defined as a point at which the quantity $\max_{m,x} |\psi_m(x)|$ becomes negligibly small.

No transformation can be made for $c < c_1$; only optimal transmission of the energy by a segment of the confocal quasioptical line is performed. The function U_0 can be expressed as a superposition of the fields $\varphi_n(x)$ of eigenwaves of this line. In the considered range of values c the losses of higher waves are large in comparison with those of the main wave. In fact, only this wave reaches the end of the transformer.

Figures 5.20–5.22 show distributions of the functions $\psi_m(x_m)$ on the correctors at fixed c for different variants of the input data; the curves are labeled by the cor-

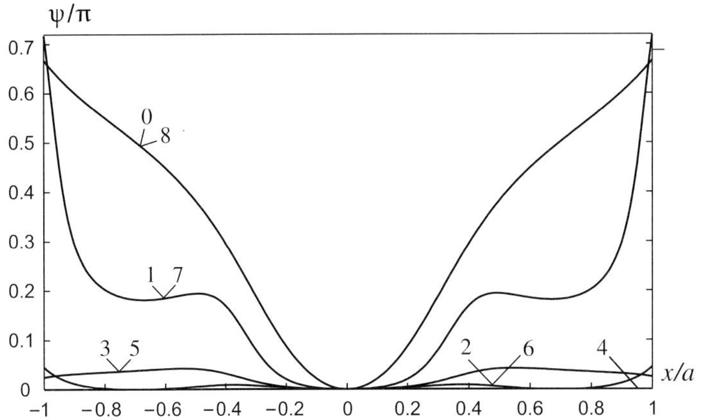

Figure 5.20 Phase corrections for $U_0(x_0) \equiv const$, $V_M(x_M) \equiv const$ at $M = 8$, $c = 7.0$.

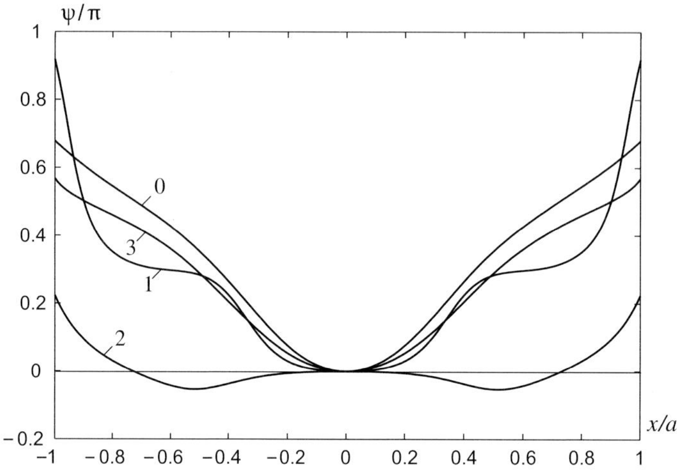

Figure 5.21 Phase corrections for $U_0(x_0) = \cos(\pi x_0/2a)$, $V_M(x_M) \equiv const$ at $M = 8$, $c = 7.0$.

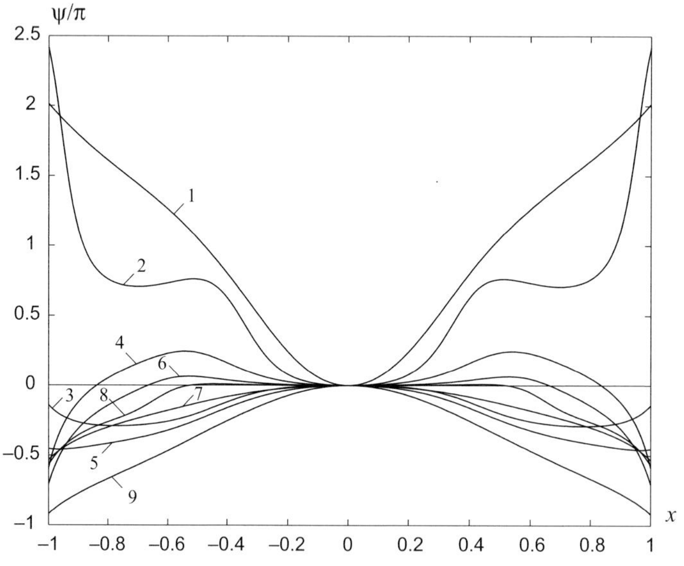

Figure 5.22 Phase corrections for $U_0(x_0) \equiv const$, $V_M(x_M) = \cos(\pi x_M/2a)$ at $M = 8$, $c = 7.0$.

rector numbers m. Recall that the phase corrections of the elements are calculated by the functions $\psi_m(x_m)$ from the relations (5.174). The functions $\psi_m(x_m)$ correspond to the optimal solutions of (5.178), that is, they provide the maximum values of the functional χ. Of course, they are only separate representatives of the corresponding equivalent groups mentioned in the previous subsection.

The physical mechanism of the field transformation may be explained in terms of the theory of regular quasi-optical lines mentioned above. The optimal transmission of the energy along such a line is attained in the case when the phase correctors of the line provide correction (5.193). In a sufficiently long transformer, the first few elements transform the field U_0 into the main eigenwave φ_1 (or into a superposition of several waves at comparatively large c when they have small losses); then, this wave propagates without distortion and the last few elements transform this eigenwave into the desired field. Such a process is observed in Figure 5.20, where the functions ψ_m are shown for the case $M = 8$, $U_0 \equiv V_M = const$ at $c = 7$. The first two correctors (with large ψ_0, ψ_1, respectively) transform the field U_0 into the wave φ_1, the correctors 2–7 transmit this wave without distortion (functions ψ_2, ψ_3, ψ_5, ψ_6 are small, $\varphi_4 \equiv 0$ due to the chosen geometrical antisymmetry $\psi_m = -\psi_{M-m}$ what is permissible in an equivalent group of solutions), and the two last correctors (functions ψ_7, ψ_8) perform the inverse transformation of φ_1 into V_M.

5.3.2
Optimization of multi-element phase beam transformers

Other problems to which the opposite directions method can be applied are the problems on the multi-element field transformer formulated in Section 2.2.3 (Problems **S**, **T** and **U**). The general theory of these problems was described in Section 3.2.2.4. An iterative algorithm was proposed for their solution. Here this algorithm is adjusted to the scheme of the opposite direction method. The two-dimensional version of Problem **Tp** is considered. The quasi-optical conditions are assumed to hold.

5.3.2.1 Problem formulation

We denote by $x_m = X_m/a$, the normed value of the X-coordinate on the mth aperture, and use it as the argument of all functions defined on the apertures (uniformly for both the vertical D_0 and inclined D_m, $m = 1, 2, \ldots, M$); $2a$ is the vertical size of the apertures.

For the two-dimensional problem, the relation (2.67a) between the fields on the apertures has the form

$$u_{m+1}(x_{m+1}) = A_m[u_m T_m] \equiv \sqrt{\frac{c}{2\pi}} \exp(-ikax_{m+1} \tan\beta_{m+1})$$
$$\times \int_{-1}^{1} u_m(x_m) T_m(x_m) \exp\left(-i\left[(c/2)(x_{m+1} - x_m)^2 - kax_m \tan\beta_m\right]\right) dx_m ,$$
$$m = 0, 1, \ldots, M-1 , \quad (5.198)$$

where $T_m(x_m)$ is the transmission coefficient of the mth element,

$$c = \frac{ka^2}{d} ; \quad (5.199)$$

$\beta_0 = 0$; $\beta_m = \beta$, $m = 1, 2, \ldots, M$. The radiation pattern of the entire system is a sum of the partial patterns of its elements

$$f(\theta) = \sum_{m=1}^{M} f_m(\theta) , \quad m = 1, 2, \ldots, M , \tag{5.200}$$

where the angle θ is measured counter-clockwise from the geometrical direction of the reflected rays making up the angle 2β with the inverse z-axis. The partial patterns $f_m(\theta)$ are defined by the fields on the mirrors and their reflection factors, as follows:

$$f_m(\theta) = B_m[u_m R_m] \equiv \exp\left(-i(m-1)kd\cos\gamma\right) \times \int_{-1}^{1} u_m(x_m) R_m(x_m) \exp(-ikax_m(\tan\beta\cos\gamma - \sin\gamma))\mathrm{d}x_m , \quad m = 1, 2, \ldots, M , \tag{5.201}$$

where $\gamma = 2\beta + \theta$; $R_m(x_m)$ is the reflection coefficient of the mth mirror (a nonessential constant factor is omitted in this formula).

There are only two independent physical parameters among the quantities ka, kd, c, β involved in (5.198) and (5.201). According to (5.199) and (2.61), ka and kd can be expressed by c and β, as follows

$$ka = \frac{2c}{\sin(2\beta)} , \tag{5.202a}$$

$$kd = \frac{4c}{\sin^2(2\beta)} . \tag{5.202b}$$

The problem consists in creating the desired amplitude radiation pattern $F(\theta)$, $\theta \in [-\theta_0, \theta_0]$, at the given amplitude $U_0(x_0)$, $x_0 \in [-1, 1]$ of the input field, by choosing appropriate complex reflection coefficients $R_m(x_m)$, $m = 1, \ldots, M$, and transmission ones $T_m(x_m)$, $m = 0, 1, \ldots, M-1$. It is assumed that these coefficients fulfill the conditions

$$|T_0(x_0)|^2 = 1 , \tag{5.203a}$$

$$|R_m(x_m)|^2 + |T_m(x_m)|^2 = 1 , \quad m = 1, \ldots, M-1 , \tag{5.203b}$$

$$|R_M(x_M)| = 1 . \tag{5.203c}$$

In [25] a system with mirrors realized as short-periodical arrays is considered. In this case the reflection and transmission coefficients depend on the physical parameters of the arrays and they are subject to other restrictions.

The functional to be maximized is

$$\chi(R_1, R_2 \ldots, R_M; T_0, T_1 \ldots, T_{M-1}) = \int_{-\theta_0}^{\theta_0} |f(\theta)|\, F(\theta)\mathrm{d}\theta . \tag{5.204}$$

Below we use the normalized coordinate $\xi = \sin\theta / \sin\theta_0$ instead of θ. It is assumed that the functions $U_0(x_0)$ and $F(\xi)$ are normed by the conditions $\int_{-1}^{1} U_0^2(x_0)\mathrm{d}x_0 = 1$, $\int_{-1}^{1} F^2(\xi)\mathrm{d}\xi = 1$.

5.3.2.2 Description of the algorithm. Numerical results

The optimization problem formulated above can be solved by the iterative procedure (3.231) proposed in Section 3.2.2.4. As in the previous subsection, this procedure can be modified to apply the opposite directions method.

First, we define the adjoint operators participating in the algorithm as

$$A^*_m[u] \equiv \sqrt{\frac{c}{2\pi}}\exp(-ikax_m\tan\beta_m)\int_{-1}^{1} u(x_{m+1})\exp\big(i\big[(c/2)(x_m-x_{m-1})^2 + kax_{m+1}\tan\beta_{m+1}\big]\big)dx_{m+1}\,,\quad m=0,1,\ldots,M-1\,; \tag{5.205}$$

$$B^*_m[\Phi] \equiv \int_{-1}^{1}\Phi(\xi)\exp\big(i\,\big[((m-1)kd + kax_m\tan\beta)\cos\gamma - \sin\gamma\big]\big)\,d\xi\,. \tag{5.206}$$

As in the algorithm (3.231), initial approximations $R^{(0)}_m$, $m=1,2,\ldots,M$, $T^{(0)}_m$, $m=0,1,\ldots,M-1$, for the functions R_m, T_m must be given, and the functions $u^{(0)}_m$, $f^{(0)}_m$, $m=1,2,\ldots,M$, and $f^{(0)}$, $\psi^{(0)}=\arg f^{(0)}$ must be calculated by (5.198), (5.201) and (5.200), respectively. The functions $u^{(p)}_0 = U_0$, $T^{(p)}_M \equiv 0$, $w^{(p)}_M \equiv 0$ are given for all iterations.

In the modified algorithm, the odd iterations (back passages) and the even ones (direct passages) are different. The odd iterations coincide with the first loop (3.231a)–(3.231j) of the algorithm (3.231), only after substitution of the iteration number p by $2p-2$ and additional calculation of the value of χ reached in the $(2p-1)$th iteration. In the even iterations the loop (3.231k), (3.231l) is modified by supplementing each step by recalculation of the functions $R_m(x_m)$, $T_m(x_m)$. As a result, the part (3.231k)–(3.231n) of algorithm (3.231) is substituted by the following steps:

$$\chi_{2p-1} = \left(|w_0^{(2p-1)}|,\,U_0\right)_0\,, \tag{5.207a}$$

$$u^{(2p)}_m = A_{m-1}\left[u^{(2p-1)}_{m-1}T^{(2p-1)}_{m-1}\right], \tag{5.207b}$$

$$\arg R^{(2p)}_m = \arg v^{(2p-1)}_m - \arg u^{(2p)}_m\,, \tag{5.207c}$$

$$\arg T^{(2p)}_m = \arg w^{(2p-1)}_m - \arg u^{(2p)}_m\,, \tag{5.207d}$$

$$R^{(2p)}_m = \left|R^{(2p-1)}_m\right|\exp\left(i\arg R^{(2p)}_m\right), \tag{5.207e}$$

$$T^{(2p)}_m = \left|T^{(2p-1)}_m\right|\exp\left(i\arg T^{(2p)}_m\right), \tag{5.207f}$$

$$f^{(2p)}_m = B_m\left[u^{(2p)}_m R^{(2p)}_m\right],\quad m=1,\ldots,M\,, \tag{5.207g}$$

$$f^{(2p)} = \sum_{m=1}^{M} f^{(2p)}_m\,,\quad \psi^{(2p)} = \arg f^{(2p)}\,, \tag{5.207h}$$

$$\chi_{2p} = \left(|f^{(2p)}|,\,F\right)_\Omega\,. \tag{5.207i}$$

Note that the functions $|R_m|$, $|T_m|$ cannot be changed in even iterations, since they depend only on the functions v_m, w_m which are not changed in these iterations. According to the method of opposite directions, these functions contain information about the right part of the system (elements from mth to Mth and the phase pattern ψ), which is not changed when passing from left to right. The function u_m containing information about the left part of the system (from the zeroth to the $(m-1)$th element) affects only the arguments of the functions R_m and T_m by (5.207c), (5.207d).

The numerical results presented below concern the two-dimensional problem consisting of 7 elements ($M = 6$) with the parameters $c = 1.5$, $\beta = 0.2$. The incident field on the input aperture is $U_0(x_0) \equiv const$, the desired amplitude pattern is $F = \cos(\pi\xi/2)$, $\theta_0 = 0.4$. These parameters correspond to $ka = 7.70$, $kL = 39.57$. The parameters of the virtual linear antenna array located on the phase front of the reflected rays are: the electrical distance between the centers of radiators $kh = kL\sin(2\beta) = 15.41$, and the main parameter of the array $c_0 = kh\sin(\theta_0) = 6.0$.

The results show that the initial approximations of the reflection and transmission coefficients can be chosen from a wide range of physically reasonable values. Two numerical experiments were made with different initial approximations. In the first case, the approximations described the system with roughly uniform distribution of the reflected energy among the elements, namely, $T_0^{(0)} \equiv 0$, $T_m^{(0)} = (M-m)^{-1/2}$, $R_m^{(0)} = (1 - T_m^{(0)2})^{1/2}$, $m = 1, 2, \ldots, M-1$; $R_M^{(0)} = 1$. In the second case, the energy of the reflected and transmitted fields at each element was approximately the same: $T_0^{(0)} \equiv 0$, $T_m^{(0)} = R_m^{(0)} = 2^{-1/2}$, $m = 1, 2, \ldots, M-1$; $R_M^{(0)} = 1$. In both cases, the iterative process led to the same results, approximately by the same number of iterations, about 20.

The optimal amplitude distributions of the reflection coefficients are shown in Figure 5.23; as above, the curves are labeled by the number m. Distributions of the transmission coefficients are calculated by the relation (5.203b). The phase distributions of these coefficients are shown in Figure 5.24. Figure 5.25 shows the desired and obtained amplitude patterns for the optimal solutions. The functional χ on this solution attains the value 0.983.

5.4
The method of generalized separation of variables[3]

The method of the generalized separation of variables is intended for the approximative solution of multi-dimensional linear and nonlinear problems. It reduces the dimensionality of the problem and represents the solution in a more compact form.

The idea of the method consists in the representation of the solution in the form of a series with terms as the products of single-variable functions. The terms are

3) This section was written by S. A. Yaroshko and contains his results.

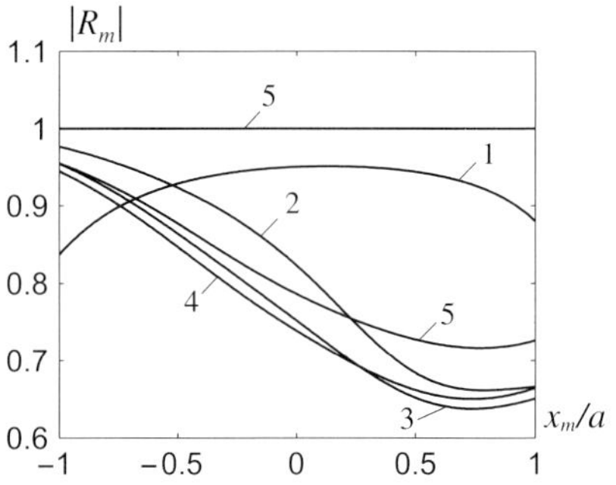

Figure 5.23 Amplitude of reflection coefficients; $U_0(x_0) \equiv const$, $F(\xi) = \cos(\pi\xi/2)$, $M = 6$, $c = 1.5$, $\beta = 0.2$, $\theta_0 = 0.4$.

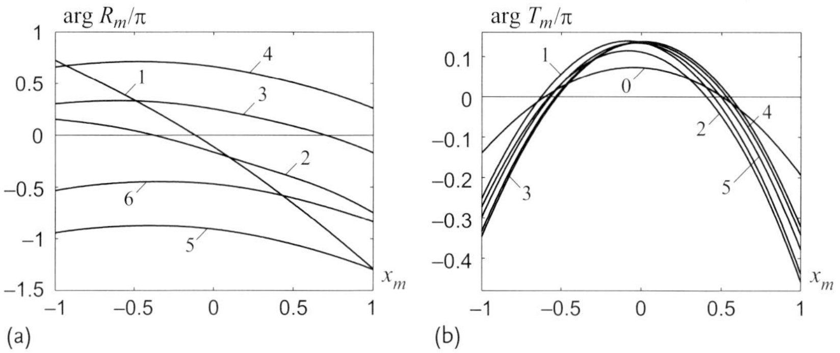

Figure 5.24 Phase of transmission and reflection coefficients; $U_0(x_0) \equiv const$, $F(\xi) = \cos(\pi\xi/2)$, $M = 6$, $c = 1.5$, $\beta = 0.2$, $\theta_0 = 0.4$.

calculated sequentially, each step determines the next term from a system of nonlinear one-dimensional equations. The system is obtained from the minimization of the appropriate functionals. The number of equations is equal to the problem dimensionality. In spite of the fact that the equations are nonlinear, the system allows solution by an iterative procedure, at each step of which a linear one-dimensional equation is solved.

The method was described in the application to the antenna array synthesis problems in [79] (see also [80]). For the linear integral equations, various versions of the method based on different variational formulations were discussed in detail in [81]. The recent results concerning its divergence are described in [82].

The method can be applied to the multi-dimensional integral and matrix equations and their variational analogs. Below, we apply it to the synthesis problem for the antenna array. For methodological reasons, we first consider the linear problem of synthesis by the complete (complex) desired radiation pattern. Next, the phase optimization problems for the two-dimensional antenna arrays will be considered.

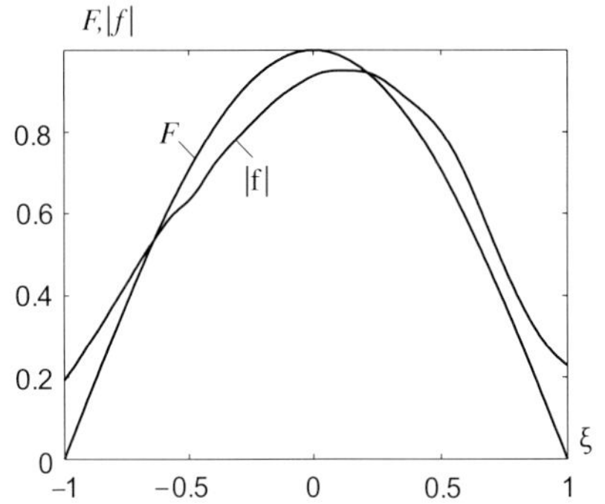

Figure 5.25 Amplitude pattern in beam transformer; $U_0(x_0) \equiv const$, $F(\xi) = \cos(\pi\xi/2)$, $M = 6$, $c = 1.5$, $\beta = 0.2$, $\theta_0 = 0.4$.

5.4.1
Description of the method for linear problems

Let us consider the aperture antenna and formulate the standard problem antenna synthesis by the desired complex radiation pattern. According to (2.28), the radiation pattern created by the field u on the aperture D is calculated as

$$f(\theta,\varphi) = Au \equiv k^2 \cos\vartheta \iint_D u(x,y) \exp(ik(x\xi + y\eta))\, dx dy \tag{5.208}$$

where

$$\xi = \sin\vartheta \cos\varphi\,, \quad \eta = \sin\vartheta \sin\varphi\,, \tag{5.209}$$

and θ, φ are the angular coordinates on the remote sphere. For simplicity, we take $D = [a, b] \times [c, d]$.

In the optimization problem similar to Problem **M** formulated in Section 2.2.2, the functional to be minimized can be written in the form

$$\sigma_t(u) = \iint_\Omega |Au - F|^2 p(\omega)\, d\omega + t \iint_D |u(x,y)|^2 dx dy\,, \tag{5.210}$$

where Ω is the angular domain in which the complex desired pattern F is given, $\omega = (\theta, \varphi)$ is the aggregate angular coordinate in Ω, $p(\omega)$ is a non-negative weight function which regulates the proximity of the created pattern to the desired one in certain directions, $d\omega = \sin\theta\, d\theta\, d\varphi$, $t > 0$ is the weight factor introduced for balancing between the degree of approximation of the pattern and minimization of the total radiated energy which is connected with minimization of the side lobes of the radiation pattern and the reactive power.

Owing to the convexity of the functional (5.210), it has a unique minimum. The function providing this minimum can be found in the form

$$u(x,y) = \sum_{k=1}^{\infty} v_k(x)\, w_k(y)\,. \tag{5.211}$$

The functions $v_k(x), w_k(y)$ are determined in turn, from the demand that the value of the functional

$$\sigma_{t,k} = \sigma_t(v_k w_k) = \iint_\Omega |A[v_k w_k] - F_k|^2 p(\omega)\,d\omega + t\iint_D \left|u^{(k-1)} + v_k w_k\right|^2 dx dy\,. \quad (5.212)$$

is minimal. Here

$$F_k = F - \sum_{i=1}^{k-1} A[v_i w_i] = F - Au^{(k-1)}\,, \quad (5.213)$$

$$u^{(k)}(x,y) = u^{(k-1)}(x,y) + v_k(x)\,w_k(y)\,, \quad k = 1,2,\ldots\,, \quad (5.214a)$$

$$u^{(0)}(x,y) \equiv 0\,. \quad (5.214b)$$

In order to obtain the Lagrange–Euler equation for the functional $\sigma_{t,k}(v_k w_k)$, we should write its partial variations $\delta_v \sigma_{t,k}$, $\delta_w \sigma_{t,k}$ with respect to the functions $v_k(x)$, $w_k(y)$. For partial variations, we understand the linear part of the increments of the functional, caused by perturbations of the function $v_k(x)$ or $w_k(y)$, respectively. Dropping terms of higher order, we obtain

$$\begin{aligned}\delta_v \sigma_{t,k} &= \sigma_{t,k}((v_k + \delta v_k)w_k) - \sigma_{t,k}(v_k w_k)\\ &= 2\,\mathrm{Re}\iint_D \left(A^*[A[v_k w_k] - F_k] + t\left(u^{(k-1)} + v_k w_k\right)\right)\\ &\quad \times \bar{w}_k(y)\cdot\overline{\delta v_k(x)}\,dx dy\\ &= 2\,\mathrm{Re}\int_a^b \left[\int_c^d \left(A^*[A[v_k w_k] - F_k] + t\left(u^{(k-1)} + v_k w_k\right)\right)\cdot\bar{w}_k(y)dy\right]\\ &\quad \times \overline{\delta v_k(x)}\,dx\,.\end{aligned} \quad (5.215a)$$

Analogously,

$$\begin{aligned}\delta_w \sigma_{t,k} &= \sigma_{t,k}(v_k(w_k + \delta w_k)) - \sigma_{t,k}(v_k w_k)\\ &= 2\,\mathrm{Re}\iint_D \left(A^*[A[v_k w_k] - F_k] + t\left(u^{(k-1)} + v_k w_k\right)\right)\\ &\quad \times \bar{v}_k(x)\cdot\overline{\delta w_k(y)}\,dx dy\\ &= 2\,\mathrm{Re}\int_c^d \left[\int_a^b \left(A^*[A[v_k w_k] - F_k] + t\left(u^{(k-1)} + v_k w_k\right)\right)\cdot\bar{v}_k(x)dx\right]\\ &\quad \times \overline{\delta w_k(y)}dy\,;\end{aligned} \quad (5.215b)$$

here A^* is the adjoint operator to A,

$$A^* g = \iint_\Omega g(\varphi,\theta)\exp(-ik(x\xi + y\eta))p(\omega)\,d\omega\,. \quad (5.216)$$

From the condition for $\delta_v \sigma_{t,k}$ and $\delta_w \sigma_{t,k}$ to be zero at arbitrary complex functions δv_k and δw_k, respectively, we get

$$\int_c^d (A^* A v_k w_k + t v_k w_k)\, w_k \mathrm{d}y = \int_c^d \left(A^* F_k - t u^{(k-1)}\right) w_k \mathrm{d}y\,, \qquad (5.217a)$$

$$\int_a^b (A^* A v_k w_k + t v_k w_k)\, v_k \mathrm{d}x = \int_a^b \left(A^* F_k - t u^{(k-1)}\right) v_k \mathrm{d}x\,. \qquad (5.217b)$$

These equalities are obtained from the necessary conditions for $\sigma_{t,k}$ to be minimal. They make up the system of Lagrange–Euler equations for the functional σ_k, being nonlinear one-dimensional integral equations with respect to the two complex functions $v_k(x)$ and $w_k(y)$ of one variable.

Owing to the special structure of (5.217), each equation is linear with respect to one of the unknown functions, $v_k(x)$ for (5.217a), and $w_k(y)$ for (5.217b). This fact allows us to apply a simple iterative procedure for their solution. At each step the following equations

$$\int_c^d \left(A^* A\left[v_k^{(p+1)} w_k^{(p)}\right] + t v_k^{(p+1)} w_k^{(p)}\right) \bar{w}_k^{(p)}(y)\mathrm{d}y = \int_c^d \left(A^* F_k - t u^{(k-1)}\right) \mathrm{d}y\,, \qquad (5.218a)$$

$$\int_a^b \left(A^* A\left[v_k^{(p+1)} w_k^{(p+1)}\right] + t v_k^{(p+1)} w_k^{(p+1)}\right) \bar{v}_k^{(p+1)}(x)\mathrm{d}x = \int_a^b \left(A^* F_k - t u^{(k-1)}\right) \mathrm{d}x \qquad (5.218b)$$

are solved as linear equations with respect to one of the functions: (5.218a) is solved with respect to $v_k^{(p+1)}$ at given $w_k^{(p)}$, after that (5.218b) is solved with respect to $w_k^{(p+1)}$ at given $v_k^{(p+1)}$. According to the derivation of (5.218a), it is the Lagrange–Euler equation for the functional $\sigma_{t,k}(v_k^{(p+1)} w_k^{(p)})$ considered as the functional of the function $v_k^{(p+1)}$ at given $w_k^{(p)}$. Its solution $v_k^{(p+1)}$ provides a minimum for this functional. Similarly, (5.218b) is the Lagrange–Euler equation for the functional $\sigma_{t,k}(v_k^{(p+1)} w_k^{(p+1)})$ considered as the functional of the function $w_k^{(p+1)}$ at the given $v_k^{(p+1)}$. Solving these equations by turns is equivalent to the minimization of the functional $\sigma_{t,k}(v_k, w_k)$ by the coordinate descent method. It decreases the value of the functional at each step, and the numeral sequence $\{\sigma_{t,k}(v_k{}^{(p)}, w_k{}^{(p)})\}$ converges with $p \to \infty$, as a monotonically decreasing sequence bounded from below.

5.4.2
Generalization for nonlinear problems. Synthesis of antenna array

The method described above can be extended to solve multi-dimensional nonlinear problems. In particular, it can be applied to three-dimensional phase optimization problems formulated for the power transmitting lines, aperture antennas and

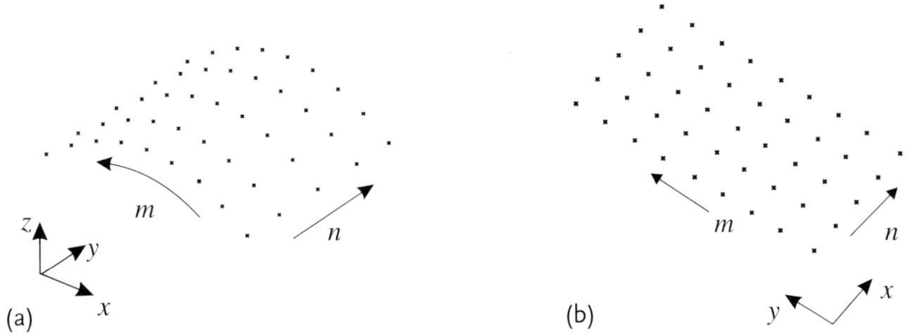

Figure 5.26 Geometry of the cylindrical (a) and plane (b) shaped arrays.

antenna arrays. Below, we illustrate it on the examples of the flat and conformal antenna arrays. We apply the method to a problem consisting in the minimization of functional (2.54) with σ_1 as the squared norm of the currents on the radiators.

In the general case the radiation pattern of the two-dimensional antenna array has the form

$$f(\omega) = Au \equiv \sum_{n=1}^{N} \sum_{m=1}^{M} u_{nm} f_{nm}(\omega) \exp(ik\vec{r}_{nm} \cdot \vec{\omega}) , \tag{5.219}$$

where u_{nm} is the current on the reflector with indices n, m; f_{nm} is the radiation pattern created by this radiator with the unit current on it in the passive presence of all other radiators, $\vec{r}_{nm}$ is the radius-vector of the point in space at which the radiator is located, $\vec{\omega}$ is the radius-vector of the point with the angular coordinates (θ, φ) on the unit sphere. In the particular cases considered here,

$$\vec{r}_{nm} \cdot \vec{\omega} = a \sin\theta \cos(\varphi - \varphi_n) + z_m \cos\theta \tag{5.220}$$

for the cylindrical array with the cylindrical coordinates φ_n, z_m of the radiators located on the surface of the cylinder of the radius a (Figure 5.26a), and

$$\vec{r}_{nm} \cdot \vec{\omega} = \sin\theta\, (x_n \cos\varphi + y_m \sin\varphi) \tag{5.221}$$

for the rectangular flat array with Cartesian coordinates x_n, y_m of the radiators located on the plane (x, y) (Figure 5.26b).

The functional to be minimized is

$$\sigma_t(u) = \iint_{\Omega} \left[|f(\omega)| - F(\omega)\right]^2 p(\omega)\mathrm{d}\omega + \frac{t}{NM} \sum_{n=1}^{N} \sum_{m=1}^{M} |u_{nm}|^2 , \tag{5.222}$$

where $u = \{u_{nm}\}$ is the matrix of currents on the radiators, $F(\omega) \geq 0$ is the desired amplitude pattern, Ω, $p(\omega)$, $\mathrm{d}\omega$, t have the same meaning as before.

According to the method of generalized separation of variables, the kth approximation to the matrix u has the form

$$u^{(k)} = \sum_{j=1}^{k} v_j^T w_j = u^{(k-1)} + v_k^T w_k , \quad k = 1, 2, \ldots , \tag{5.223}$$

where $u^{(0)} \equiv 0$; $v_j = (v_1^{(j)}, v_2^{(j)}, \dots, v_N^{(j)})$, $w_j = (w_1^{(j)}, w_2^{(j)}, \dots, w_M^{(j)})$ are the vector-rows, the superscript T denotes transposition of the vector. Expression (5.223) is the discrete analog of (5.214a).

The functional minimized at the kth step is

$$\sigma_{t,k}(v_k w_k) = \iint_\Omega \left(\left|A\left[u^{(k-1)} + v_k w_k\right]\right| - F\right)^2 p(\omega)\mathrm{d}\omega + \frac{t}{NM}\sum_{n=1}^{N}\sum_{m=1}^{M}\left|u_{nm}^{(k-1)} + v_n^{(k)} w_m^{(k)}\right|^2 . \tag{5.224}$$

Here the matrix $u^{(k-1)}$ is known, calculated in the previous steps of the method, the vectors v_k, w_k are to be found. We calculate the partial variations of (5.224) with respect to the vectors v_k, w_k. Taking into account that the first variation of the modulus of a complex function f is $\delta|f| = \mathrm{Re}(\delta f \exp(-i \arg f))$ (see 3.13), we obtain

$$\delta_v \sigma_{t,k} = 2\,\mathrm{Re}\left(A^* A u^{(k)} - A^*\left[F \exp\left(i \arg A u^{(k)}\right)\right] + \frac{t}{NM} u^{(k)}, \delta v_k \cdot w_k\right)_U , \tag{5.225a}$$

$$\delta_w \sigma_{t,k} = 2\,\mathrm{Re}\left(A^* A u^{(k)} - A^*\left[F \exp\left(i \arg A u^{(k)}\right)\right] + \frac{t}{NM} u^{(k)}, v_k \cdot \delta w_k\right)_U , \tag{5.225b}$$

where $(.,.)_U$ is the inner product of the $(N \times M)$ matrices, defined as

$$(u, v)_U = \sum_{n=1}^{N}\sum_{m=1}^{M}\sum_{n'=1}^{N}\sum_{m'=1}^{M} u_{nm}\bar{v}_{n'm'} . \tag{5.226}$$

The adjoint operator A^* is defined as $A^* g = \{C_{nm}[g]\}$, $n = 1, 2, \dots, N$, $m = 1, 2, \dots, M$, where

$$C_{nm}[g] = \iint_\Omega g(\omega)\bar{f}_{nm}(\omega)\exp\left(-ik\vec{r}_{nm}\cdot\vec{\omega}\right) p(\omega)\mathrm{d}\omega . \tag{5.227}$$

The operator A^*A has the form $\{A^*Au\}_{nm} = \sum_{n'=1}^{N}\sum_{m'=1}^{M} D_{nmn'm'} u_{n'm'}$, $n = 1, 2, \dots, N$, $m = 1, 2, \dots, M$, where

$$D_{nmn'm'} = \iint_\Omega f_{n'm'}(\omega)\bar{f}_{nm}(\omega)\exp\left(ik(\vec{r}_{n'm'} - \vec{r}_{nm})\cdot\vec{\omega}\right) p(\omega)\mathrm{d}\omega . \tag{5.228}$$

The Lagrange–Euler equation is obtained from the demand that the right-hand sides of (5.225a) and (5.225b) are zero for any complex vectors δv_k and δw_k, respectively. Transferring to the right-hand side those terms with the phase pattern $\exp(i \arg A u^{(k)})$ and with the known current distribution $u^{(k-1)}$, and interchanging the summation order in the rest of the terms, we obtain

$$\sum_{n'=1}^{N} v_{n'}^{(k)} \sum_{m=1}^{M} \bar{w}_m^{(k)} \sum_{m'=1}^{M} w_{m'}^{(k)} D_{nmn'm'} + \frac{t}{NM} v_n^{(k)} \sum_{m=1}^{M} \bar{w}_m^{(k)} w_m^{(k)}$$
$$= \sum_{m=1}^{M} \bar{w}_m^{(k)} \left(C_{nm} \left[F \exp\left[i \arg f^{(k)} \right] - f^{(k-1)} \right] - \frac{t}{NM} u_{nm}^{(k-1)} \right),$$
$$n = 1, 2, \ldots, N, \quad (5.229a)$$

$$\sum_{m'=1}^{M} w_{m'}^{(k)} \sum_{n=1}^{N} \bar{v}_n^{(k)} \sum_{n'=1}^{N} v_{n'}^{(k)} D_{nmn'm'} + \frac{t}{NM} w_m^{(k)} \sum_{n=1}^{N} \bar{v}_n^{(k)} v_n^{(k)}$$
$$= \sum_{n=1}^{N} \bar{v}_n^{(k)} \left(C_{nm} \left[F \exp\left[i \arg f^{(k)} \right] - f^{(k-1)} \right] - \frac{t}{NM} u_{nm}^{(k-1)} \right),$$
$$m = 1, 2, \ldots, M. \quad (5.229b)$$

As before, it is denoted $Au^{(k)} = f^k$.

Similar to (5.217), the equation system (5.229) is nonlinear. However, it is more complicated, because it has a nonlinearity in the phase pattern on the right-hand side. This nonlinearity is caused by the nonlinearity of the original problem of the antenna synthesis by the amplitude radiation pattern. Transferring the terms with the phase pattern to the right-hand side, we keep the main property of the left-hand side: it is linear with respect to v_k and w_k for subsystems (5.229a) and (5.229b), respectively. As before, this property permits us to construct the iterative procedure, which solves linear problems with respect to the mentioned vectors at each step. In these problems the phase pattern on the right-hand side is taken from the previous iteration. It can be shown that such a procedure decreases the value of the functional $\sigma_t(u)$ at each step, and hence converges.

Note that system (5.229) has a collateral 'trivial' solution $v_k = w_k \equiv 0$, which obviously does not provide the minimum for (5.224). In order that the iterative procedure does not lead to this solution, it is convenient to take the initial approximation in such a way that the functional $\sigma_{t,k}$ on it is smaller than on the trivial solution. To this end, it is sufficient to write the initial approximation at each step in the form $C_k v_k^{(0)T} w_k^{(0)}$ with a constant C_k, and determine this constant in such a way that the value of $\sigma_{t,k}$ considered as a function of $\sigma_{t,k}(C_k)$, is smaller than $\sigma_{t,k}(0)$. It can be easily checked that this inequality is obtained at

$$C_k = \left\{ \iint_\Omega \left[F(\varphi, \theta) \exp\left(i \arg f^{(k-1)} \right) - f^{(k-1)}(\varphi, \theta) \right] \right.$$
$$\left. \times \overline{A\left[v_k^{(0)T} w_k^{(0)} \right]} p(\omega) \mathrm{d}\omega - \frac{t}{NM} \sum_{n=1}^{N} \sum_{m=1}^{M} u_{nm}^{(k-1)} \overline{v_n^{(k,0)} w_m^{(k,0)}} \right\}$$
$$\times \left\{ \iint_\Omega \left| A\left[v_k^{(0)T} w_k^{(0)} \right] \right|^2 p(\omega) \mathrm{d}\omega + \frac{t}{NM} \sum_{n=1}^{N} \sum_{m=1}^{M} \left| v_n^{(k,0)} w_m^{(k,0)} \right|^2 \right\}^{-1}.$$
$$(5.230)$$

Such an optimization is not needed at $k = 1$, when this constant factor is included in the vector which is first determined in the algorithm.

5.4.3 Numerical results

The method was applied to the two problems with the geometry of the arrays as shown in Figure 5.26. In the first case, the array of 9×9 elements is considered, which occupies a part of the circular cylinder surface with the radius $ka = 15$. The coordinates of the radiators are $\varphi_n = (n-5)\pi/(ka)$, $kz_m = (m-5)\pi$. The radiation pattern of the mnth radiator is

$$f_{nm}(\varphi,\theta) = \begin{cases} \sin\theta\cos^2(\varphi-\varphi_n)\,, & |\varphi-\varphi_n| \le \pi/2\,; \\ 0\,, & |\varphi-\varphi_n| > \pi/2\,. \end{cases} \tag{5.231}$$

The desired amplitude pattern $F(\varphi,\theta) = \sum_{j=1}^{4} f_{nm}^{(j)}(\varphi,\theta)$ is shown in Figure 5.27 in the generalized angular coordinates $\xi = \cos\theta$, $\eta = \sin\varphi\sin\theta$, $(\theta,\varphi) \in \Omega$, $\Omega = [-\pi/2; \pi/2] \times [0; \pi]$. It is a sum of four lobes of the form

$$f_{nm}^{(j)}(\varphi,\theta) = \begin{cases} \cos^2\left(4g^{(j)}(\varphi,\theta)\right), & g^{(j)}(\varphi,\theta) \le \pi/8\,; \\ 0\,, & g^{(j)}(\varphi,\theta) > \pi/8\,, \end{cases} \tag{5.232}$$

where $g^{(j)}(\varphi,\theta) = \arccos(\cos(\varphi-\varphi_j)\sin(\theta-\theta_j))$, $\varphi_1 = \varphi_3 = 0$; $\varphi_2 = -\varphi_4 = \pi/(4\sqrt{2})$; $\theta_1 = -\theta_3 = \pi/(4\sqrt{2})$; $\theta_2 = \theta_4 = 0$. The squared norm of this pattern is $\|F\|^2 = 0.30773$. The weight factors are $p_{nm}(\varphi,\theta) \equiv 1$, $t = 0$.

The five steps of the method were made. The amplitude pattern obtained is shown in Figure 5.28. The step of isolines of the function $|f(\xi,\eta|$ in Figure 5.28b

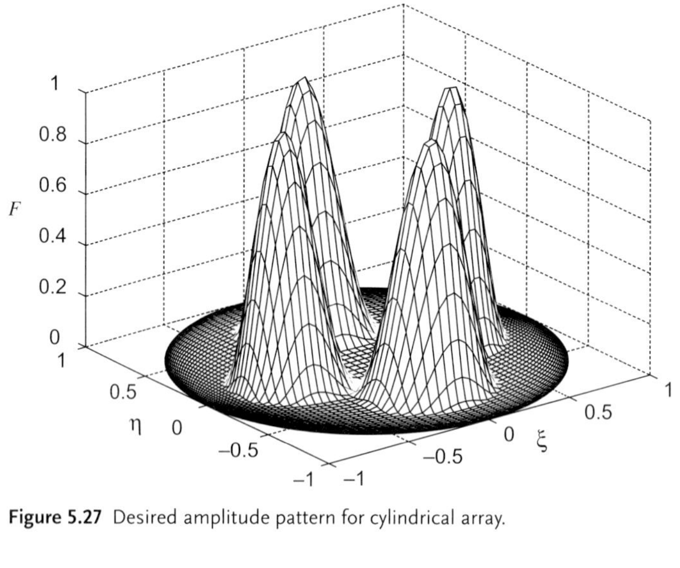

Figure 5.27 Desired amplitude pattern for cylindrical array.

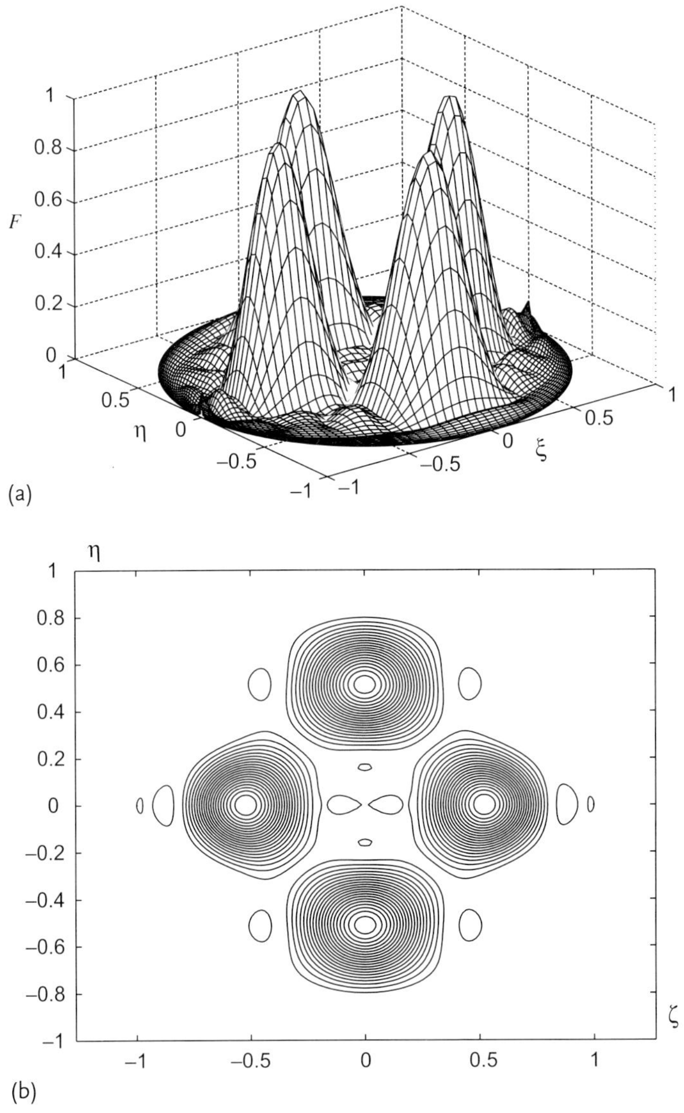

Figure 5.28 Amplitude pattern obtained for cylindrical array.

equals 0.05. The sequence of values of the functional σ obtained at different steps of the method is $\{\sigma_p\} = \{0.13717, 0.00809, 0.00565, 0.00529, 0.00525\}$. It can be seen that the four terms are sufficient to take in the sum (5.223) to obtain a good result. The approximate amplitude patterns obtained at different steps of the method are shown in Figure 5.29 in the form of perpendicular cross-sections of the angular domain $\eta = 0$ (a) and $\xi = 0$ (b).

The second problem consists in obtaining a so-called *contoured radiation pattern* which is constant inside a complicated angular domain $T \subset \Omega$, $\Omega = [0; \pi/2] \times$

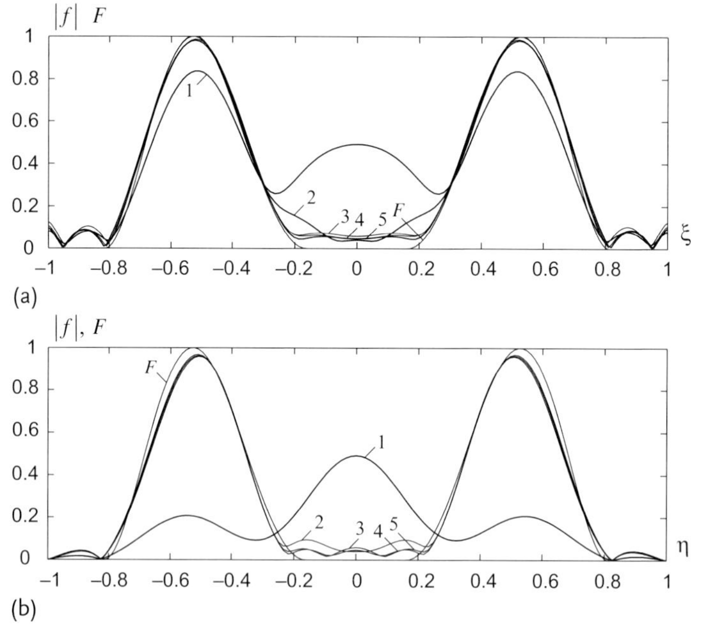

Figure 5.29 Cross-sections of pattern obtained for cylindrical array.

$[0; 2\pi]$, and which decreases quickly outside it. Such a problem can arise when uniformly illuminating certain areas of the Earth (only) from the satellite for the purposes of communication.

The rectangular antenna array of 17×17 elements is used. Their coordinates are $kx_n = (n-9)\pi$, $ky_m = (m-9)\pi$. The radiation patterns of all radiators are uniform, $f_{nm} \equiv 1$.

The generalized angular coordinates ξ and η are given in (5.209). The contour imitates the borders of a country with the angular sizes about $3\pi/5$ in the direction $\eta = 0$ and $4\pi/9$ in the direction $\xi = 0$. The desired amplitude pattern $F(\xi, \eta) \equiv 1$ for $(\xi, \eta) \in T$, and decreases from one to zero on the layer $h_\omega = \pi/16$ in all directions away from its contour; $\|F\|^2 = 1.60716$. Its contours with the step 0.1 are shown in Figure 5.30a.

The twelve steps of the method were made. The sequence of the obtained values of σ is $\{\sigma_p\} = \{0.28611, 0.12505, 0.07511, 0.05485, 0.04045, 0.03372, 0.02994, 0.02706, 0.02511, 0.02330, 0.02226\}$. The uniformly distanced isolines of the obtained radiation pattern $|f(\varphi, \psi)|$ are shown in Figure 5.30b with the step 0.1. The contour of T is shown by the bold line. The angular cross-sections at $\eta = 0$ and $\xi = 0$ of the approximate amplitude patterns $|f^{(p)}|$, $p = 3, 6, 8$, obtained at different steps, are shown in Figure 5.31a,b, respectively.

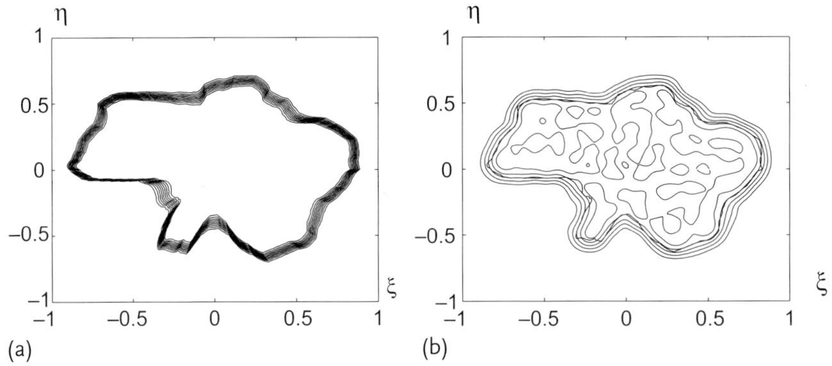

Figure 5.30 Contours of desired (a) and obtained (b) amplitude patterns for rectangular array.

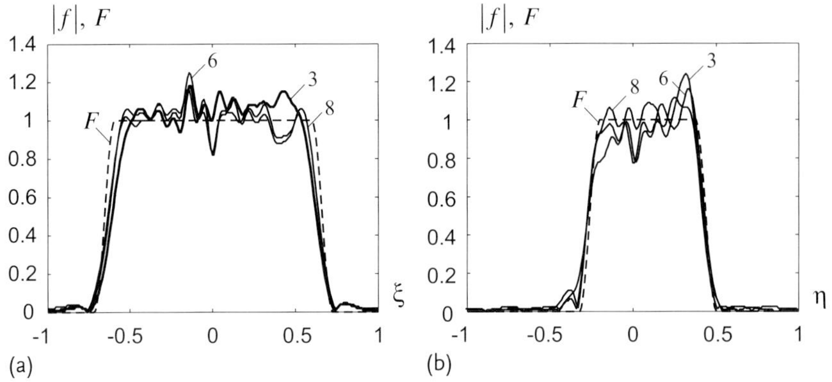

Figure 5.31 Cross-sections of pattern obtained for rectangular array.

5.5
Homogeneous problems

In this section phase optimization problems are considered, in which no desired functions are given, but specific properties (as a rule, undesirable losses) of the eigenwaves or eigenoscillations in the waveguides or resonators are optimized. Such problems were formulated in Section 2.3 and partially described in Section 3.3.

5.5.1
Minimization of losses in waveguide walls

5.5.1.1 Problem formulation

Consider Problem **X** from Section 2.3.2 in the particular case of the circular waveguide with perfectly conducting walls of radius a (this problem was first considered in [83]). In this problem the lenses are periodically located inside the waveguide ($d_n = L$ in Figure 5.32) to create a 'synthesized' eigenwave with minimal current

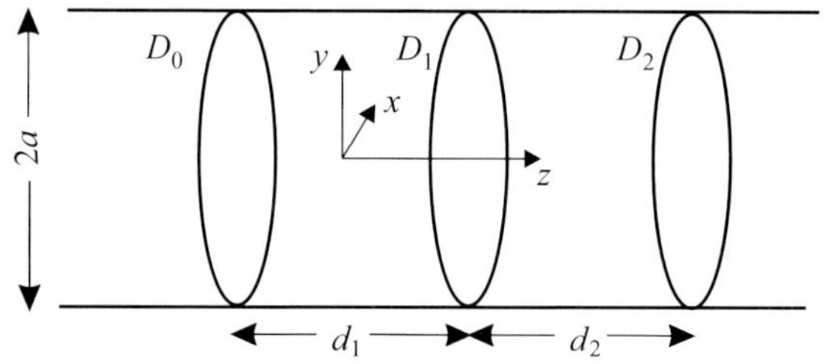

Figure 5.32 Lens line in the waveguide.

in the walls. The fields $\vec{E}$ and $\vec{H}$ of the eigenwave depend on the shape of the lenses, which are described by the function $\psi(r,\varphi)$. It is the optimization function in the problem. In our case, functional (2.110), minimized in the problem, has the form

$$j(\psi) = \frac{\frac{a}{L}\int_{-L/2}^{L/2}\int_0^{2\pi}|H_t(a,\varphi,z)|^2 d\varphi dz}{\int_0^{2\pi}\int_0^a \operatorname{Re}\left[\vec{E}(r,\varphi,0)\times\overline{\vec{H}(r,\varphi,0)}\right]_z r dr d\varphi}, \tag{5.233}$$

where $H_t(a,\varphi,z)$ is the projection of the vector $\vec{H}$ onto the tangent plane to the cylinder wall, $\vec{E}(r,\varphi,0)$, $\vec{H}(r,\varphi,0)$ are the electric and magnetic fields at $z = 0$ (in the middle between the lenses.) After applying the Lagrange theorem to (5.233), the problem is reduced to minimization of the functional

$$\Lambda(\psi) = \frac{a}{L}\int_{-L/2}^{L/2}\int_0^{2\pi}|H_t(a,\varphi,z)|^2 d\varphi dz - j\int_0^{2\pi}\int_0^a \operatorname{Re}\left[\vec{E}(r,\varphi,0)\times\overline{\vec{H}(r,\varphi,0)}\right]_z r dr d\varphi\,, \tag{5.234}$$

where j is an indeterminate factor; its value in the stationary (minimal) point is equal to the minimum of $j(\psi)$.

5.5.1.2 Description of the algorithm

In order to determine the function $\psi(r,\varphi)$ minimizing the functional $j(\psi)$, we should first minimize it with respect to the fields $\vec{E}$ and $\vec{H}$ satisfying the Maxwell equations and the boundary conditions in the waveguide segment $-L/2 < z < L/2$ with the periodicity conditions (2.109).

After the fields are found, the function $\psi(r,\varphi)$ is calculated from (2.108), which together with (2.109) gives

$$\psi(\vec{R}) = \arg\left(\{\vec{E},\vec{H}\}\Big|_{z=-L/2}\right) - \arg\left(\{\vec{E}^-,\vec{H}^-\}\Big|_{z=L/2}\right)\ ; \tag{5.235}$$

the nonessential constant phase summand hL is omitted here. In the general case, different components of the fields can have different phase distributions over the cross-section, and $\psi(\vec{R})$ defined in such a way is not the scalar function.

The fields $\vec{E}$ and $\vec{H}$ can be expressed as a superposition of the fields of traveling modes of the same type as the synthesized wave which we want to construct. For instance, if the desired wave belongs to the TE type, that is, its component E_z equals zero, then the base functions for $\vec{E}$ and $\vec{H}$ are the fields of the magnetic modes having the same property. The dependencies of their nonzero components on the transversal coordinates can be expressed by the magnetic potentials $\Psi_n(r,\varphi)$ which satisfy the two-dimensional Helmholtz equation

$$\Delta \Psi_n + \left(k^2 - h_n^2\right) \Psi_n = 0 \tag{5.236}$$

and the boundary condition

$$\left.\frac{\partial \Psi_n}{\partial r}\right|_{r=a} = 0\,. \tag{5.237}$$

As for any mode, the z-dependence of the components is described by the factor $\exp(-i h_n z)$, where h_n is the propagation constant of the nth mode. For traveling modes, $h_n^2 > 0$.

We confine ourselves to the symmetrical modes H_{0n} of this type ($\partial \Psi_n / \partial \varphi \equiv 0$). In this case, the only functions solving the boundary value problem (5.236), (5.237) are

$$\Psi_n(r) = J_0\left(\frac{\mu_n}{a} r\right)\,, \tag{5.238}$$

$n = 1, 2, \ldots$, where μ_n are the roots of the equation $J_1(\mu_n) = 0$. These roots are connected with h_n as

$$h_n^2 = k^2 - \frac{\mu_n^2}{a^2}\,. \tag{5.239}$$

The nonzero components of the considered modes are calculated by Ψ_n as

$$H_{n,z} = (k^2 - h_n^2)\Psi_n(r)\exp(-i h_n z)\,, \tag{5.240a}$$

$$E_{n,\varphi} = i k \Psi_n'(r)\exp(-i h_n z)\,, \tag{5.240b}$$

$$H_{n,r} = -i h_n \Psi_n'(r)\exp(-i h_n z)\,, \tag{5.240c}$$

where $\Psi_n'(r)$ denotes the usual derivative with respect to r.

Express the fields $\vec{E}$, $\vec{H}$ in the form

$$\begin{Bmatrix} \vec{E} \\ \vec{H} \end{Bmatrix} = \sum_{n=1}^{N} \frac{c_n}{\mu_n J_0(\mu_n)} \begin{Bmatrix} \vec{E}_n \\ \vec{H}_n \end{Bmatrix}\,, \tag{5.241}$$

where N is the number of traveling modes for which $\mu_n < ka$. The components of the fields $\vec{E}_n$, $\vec{H}_n$ are expressed by the functions Ψ_n according to (5.240). If there are no losses in the walls, then, due to symmetry, the fields $\vec{E}$, $\vec{H}$ are in-phase at $z = 0$ (in the middle between the lenses). Hence, the coefficients c_n are real. The factors $[\mu_n J_0(\mu_n)]^{-1}$ are introduced into the coefficients in order to simplify the elements of the matrix in the equation system for c_n.

According to (5.241), (5.240) and (5.238), the expressions participating in the integrands of (5.234) are

$$H_t(a,\varphi,z) \equiv H_z(a,z) = \frac{1}{a^2}\sum_{n=1}^{N} c_n\mu_n \exp(-ih_n z)\,, \tag{5.242a}$$

$$\begin{aligned}\left[\vec{E}(r,\varphi,0)\times\overline{\vec{H}}(r,\varphi,0)\right]_z &\equiv -E_\varphi(r,\varphi,0)\bar{H}_r(r,\varphi,0)\\ &= \frac{k}{a^2}\sum_{n=1}^{N}\frac{c_n}{J_0(\mu_n)}J_1\left(\frac{\mu_n}{a}r\right)\sum_{m=1}^{N}\frac{\bar{c}_m h_m}{J_0(\mu_m)}J_1\left(\frac{\mu_m}{a}r\right).\end{aligned} \tag{5.242b}$$

Substituting (5.242) into (5.234) and calculating the integrals yields

$$\Lambda(\psi) = \frac{2\pi}{a^4}\frac{2a}{L}\sum_{n=1}^{N} c_n\mu_n\sum_{m=1}^{N} c_m\mu_m\frac{\sin((h_m-h_n)L/2)}{h_m-h_n} - 2\pi j\frac{k}{2}\sum_{n=1}^{N}|c|_n^2\,h_n\,; \tag{5.243}$$

here the orthogonality of the functions $\Psi_n(r)$

$$\int_0^a J_0\left(\frac{\mu_n}{a}r\right)J_0\left(\frac{\mu_m}{a}r\right)r\mathrm{d}r = \delta_{nm}\frac{a^2}{2}J_0^2(\mu_n) \tag{5.244}$$

(see [65]) is used. Differentiating with respect to c_n, we obtain the algebraic eigenvalue problem

$$AC = jBC \tag{5.245}$$

where $C = \{c_n\}^T$ is the transposed vector of the unknown coefficients, and the elements of the matrices $A = \{a_{nm})$ and $B = \{b_{nm}\}$ are

$$a_{nm} = \frac{2}{aL}\mu_n\mu_m\frac{\sin((h_m-h_n)L/2)}{h_m-h_n}\,, \tag{5.246}$$

$$b_{nm} = \delta_{nm}\frac{ka^2h_n}{2}\,, \tag{5.247}$$

j is the spectral parameter. All eigenvalues j_n of the problem are positive. They are the values of the average currents (5.233) of the synthesized modes in the wave-lens guide. The minimum of j_n is the required minimum of functional (5.233).

5.5.1.3 Numerical results

The numerical results concerning waveguides with $ka = 10.5$ and $ka = 28,56$ are presented below. In these waveguides three and eight modes of the H_{0n} type can propagate, respectively. The distance between the lenses is normed to $L_0 = ka^2/(2\pi)$; it is a characteristic distance between the lenses in the open quasi-optical line. The minimum value of j is normed to $j_0 = k^2\mu_1^2/(2h_1^2)$; it is the average

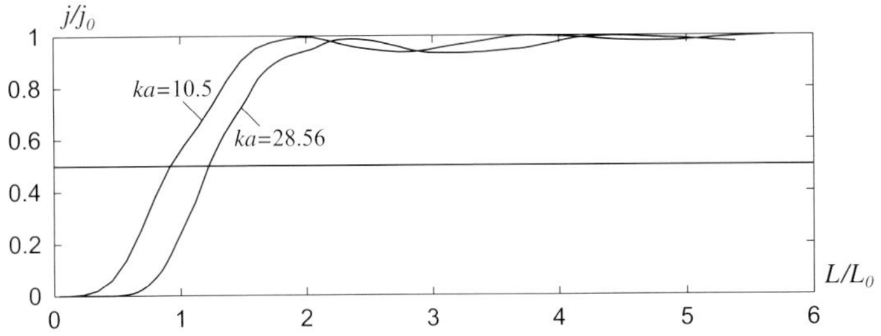

Figure 5.33 Minimized current in synthesized waveguide.

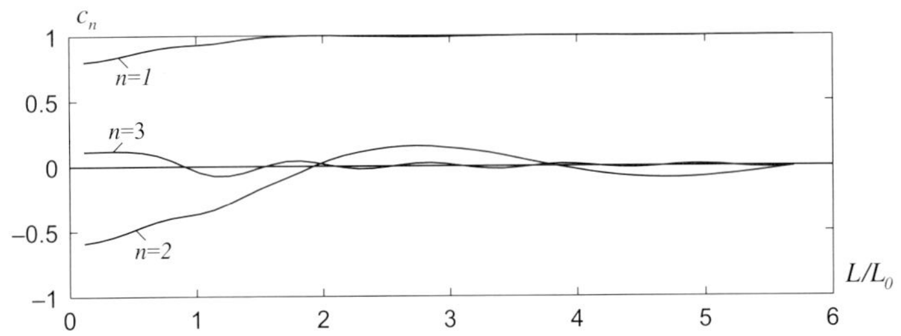

Figure 5.34 Amlitudes of traveling modes in synthesized eigenwave, $ka = 10.50$.

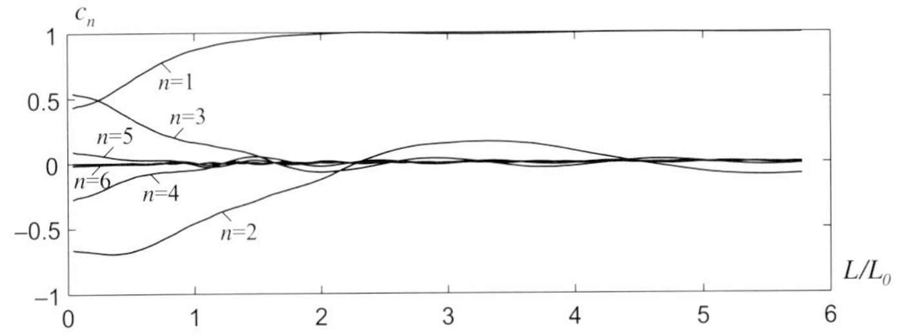

Figure 5.35 Amlitudes of traveling modes in synthesized eigenwave, $ka = 28.56$.

squared current of the H_{01} mode of the unit amplitude $c_1 = 1$. This value is shown in Figure 5.33. In both cases, it monotonically grows at small L, and then asymptotically tends to unity. It can be seen that the system considered can be effective for symmetrical magnetic modes, only for relatively small L. The limiting distance between the lenses, at which the squared current can be reduced by half, equals about $0.9L_0$ and $1.3L_0$ for $ka = 10.5$ and $ka = 28.56$, respectively.

The amplitudes of the traveling H_{0n} modes in the synthesized eigenwave are presented in Figures 5.34 and 5.35. Except for the first mode, amplitudes oscillate with kL varying. This fact causes oscillations of the current in Figure 5.33. In fact, only some lower modes participate in construction of the synthesized wave. As was expected, the lenses providing the synthesized wave are anisotropic; their phase corrections are different for different components of the fields. This can be

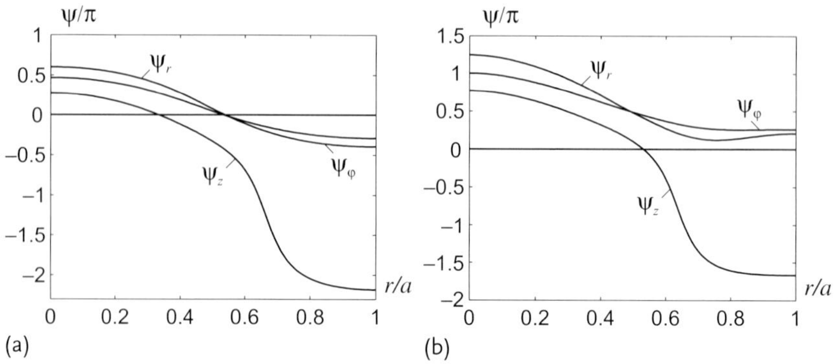

Figure 5.36 Shape of lenses in the waveguide; $ka = 10.50$ (a), $ka = 28.56$ (b); $L/L_0 = 0.86$.

seen from Figure 5.36a,b, in which the functions ψ_r, ψ_φ, ψ_z are presented for the two considered cases, respectively. These functions are calculated by (5.235) for the components H_r, E_φ, H_z at $L/L_0 = 0.86$. The phase corrections of the transversal components H_r and E_φ are close, whereas the component H_z needs to be corrected in a more complicated way.

5.5.2 Open resonators with shifted mirrors

In this subsection we consider the optimization problem of the mirror shape in an open resonator with shifted mirrors (Problem **Z**). This problem is one of maximization of the spectral radius of compact operators, formulated in Section 3.3.2, and the results obtained there will be used below.

5.5.2.1 Problem formulation

In the two-dimensional case considered here, the linear homogeneous integral equation related to the problem has the form (see 3.292)

$$\lambda w(y) = \int_{-a}^{a} K(x, y; \delta) \exp\left(i[\varphi(x) + \varphi(y)]\right) w(x)\, dx\,, \tag{5.248}$$

where

$$K(x, y; \delta) = \frac{1}{\sqrt{2\pi}} \exp\left(\frac{ik}{2a}(x + y + \delta)^2\right), \tag{5.249}$$

$2a$ is the width of the mirrors, d is the distance between them, δ is the mutual shift of the resonators; $\varphi(x)$ is the phase correction provided by the mirrors. In order to keep the symmetry of the equation, the specific local coordinates x and y are introduced on the mirrors (see Figure 5.37). Note that their origins are shifted with respect to each other in the vertical direction by a distance δ.

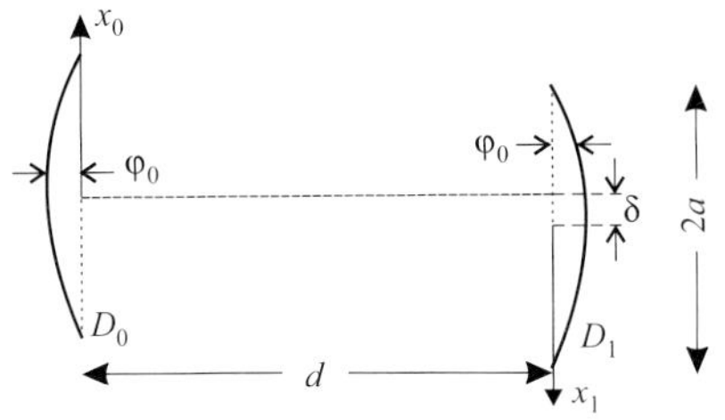

Figure 5.37 Resonator with shifted mirrors.

Since the direction of the shift is assumed to be unknown, the phase correction $\varphi(x)$ should be an even function,

$$\varphi(x) = \varphi(-x)\,. \tag{5.250}$$

The problem is to find $\varphi(x)$ providing the maximum modulus $|\lambda_1|$ of the first eigenvalue of (5.248). For simplicity, below we drop the index 1 in λ.

After denoting

$$u(x) = w(x)\exp\left(-i\varphi(x)\right)\,, \tag{5.251a}$$

$$\psi(x) = 2\varphi(x)\,, \tag{5.251b}$$

this equation obtains the form (3.282)

$$\lambda u(y) = \int_{-a}^{a} K(x, y; \delta)\, u(x) \exp\left(i\psi(x)\right) \mathrm{d}x\,. \tag{5.252}$$

The adjoint equation is

$$\bar{\lambda} v(x) = \exp\left(-i\psi(x)\right) \int_{-a}^{a} \bar{K}(y, x; \delta)\, v(y)\, \mathrm{d}y\,. \tag{5.253}$$

It can be seen, that the solutions to these equations are connected as

$$v(x) = \bar{u}(x)\exp(-i\psi(x))\,. \tag{5.254}$$

Now we can use expression (3.289) for the first variation of the squared modulus of λ, which in our case is

$$\int_{-a}^{a} \delta\psi(x)\, \mathrm{Im}\, \frac{\bar{u}(x)v(x)}{(u, v)} \mathrm{d}x = 0\,; \tag{5.255}$$

the constant factor $|\lambda|^2$ is omitted here. Due to the evenness of $\varphi(x)$, the function $\delta\psi$ in (5.255) is an arbitrary even function,

$$\delta\psi(-x) = \delta\psi(-x)\,, \tag{5.256}$$

and (3.289) is rewritten as

$$\int_0^a \delta\psi(x)\,\mathrm{Im}\,\frac{\bar{u}(x)v(x)+\bar{u}(-x)v(-x)}{(u,v)}\mathrm{d}x\,. \tag{5.257}$$

Equating this expression to zero for arbitrary $\delta\psi(x)$ with $x \in [0, a]$ leads to

$$\mathrm{Im}\,\frac{\bar{u}(x)v(x)+\bar{u}(-x)v(-x)}{(u,v)} = 0\,, \tag{5.258}$$

or, after using (5.254), to

$$\mathrm{Im}\,\frac{\left[\bar{u}^2(x)+\bar{u}^2(-x)\right]\exp(-i\psi(x))}{(u,\bar{u}(x)\exp(-i\psi(x)))} = 0\,. \tag{5.259}$$

This equality holds if

$$\psi(x) = -\arg\left[u^2(x)+u^2(-x)\right] - t\,, \tag{5.260}$$

where

$$t = (u,\bar{u}(x)\exp(-i\psi(x))) + \begin{Bmatrix} 0 \\ \pi \end{Bmatrix} = const\,. \tag{5.261}$$

Substituting (5.260) into (5.252) gives

$$\hat{\lambda}u(y) = \int_{-a}^{a} K(x,y;\delta)u(x)\exp\left\{-i\arg\left[u^2(x)+u^2(-x)\right]\right\}\mathrm{d}x \tag{5.262}$$

with $\hat{\lambda} = \lambda\exp(it)$. This equation represents a nonlinear eigenvalue problem. The problem consists in determining the maximum modulus eigenvalue λ and the corresponding eigenfunction $u(x)$. After $u(x)$ has been found, the required phase correction $\varphi(x)$ is calculated by relations (5.262) and (5.251b).

5.5.2.2 Description of the algorithm. Numerical results

The simplest way to find the maximum modulus eigenvalue of the nonlinear integral equation (5.262) and the corresponding eigenfunction is the iteration method generalizing the power method for similar linear problems (see 3.267):

$$\hat{u}^{(p+1)}(y) = \int_{-a}^{a} K(x,y;\delta)u^{(p)}(x)\exp(i\psi^{(p)}(x))\mathrm{d}x\,, \tag{5.263a}$$

$$\left|\lambda^{(p+1)}\right| = \left\|\hat{u}^{(p+1)}\right\|\,,\quad u^{(p+1)} = \hat{u}^{(p+1)}/\left|\lambda^{(p+1)}\right|\,, \tag{5.263b}$$

$$\psi^{(p+1)}(x) = -\arg\left\{\left[u^{(p+1)}(x)\right]^2 + \left[u^{(p+1)}(-x)\right]^2\right\}\,; \tag{5.263c}$$

$\|u^{(0)}\| = 1$ is assumed. The convergence ranges of the method have not yet been investigated. Similar to the linear case, the convergence essentially slows down as c grows.

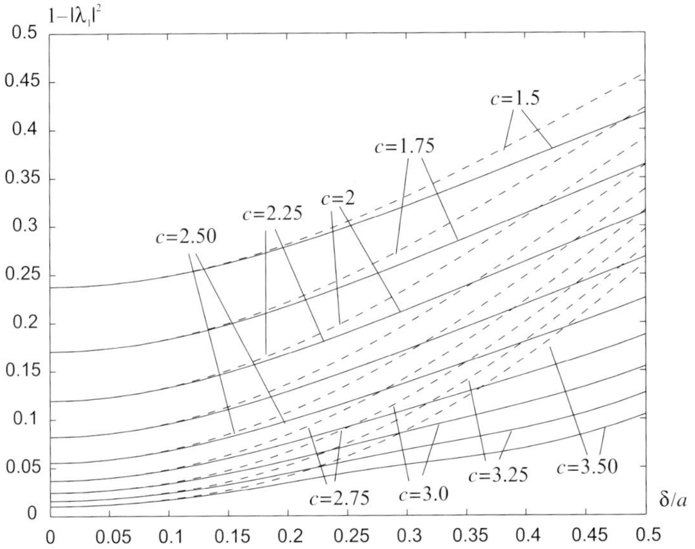

Figure 5.38 Optimized losses of the main eigenoscillation in the resonator with shifted mirrors.

The numerical results for the resonators with different values of the geometrical parameter c and shift d are presented in Figures 5.38–5.42. The losses of the main eigenoscillation in the resonator versus d for different c are shown in Figure 5.38. For comparison, the losses in the confocal resonators $\varphi(x) = \varphi_c(x) = -c/2 \cdot (x/a)^2$, $c = ka^2/d$ of the same geometry are shown by the dashed lines. It can be seen

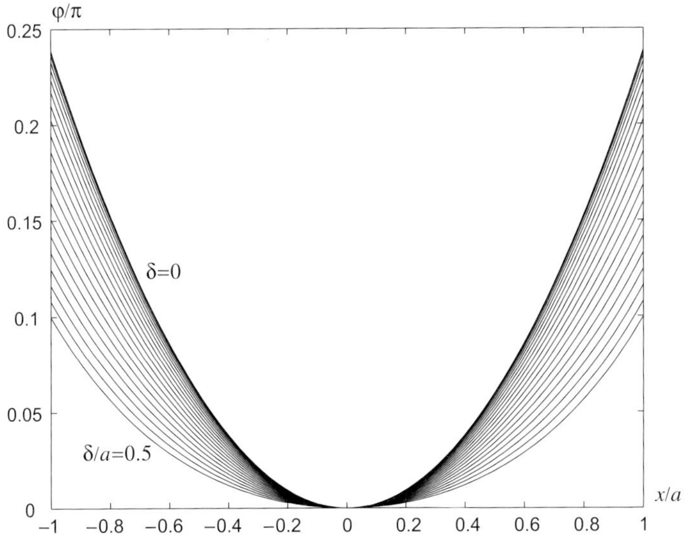

Figure 5.39 Optimal phase corrections in the resonator with shifted mirrors; $c = 1.50$.

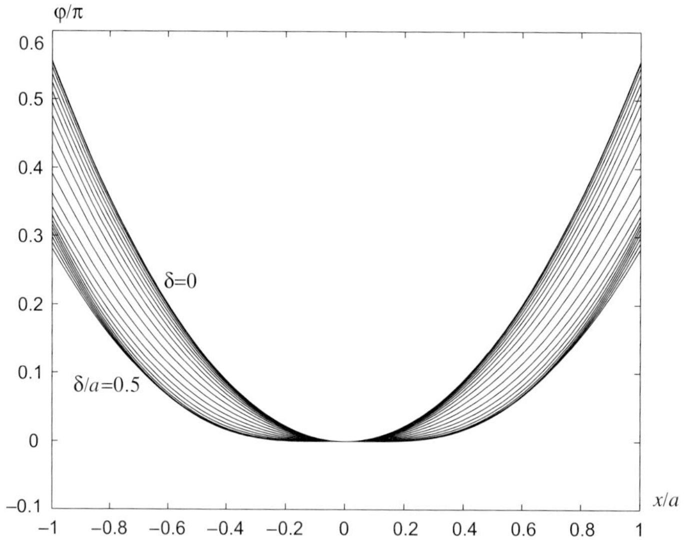

Figure 5.40 Optimal phase corrections in the resonator with shifted mirrors; $c = 3.50$.

that, for small shifts, the losses in the optimized resonators do not differ a great deal from those in the confocal resonators with the same shift. Efficiency of the optimization becomes noticeable for shifts larger than $0.1a$. Its relative influence increases with increasing c when the losses themselves decrease.

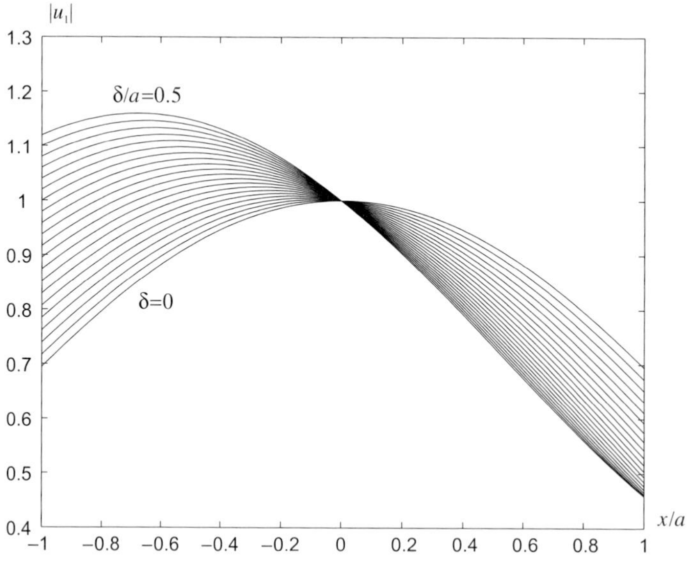

Figure 5.41 Field distribution of the main eigenoscillation in the resonator with optimal shifted mirrors; $c = 1.5$.

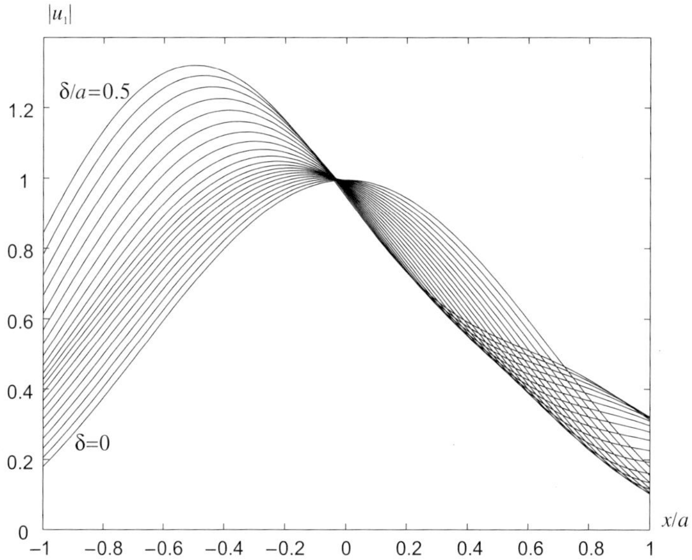

Figure 5.42 Field distribution of the main eigenoscillation in the resonator with optimal shifted mirrors; $c = 3.5$.

The optimal phase correction of the mirrors is shown in Figures 5.39 and 5.40 for the values $c = 1.50$ and $c = 3.50$, respectively. The values of the shift are varied from $d = 0$ to $d = a/2$ with the step $0.025a$; only the curves corresponding to the limiting values of d are marked. As was expected, the optimal phase correction weakens if the value of the shift grows. For small c it remains almost quadratic, $\varphi(x) \approx \nu\varphi_c(x)$, with ν decreasing, starting at $\nu = 1$ when d increases (Figure 5.39). For larger c it ceases to be quadratic, $\varphi(x)$ becomes more slanting in its central part (Figure 5.40). This part of the mirrors becomes nonfocussing and even somewhat scattering.

The field distributions $u_1(x)$ of the main eigenoscillation in the resonators with optimized mirrors are shown in Figures 5.41 and 5.42. The parameters correspond to those of Figures 5.39 and 5.40. For best visualization, the functions are normed by the condition $u_1(0) = 1$. In all cases the maximum of the field is shifted from the mirror centers in a direction opposite to the direction of the shift. This effect can easily be explained in terms of geometrical optics. For larger c and d the shape of the distribution becomes more complicated and a weak oscillation appears on it.

6
Nonstandard Inverse Problems of Diffraction Theory

6.1
Introduction. Outline of the chapter

6.1.1
Object of investigation

In this chapter those problems are considered in which the required values are the parameters of the body on which the field is diffracted. The incident field (or the currents creating it) is given and the field that arises as a result of diffraction, is given also.

The three following objects participate in the diffraction process: the incident field, the body on which it diffracts, and the desired field to be obtained. In direct problems, the incident field and the body are given, and the diffraction field is to be found. With standard inverse problems we understand the problems in which the body and the desired field are given, whereas the incident field or the currents creating it are to be found. The *nonstandard inverse problems* are the problems in which the parameters of the body are unknown; they must be found from the requirement that the diffraction on the body transforms the incident field into the desired one.

Problems of this type arise in different technical applications. They are the basis of the synthesis and design of the most passive radio engineering elements. These elements are constructed in such a way that the field diffracted on them has a given structure. The diffraction should provide the desired transformation of the field in the waveguide transformers of mode type, in the transformers of polarization, etc. An antiradar coating should provide the smallness of the diffracted field in certain directions. The determination of the shape of certain types of antenna, such as mirror or cavity ones, is also an inverse problem of this type. These antennas should transform the field of the feeder into the field with the desired properties in the near or far zones.

There are different types of nonstandard inverse problems. For instance, some parameters of the body may be known, whereas the other parameters must be found. The incident field may not be given completely, for instance, its phase may be unknown; the same relates to the field arising after diffraction. In different vari-

Phase Optimization Problems. O. O. Bulatsyk, B. Z. Katsenelenbaum, Y. P. Topolyuk, and N. N. Voitovich

ISBN: 978-3-527-40799-6

ants of the problems, the questions about the existence and uniqueness of solutions are formulated and answered in different ways.

The solutions of the inverse problems considered here can be found in the following way: first find a solution of the direct problem valid for any values of the required parameters of the body, and then find these parameters from the requirements formulated in the inverse problem. However, such a way of solving an inverse problem is rarely realizable in the pure form. The problems considered here are inverse to the problems that are related to different parts of diffraction theory. Various methods of this theory are used for solving inverse problems.

If the inverse problem has no solution in its rigorous formulation, then it can be substituted by an optimization problem. This problem consists in finding parameters of the body so that the field arising as the result of diffraction on this body is as close as possible to the desired scattered field. The 'distance' between the fields is usually measured as their mean square difference, that is, the appropriately normed integral of the squared modulus of the difference between them. The values of the parameters at which this difference is minimal under physically realizable conditions are the solution to the optimization problem. This minimum characterizes the degree of reality of the formulated inverse problem. If the phases of one or both fields are not fixed, then their appropriate choice can decrease the minimum, that is, decrease the distinction between the rigorous and optimization statements of the inverse problem. Different examples of these optimization problems have been investigated in previous chapters of the book.

6.1.2
Content of the chapter

Various inverse problems which differ from each other in the required physical parameters are considered in the next sections of this chapter. In the inverse problems considered in Section 6.2 the impedance of the body surface is sought. The impedance must be found such that the diffraction on this body leads to the desired transformation of the given field.

The surface is called the *impedance surface* if the relation between the tangential components of the electric and magnetic fields on it is linear and independent of the exterior field [36]. The isotropic impedance is described by a scalar number, the anisotropic one is described by two numbers and a direction in the tangential plane. These numbers are complex; their real parts define the absorbing ability of the surface, whereas the imaginary parts describe the ability to support the propagation of the surface waves. Different constructive possibilities for creation of the desired impedance are considered and some methods of manipulation of its frequency dependence are described [84].

The main objective of the investigation is the determination of the impedance distribution along the surface which provides the radar invisibility of the body, that is, the equality to zero (or, at least, the smallness) of the amplitude of the scattered wave, which propagates in the opposite direction to the incident plane wave. The simplest way to determine such an impedance consists in equating to zero the expression for the amplitude valid for any impedance and then considering this condition as an equation for the required impedance.

For nonsymmetrical bodies the 'invisibility' condition should be fulfilled at any orientation of the body with respect to the incident wave. This is possible only if the impedance is variable, that is, different at different points on the surface. In order to find the function describing the distribution of the impedance, it is necessary to solve a system of two nonlinear integral equations. The numerical results are presented for the body in the form of an elliptical cylinder with any eccentricity of the cross-section [85].

The 'invisibility' is provided by simultaneous actions of the two mechanisms: absorption of the energy and creation of surface waves that take away a part of the energy and radiate it in other directions. For the circular impedance cylinder, the amplitudes of the surface waves are calculated by means of the technique used in the direct diffraction problems and based on the Watson series [86]. If the impedance allows existing surface waves, then each term of this series describes such a wave.

If the body orientation with respect to the incident wave is fixed, then the protection (at least incomplete) can be realized if the impedance of the illuminated part of the body is matched with the incident wave. This means that the impedance should provide the same relation between the components of the electric and magnetic fields, as it is in the incident wave [87]. The conditions under which the matched impedance can be isotropic are considered. This type of protection can also be effective against multi-positional radiolocation.

The impedance surfaces are also used in interior problems, for instance, when constructing the waveguides and cavity resonators. Substitution of the smooth metallic surfaces by impedance ones changes the phase velocities of the modes in waveguides or the eigenfrequencies of resonators. In multi-mode waveguides, the choice of the wall impedance may increase the difference between the phase velocities of two modes and, in this way, weaken the process of their mutual transformation. Vice versa, the desired transformation of a mode into another one can be provided by creating a segment with variable wall impedance in the multi-mode waveguide [88]. The difference between eigenfrequencies of two oscillations (and in this way, the one-mode range) can be increased by an appropriate choice of the wall impedance.

A short-periodical array with metallic wires is considered in Section 6.3. It is a surface with an anisotropic transparency [89]. This property may be used for creating the polarization transformers. When a linear polarized plane wave falls onto the array, then the $\vec{E}$-field component perpendicular to the wires passes through the array almost without decreasing, whereas the transmitted component parallel to the wires strongly decreases and, moreover, obtains an additional phase shift $\pi/2$. This decrease depends on the density of the array and on the shape of the wires. If the vector $\vec{E}$ makes up a small angle with the wires, then it is possible to provide the equality of the amplitudes of both components and, at the same time, the $\pi/2$-difference of their phases, that is, to create the wave of circular polarization [90]. The transmission resonator having such an array (with appropriately chosen parameters) transforms linear into circular polarization [91].

It is worth mentioning more complicated elements that are not described here, but which deserve detailed analysis. In the literature short-periodical arrays with the wires in the form not of uniform solid cylinders, but of short-periodical helixes have been described. Such arrays have complicated frequency and polarization properties depending on the parameters of the helixes and the array itself. They can be used to construct different passive radio engineering devices [92].

6.1.3 Other inverse problems

There is another class of nonstandard inverse diffraction problem consisting of problems about the shape of a semi-transparent layer and the distribution of its transparency, such that a given field falling onto such a layer undergoes the desired transformation. These problems are not discussed here. They are described in detail in the literature; some part of the results are summarized in [24]. We confine ourselves only to a short overview of the possibilities that the use of semi-transparent layers provides. These layers can be realized in the form of thin dielectric bands with $|\varepsilon| \gg 1$, perforated metallic sheets, a layer with $|\varepsilon| \approx 1$ located between metallic plates [93], etc.

Semi-transparent layers are used when constructing the cavity antennas. They are the open resonators with quality factor and radiation patterns which depend on the wall transparency. By appropriate choice of the parameters of the walls, it is possible to construct the antennas with a practically arbitrary radiation pattern. Simultaneously, in some cases, it is possible to provide the existence of a domain near the antenna with very small field. This property is very important for electromagnetic compatibility and safety (e.g. mobile phones). The existence of such a 'black spot' does not affect the pattern almost; in particular, it does not decrease the radiation in the direction in which it is located itself: the energy as if flows around it.

The cavity antennas are not the only devices in which closed semi-transparent surfaces can be used. The field of any sources can be transformed into any other field by surrounding it with a surface of the appropriate shape and transparency.

Semi-transparency may also be used for compensation of distortion of the antenna pattern caused by an obstacle located near the antenna. If the locations of the antenna and obstacle are given, then a semi-transparent shell can be found, such that the field which arises when diffracting the field of the antenna on the 'body' consisting of the obstacle and the shell surrounding it, has the same pattern as the pattern created by the antenna in the absence of the obstacle [94], [95].

6.2 Diffraction on impedance surfaces. 'Invisible' bodies and screens

6.2.1 Impedance plane

6.2.1.1 Impedance conditions

We use the following notation. As before, the electric and magnetic fields in a medium are denoted by $\vec{E}$ and $\vec{H}$. The components of these fields, tangential to a certain surface, are denoted by $\vec{e}$ and $\vec{h}$, respectively. These two-component vectors are also called the electric and magnetic fields (omitting the words 'tangential components'), except for cases when this leads to ambiguity.

On the boundary of the dielectric body, the conditions are fulfilled that contain values of the fields $\vec{e}$ and $\vec{h}$ on both sides of the boundary. On the boundaries of certain bodies the conditions contain the values of these fields only on one side of the boundary. The fields in the body interior (i.e. in the dielectric) do not participate in these conditions. Such conditions hold in the media in which the structure of the interior field near the boundary does not depend on the field exterior to the body. The relations between the fields $\vec{e}$ and $\vec{h}$ do not depend on the exterior fields. These relations are the *boundary conditions* for the fields $\vec{E}$ and $\vec{H}$ outside the body. Such boundaries are called the *impedance boundaries* and the bodies having such boundaries are called the *impedance bodies*. The coefficients involved in these conditions are the *boundary impedances*.

The simplest impedance boundary is the boundary of the dielectric body with material constants ε and μ such that

$$\sqrt{|\varepsilon\mu|} \gg 1 . \tag{6.1}$$

In such a body the field near the boundary is close to that of the plane wave normally falling onto the body. In this wave the fields $\vec{E}$ and $\vec{H}$ are perpendicular, and the ratio of their amplitudes is $\sqrt{\mu/\varepsilon}$. If the z-axis of the Cartesian coordinate system is directed along the exterior normal to the boundary, then the components of these fields are

$$H_x = A\exp\left(ik\sqrt{\varepsilon\mu}\cdot z\right) , \quad E_y = A\sqrt{\mu/\varepsilon}\exp\left(ik\sqrt{\varepsilon\mu}\cdot z\right) , \tag{6.2a}$$

$$H_y = B\exp\left(ik\sqrt{\varepsilon\mu}\cdot z\right) , \quad E_x = -B\sqrt{\mu/\varepsilon}\exp\left(ik\sqrt{\varepsilon\mu}\cdot z\right) . \tag{6.2b}$$

The coefficients A and B depend on the value and structure of the exterior field, whereas the ratios E_y/H_x and E_x/H_y do not depend on them. From the continuity of the tangential components of all fields on the boundary, it follows that the conditions

$$e_x = -wh_y , \quad e_y = wh_x \tag{6.3}$$

hold on the exterior side (at $z = +0$); here

$$w = \sqrt{\frac{\varepsilon}{\mu}} . \tag{6.4}$$

These impedance conditions hold on boundaries of the media in which the inequality (6.1) is valid. The stronger the inequality (6.1), the more precisely conditions (6.3) are fulfilled. The Shchukin–Leontovich boundary condition [97], which is true on the boundary of a 'perfect' dielectric (i.e. with $|\mu| \approx 1$, $|\varepsilon| \gg 1$), has the form (6.3) with $|w| \ll 1$. The limiting case of such a dielectric is the perfect conductor ('metal') that is, the medium in which $|\varepsilon| = \infty$. On its boundary,

$$\vec{e} = 0\,, \tag{6.5}$$

that is, the impedance is zero.

In the general case, the impedance is complex:

$$w = w' + i w''\,. \tag{6.6}$$

Only surfaces with

$$w' \geq 0 \tag{6.7}$$

are realized. This follows from the fact that the flux of the Poynting vector $\vec{S} = c/(8\pi)\,\mathrm{Re}[\vec{E}\vec{H}^*]$ is directed into the body, that is, $S_x < 0$. In order to avoid misunderstanding, we denote the complex-conjugated values by an asterisk (not by dashed symbol, as in the previous sections). According to (6.3),

$$S_z = -\frac{c}{8\pi}\,\mathrm{Re}\,w \cdot |\vec{h}|^2\,. \tag{6.8}$$

If condition (6.7) is violated, then $S_z > 0$, that is, the energy flows out of the body; the body does not absorb the energy, but, oppositely, radiates it. This fact takes place only for the laser media, that is, for the media with inverse distribution of the quantum levels. Such media are not considered here.

Media with any sign of w'' are realizable. If the plane wave normally falls onto the plane with condition (6.3), a reflected wave arises with the squared reflection coefficient

$$R^2 = \frac{(1-w')^2 + (w'')^2}{(1+w')^2 + (w'')^2}\,. \tag{6.9}$$

There is no reflection when

$$w' = 1\,, \quad w'' = 0\,. \tag{6.10}$$

In this case, the wave normally falling onto the plane is fully absorbed into the material located under the surface.

The mechanism of protection of the flat surface, based on conditions (6.10), requires that w' is maximally close to unity, whereas w'' is the smallest possible. The reflection coefficient does not depend on the sign of w''. As will be shown later, these assertions are not true for nonflat surfaces.

6.2.1.2 Surface waves on the plane

If $w'' \neq 0$, then the surface waves can propagate along the plane; the fields of these waves decrease quickly in a direction normal to the plane.

The field of the wave propagating along the x-axis does not depend on the y-coordinate, whereas its dependence on x and z is described by the factor

$$\Psi(x, z) = \exp(-ihz)\exp(-\alpha z)\,, \tag{6.11}$$

where $h = h' + h''$, $\alpha = \alpha' + \alpha''$ (do not confuse the vector $\vec{h}$ with the scalar h). The surface waves exist if $\alpha' > 0$.

There are waves of the two following types: $E(TM)$ waves having only the components E_x, E_z, H_y, and $H(TE)$ waves having only the components H_x, H_z, E_y. First we consider the $E(TM)$ waves. In them,

$$E_x = \frac{i}{k}\frac{\partial H_y}{\partial z}\,. \tag{6.12}$$

It follows from (6.11) and (6.12) that $e_x = -(i\alpha/k)h_y$ at $z = 0$. Comparing this relation with the first impedance condition (6.3) gives

$$\alpha' = kw''\,, \quad \alpha'' = -kw'\,. \tag{6.13}$$

Consequently, the surface wave of $H(TE)$ type can propagate along the impedance plane only if $w'' > 0$. The larger w'', the smaller the effective height $d = (\alpha')^{-1}$ of this wave. The amplitude of the field decreases in the direction from the surface as $\exp(-z/d)$. For the $E(TM)$ wave, $d = \lambda/(2\pi w'')$, where λ is the wavelength.

The complex propagation constant h is found from the equation

$$\frac{\partial^2\Psi}{\partial x^2} + \frac{\partial^2\Psi}{\partial z^2} + k^2\Psi = 0\,, \tag{6.14}$$

which is satisfied by $\Psi(x, z)$. According to (6.11), $h^2 = k^2 + \alpha^2$. The imaginary part of the propagation constant is negative, $h''h' = -k^2 w'' w'$. The larger w' is, that is, the larger the part of the energy which penetrates into the medium, the faster the wave decreases by the law $\exp(h''x)$.

The phase velocity of the wave is proportional to $(h')^{-1}$. The value h' satisfies the equation

$$(h')^4 - k^2\left[1 + (w'')^2 - (w')^2\right](h')^2 - k^4(w')^2(w'')^2 = 0\,. \tag{6.15}$$

If there are no losses in the surface ($w' = 0$), then $h'/k = \sqrt{1 + (w'')^2}$, that is, $h' > k$. The wave velocity kh'/c is smaller than c, that is, the wave is slow. It is slow also at $w' \neq 0$ if $w'' > 1$. However, if $w'' < 1$, that is, $d > \lambda/(2\pi)$, and simultaneously $w' > w''/\sqrt{1 - (w'')^2}$, then it follows from (6.15) that $h' < k$. If the field decreases weakly in the direction of the surface and the losses are large, then the surface wave is fast.

In the waves of $H(TE)$ type, the coordinate dependence of the fields is described by the same expression (6.11), but only with other values of the parameters α and h.

The component H_x is expressed by E_y, as follows

$$H_x = -\frac{i}{k}\frac{\partial E_y}{\partial z}\,, \tag{6.16}$$

which gives $h_x = (i\alpha/k)e_y$. Comparing this expression with the second condition (6.3) gives $\alpha = -ik/w$, that is,

$$\alpha' = -\frac{kw''}{|w|^2}\,, \quad \alpha'' = -\frac{kw'}{|w|^2}\,. \tag{6.17}$$

The surface wave of $H(TE)$ type propagates only along the impedance surface with $w'' < 0$. Its effective height is $d = \lambda/(2\pi) \cdot |w|^2/(-w'')^2$.

Consequently, if the impedance of a plane is complex ($w'' \neq 0$), then a surface wave can propagate along it in any direction. If $w'' > 0$, then this wave is the $E(TM)$ wave. The field $\vec{E}$ is directed along the propagation direction, the component H_z, normal to the plane, equals zero. If $w'' < 0$, then the wave is of $H(TE)$ type. The field $\vec{H}$ is directed along the propagation direction and the component E_z, normal to the plane, equals zero.

The surface waves are not excited on the impedance plane by the plane wave falling at any angle. In order for these waves to exist, the field of the incident wave must have a complicated structure. However, waves with similar properties can be excited by the plane wave on nonflat surfaces. Such waves play a significant role in the diffraction process on nonflat impedance surfaces.

6.2.1.3 **Corrugated surface (goffer)**

Both equalities in the boundary conditions (6.3) contain the same number w. These conditions describe the impedance surface having no separate direction, an *isotropic* surface. The more general form of the conditions is

$$e_x = -w_1 h_y\,, \quad e_y = w_2 h_x\,, \tag{6.18}$$

where $w_1 \neq w_2$. Such conditions are fulfilled only for a certain definition of the Cartesian axes x, y. For other orientations of the coordinate system the vector $\vec{e}$ is obtained as a product of the vector $\vec{h}$ by an impedance tensor. The simple relations (6.18) take place only if the x- and y-axes coincide with the principal axes of the tensor. The numbers w_1, w_2 equal its principal values. If the impedance is anisotropic ($w_1 \neq w_2$) and the vector $\vec{h}$ is not directed along a principal axis, then the angle between $\vec{e}$ and $\vec{h}$ differs from $\pi/2$.

Condition (6.7) of the impedance realizability remains also for the coefficients w_1, w_2 in (6.18). The normal component of the Poynting vector is

$$S_z = -\frac{c}{8\pi}\,\mathrm{Re}\left(w_1 \cdot |h_y|^2 + w_2 \cdot |h_x|^2\right)\,. \tag{6.19}$$

This value should be nonpositive for any vector $\vec{h}$, which gives $w_1' \geq 0$, $w_2' \geq 0$.

The simplest example of an anisotropic surface is the *goffer* (*corrugated* surface, Figure 6.1). If its period p is small in comparison with the wavelength λ, then at

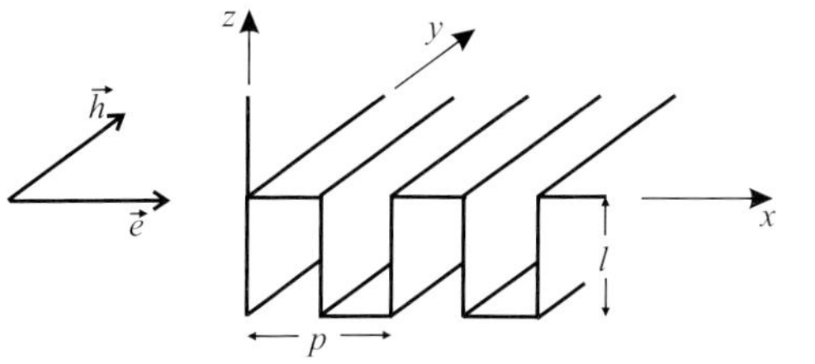

Figure 6.1 Goffer with rectangular grooves.

a distance from its boundary, large in comparison with p and small in comparison with λ, the fields are 'smoothed'; they are averaged over the period. Although the fields $\vec{E}$, $\vec{H}$ satisfy the 'metallic' conditions on the goffer surface, in the domain $z \gg p$ these fields are approximately the same as the fields on the plane $z = 0$ which fulfill the boundary conditions (6.18) for the averaged fields. These conditions are determined when considering the fields near the plane $z = 0$. They depend on the shape and the parameters of the goffer.

Consider a goffer formed by rectangular grooves filled by a material with parameters ε, μ. The fields of the higher waveguide modes arising in the grooves cancel after averaging over the period. In the grooves the field is approximately the same as the field of the cable wave, that is, it does not depend on the x-coordinate. On the groove bottom (at $z = -l$), $E_x = 0$. The field in the groove is the standing wave, in which

$$E_x = A \sin\left[k\sqrt{\varepsilon\mu}(z+l)\right], \quad H_y = i\sqrt{\frac{\varepsilon}{\mu}} A \cos\left[k\sqrt{\varepsilon\mu}(z+l)\right]. \tag{6.20}$$

The coefficient A depends on the exterior (i.e. at $z \gg p$) field; however, the ratio of the fields does not depend on A. In the plane $z = 0$, we have

$$E_x = A \sin\left(k\sqrt{\varepsilon\mu}\cdot l\right), \quad H_y = i\sqrt{\frac{\varepsilon}{\mu}} A \cos\left(k\sqrt{\varepsilon\mu}\cdot l\right). \tag{6.21}$$

On the metallic partitions between the grooves, we have $E_z = 0$ at $z = 0$, so that in the plane $z = 0$ the average value of the component E_x is $e_x = Aq \sin(k\sqrt{\varepsilon\mu}\cdot l)$. Here q is the ratio of the groove width to the period p. On the entire plane $z = 0$, the magnetic field is approximately the same as in the mouth of the groove, that is, $h_y = i\sqrt{\varepsilon/\mu}A\cos(k\sqrt{\varepsilon\mu}\cdot l)$. Substituting these fields into the first relation of (6.18) gives

$$w_1 = iq\sqrt{\frac{\varepsilon}{\mu}} \tan\left(k\sqrt{\varepsilon\mu}\cdot l\right). \tag{6.22}$$

On the metallic partitions, $E_y = 0$ and there are neither cable nor waveguide waves for this polarization. Therefore, e_y is very small, and we can approximately put $w_2 = 0$.

The surface wave of both TM and TE type can propagate along the impedance plane with a large anisotropy ($w_1 \neq 0$, $w_2 = 0$). The components e_x and h_y are not

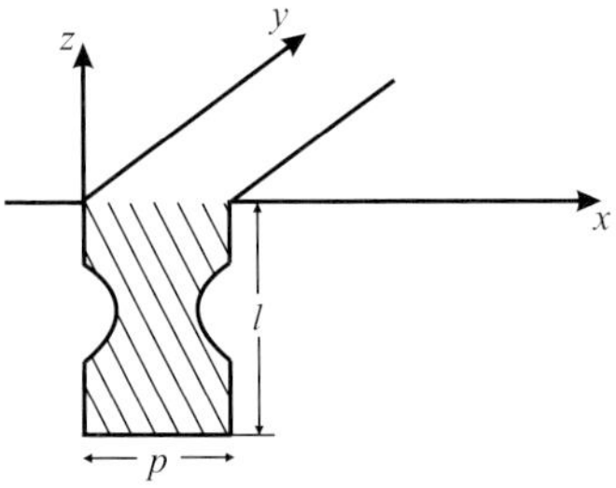

Figure 6.2 Goffer with narrowing.

zero in waves of both types. If $w_1'' > 0$, then the TE wave can propagate, that is, $E_z \neq 0$ in it, and this wave propagates across the grooves. If $w_1'' < 0$, then the TM wave can propagate, $H_z \neq 0$ in it, and this wave propagates along the grooves. The damping of the waves is determined by the value w', that is, according to (6.22), by the imaginary parts of the constant ε and μ (absorption in the medium).

If two sets of perpendicular grooves are created on the plane, then both coefficients w_1 and w_2 differ from zero. Their dependences on the geometrical parameters l and q are more complicated than (6.22). Any anisotropic impedance can be created on the plane by choosing these parameters to be different in both sets.

The frequency dependence of the impedance (6.22) is caused by both frequency dependences of the parameters ε and μ, and mostly by the fact that the argument of the tan function is proportional to the frequency. It is possible to change this dependence by substituting the rectangular grooves with those of more complicated profile (Figure 6.2), and in this way to manipulate this dependence in a certain range.

Consider a groove of this type, having a narrowing which divides it into two parts. The electrodynamic coupling of these parts depends on the frequency. The larger this coupling, the smaller the influence of the narrowing, so that, for strong coupling, the corrugated surface has approximately the same impedance as the goffer with the rectangular grooves of the same depth. If the coupling between the parts of the groove is weak, then the impedance is the same as in the case when the grooves were rectangular but had a smaller depth.

For the groove with narrowing in the form of a diaphragm with a slot parallel to the y-axis (Figure 6.3), the impedance can be found explicitly as a function of the geometrical parameters and the frequency. To find it, we assume that the width b of the slot is much smaller than the distance l_1 from the slot to the groove mouth and than the distance l_2 from the slot to the groove bottom. The local waves, which arise near the slot owing to the nonhomogeneity of the diaphragm in the x-direction, are damped at a distance of the order b and do not reach the planes $z = 0$ and $z = -l$. The fields on these planes are the same as in the system in which the diaphragm with the slot is substituted by a semi-transparent solid film. The fields on both sides of the film are connected as follows:

$$E_x^+ = E_x^- , \quad E_x^+ = -iQ\left(H_y^+ - H_y^-\right) , \tag{6.23}$$

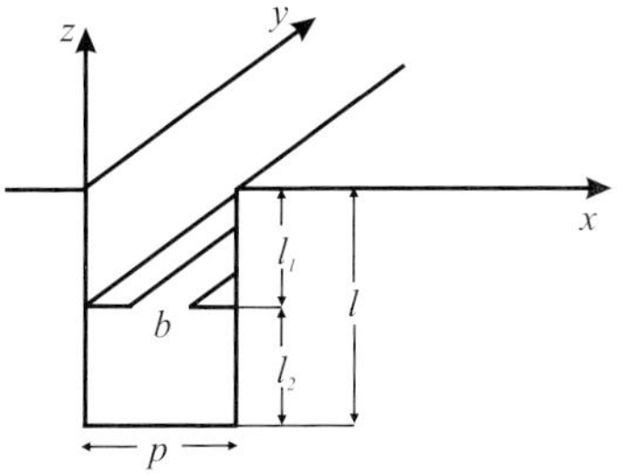

Figure 6.3 Goffer with the diaphragm.

where the values of the components on the upper side of the film ($z = -l_1 + 0$) are marked by the upper index '+', and on the lower side ($z = -l_1 - 0$) by the index '−'. The quantity Q in (6.23) is the transparency coefficient for the diaphragm. It is equal to the coefficient for the short-periodical array consisting of bands, which would arise in sequential reflecting of the diaphragm in the two metallic planes between which it is located. This coefficient is determined by solving the electrostatic problem (because $kp \ll 1$) on the field near to the array. The coefficient Q connects the values of the fields on both sides of the array at distances large in comparison with p and small in comparison with λ. Its value is

$$Q = -\frac{\pi}{2}\frac{1}{kp}\frac{1}{\ln\sin(\pi b/(2p))} \tag{6.24}$$

(see (6.111b) in [36] substituting $q = (p - b)/p$).

The value $|w_1|$ increases as the frequency increases. This effect can be weakened by using the diaphragm with a slot, due to the fact that at $kp \ll 1$, Q decreases as the frequency increases. The larger the frequency, the smaller the transparency, that is, the smaller the effective depth of the groove with the diaphragm. If the frequency increases, then $|w_1|$ decreases more slowly than in the groove without the narrowing, in particular, without the diaphragm with a slot.

The transparency decrease with increasing frequency can be explained in terms of quasi-statics. The slot plays the role of capacity for the field $\vec{E}$ perpendicular to it. The current flowing between its edges is the displacement current. It is proportional to the frequency and equal to the difference of the components H_y in the problem about the uniform semi-transparent diaphragm equivalent to the film with the slot. The current excited by the electric field is proportional to the product of the frequency and the capacity between two bodies where this current flows. The value $(kQ)^{-1}$ is proportional to the capacity of the condenser consisting of two bands located between the two infinite plates perpendicular to the field $\vec{E}$. The capacity between these plates is connected with the electric field, uniformly distributed along the z-axis, which does not participate in forming the difference in the magnetic fields (i.e. the displacement current) on both sides of the condenser formed by the bands. According to (6.24), the capacity between two infinitely thin bands is large only if the slot width b is so small, that $\ln(p/b) \gg 1$. In order to increase the capacity, it is possible to place two plates on the slot edges perpendicular to the diaphragm; these plates make up the condenser. The halves of the diaphragm can be placed in two mutually shifted planes, overlapping each other, as in [93].

Another interpretation of the process in the goffer grooves is possible. In this interpretation the slot is a load at $z = -l_1$, whereas the groove is a two-wire circuit of length l_1 occupying the domain $-l_1 < z < 0$. The domain is coupled with the resonator occupying the domain $-l < z < -l_1$ by the slot at $z = -l_1$. Other forms of resonator as well as other coupling elements, in particular, the slots parallel to the field $\vec{E}$ (inductive coupling elements) can be used. This allows us to create an arbitrary (in certain limits) combination of oppositely-directed cable waves in the line $-l_1 < z < 0$, and, consequently, an arbitrary frequency dependence of the impedance.

In order to calculate the impedance of the goffer shown in Figure 6.3, one should express the fields E_x and H_y in both domains as oppositely-directed cable waves, similar to (6.2), but with four coefficients. These coefficients are determined from the requirement that $E_x = 0$ at $z = -l$ and conditions (6.23) are fulfilled at $z = -l_1$. The required impedance equals the ratio E_x/H_y in the plane $z = 0$, multiplied by q, which gives

$$w_1 = -i\frac{\tan(kl) - U}{1 + U\tan(kl)}q\,, \tag{6.25}$$

where

$$U = \frac{\tan(kl_2)}{1 + Q/(\sin(kl_2)\cos(kl_2))}\,. \tag{6.26}$$

At $Q = 0$ the diaphragm completely separates both parts of the groove, $w_1 = -iq\tan(kl_1)$. At $Q = \infty$ the screening is absent, $w_1 = -iq\tan(kl)$. The first case corresponds to $b = 0$ and the second to $b = p$.

6.2.2
Diffraction on the impedance circular cylinder and on the impedance sphere

6.2.2.1 Surface waves on the cylinder

In order to derive the boundary conditions on the surface of the circular cylinder from those on the impedance plane, we should substitute the right Cartesian coordinates system (x, y, z) with the right cylindrical one (φ, z, r). According to (6.3), the required conditions on the impedance surface of the cylinder are

$$e_\varphi = -w_1 h_z\,, \quad e_z = w_2 h_\varphi\,. \tag{6.27}$$

The Poynting vector is directed inside the cylinder, and conditions (6.7) follow for both impedances w_1 and w_2. Conditions (6.27) are valid for a cylinder of arbitrary cross-section if φ is directed perpendicular to the vectors $\vec{z}$ and $\vec{N}$.

The isotropic impedance ($w_1 = w_2$) can be realized by a dielectric layer, and the anisotropic one ($w_1 \neq w_2$) by a goffer (i.e. by grooves). If the grooves are directed either axially (along the z-axis) or azimuthally (along the φ-axis), then the compo-

Table 6.1 Surface waves and conditions for E and H polarization.

Polarization of incident wave	Grooves direction	Type of surface wave	Condition for w''
E	Azimuthal	H	$w_2'' < 0$
H	Axial	E	$w_1'' > 0$

nents e_φ, e_z are not coupled with h_φ, h_z, respectively. In other cases (e.g. for the helical grooves) the boundary conditions are more complicated than (6.27). The axial goffer couples e_φ with h_z by the impedance w_1, whereas the azimuthal one couples e_φ with h_z by w_2.

If the plane wave with $H_z \equiv 0$, $\partial/\partial z \equiv 0$ falls onto the cylinder, then the total field also has this property. This is true also for the wave with $E_z \equiv 0$, $\partial/\partial z \equiv 0$. The first case is called $E(TM)$ polarization and the second $H(TE)$ polarization. This separation takes place only for the cylinder with conditions (6.27) but not for the 'tensor' ones.

The azimuthal grooves violate the invariance of the surface with respect to displacements along the z-axis, hence the total field depends on z. However, for the short-periodical goffer ($kp \ll 1$), it is possible to consider the fields $\vec{e}$, $\vec{h}$, $\vec{E}$, $\vec{H}$ averaged over the period, or, which is equivalent, the fields at distances from the surface which are large in comparison with the period. At such distances, these fields do not depend on z and they satisfy the averaged conditions (6.27).

If the surface waves can propagate along a nonflat surface, then these waves can be excited by the plane wave. Only the azimuthal surface waves exist on the impedance cylindrical surface with either azimuthal or axial grooves. The axial waves are not generated, because the field of the incident wave, as well as the properties of the surface, do not depend on z. Similar to the plane surface, these waves are separated onto the waves of E type (TM waves) and H type (TE waves). In the waves of E type, the component E_φ differs from zero, and $H_\varphi \equiv 0$; such waves arise at the H polarization of the incident field. On the other hand, in the waves of the H type, arising at the E polarization of the incident field, the component H_φ differs from zero, and $E_\varphi \equiv 0$.

As it will be shown below, the demands on the sign of the imaginary part of the impedance, imposed for the waves on the plane, also remain the same for the waves on the cylindrical surface (see Table 6.1).

If the impedance is isotropic, then the waves arise at $w'' < 0$ in the E polarization case, and at $w'' > 0$ in the H polarization case. In the first case the waves arise if the goffer is azimuthal and $w_2'' < 0$, in the second one they arise if the goffer is axial and $w_1'' > 0$.

If the polarization of the incident wave does not coincide with the cylinder axis and is not perpendicular to it, then the azimuthal waves of both types can arise depending on the sign of the imaginary part of the impedance.

We find the structure of the surface waves on the cylinder. We assume that the z-axis coincides with the cylinder axis, and the angle φ is counted from the direction from which the incident wave arrives.

In the E polarization case, considered first, the field contains the components E_z, H_φ, H_r. Omitting the index z in the function E_z, we write the Maxwell equations in the form

$$H_\varphi = -\frac{i}{k}\frac{\partial E}{\partial r}\,, \qquad H_r = \frac{i}{kr}\frac{\partial E}{\partial \varphi}\,. \tag{6.28}$$

Let the currents creating the incident field be in the axial direction and be located in the half-plane $\varphi = 0$ at a finite distance, near the line $r = r_0$, where r_0 is a finite number. Then the total field satisfies the radiation condition. In order that the incident wave near the cylinder be the plane wave, that is,

$$E^{\text{in}} = \exp(ikr\cos\varphi)\,, \tag{6.29}$$

we will pass to the limit $r_0 \to \infty$ later. Passing to the limit in this order (the distance to the source is small in comparison with the distance where the radiation conditions hold) facilitates the derivations.

On the cylinder boundary, at $r = a$ (where a is the cylinder radius), we have

$$e = wh_\varphi\,; \tag{6.30}$$

here the index '2' is omitted for w.

We express the solution of the problem about diffraction of the plane wave on the impedance cylinder as a series of functions satisfying condition (6.30). This technique, developed for the metallic cylinder and sphere (see, e.g. [36], Section 7.1.3), is extended to the impedance bodies in an obvious way.

The function $E(r, \varphi)$, as a Cartesian component of the electromagnetic field, satisfies the wave equation

$$\frac{\partial^2 E}{\partial r^2} + \frac{1}{r}\frac{\partial E}{\partial r} + \frac{1}{r^2}\frac{\partial^2 E}{\partial \varphi^2} = 0\,. \tag{6.31}$$

The existence of axial electric currents on the half-plane $\varphi = 0$ can be described by the discontinuity of the component H_r on this half-plane. According to (6.28), this component is proportional to $\partial E/\partial\varphi$, so that the currents can be given as a jump of this derivative. The field E itself is continuous on this half-plane; its discontinuity would mean the existence of the ring magnetic fields.

Introduce a function $F(r)$ proportional to the currents exciting the incident wave (6.29); its value consistent with the normalization used in (6.29) will be found further. The function $E(r, \varphi)$ satisfies the condition

$$\left.\frac{\partial E}{\partial \varphi}\right|_{\varphi=2\pi} - \left.\frac{\partial E}{\partial \varphi}\right|_{\varphi=0} = F(r)\,. \tag{6.32}$$

The problem consists in finding the total field which satisfies the radiation condition at infinity, condition (6.30) on the surface $r = a$, (6.31) in the domain $a < r < \infty$, and the jump condition (6.32) with a given function $F(r)$ on the half-plane $\varphi = 0$.

The solution can be found by the separation of variables method, as a series of products of the functions of r and φ. We subordinate each of the terms in the series to (6.31) at $a < r < \infty$, the radiation condition at $r = \infty$, and the boundary condition at $r = a$. The series with such properties is called the *Watson series*.

As will be shown further, in the case of the impedance cylinder, the terms of this series are the surface waves if they can propagate at a given impedance. In contrast to the surface waves on the plane, the set of waves on the nonflat surface (if they exist) are countable. The functions describing them apparently make up a complete system.

After separating the multiplier dependent on φ, (6.31) is reduced to the Bessel equation. The only functions satisfying this equation and the condition at infinity are the Hankel functions $H^{(2)}_{\nu_n}(kr)$, where $\nu_n = \nu'_n + i\nu''_n$, $n = 0, 1, \ldots,$ are complex separation constants in (6.31). After multiplying by the angular factor $\exp(-i\nu_n\varphi)$ according to the corresponding differential equation, the terms of the series have the form

$$H^{(2)}_{\nu_n}(kr)\exp(-i\nu_n\varphi)\,. \tag{6.33}$$

This function describes the field of the nth surface wave. The separation constants are determined from the equation

$$H^{(2)}_{\nu_n}(ka) + iwH^{(2)\prime}_{\nu_n}(ka) = 0\,, \tag{6.34}$$

where the prime means differentiation with respect to the argument. This equation is investigated in detail in the literature (see, e.g. references in [85, 86]). Further, functions (6.33) with the respective values ν_n will be connected with the diffraction problem on the cylinder, and the amplitudes of the surface waves will be found.

The roots of (6.34) lie in the quadrant

$$\nu'_n > 0\,, \quad \nu''_n < 0\,. \tag{6.35}$$

The angular dependence

$$\exp(\nu''_n\varphi)\exp(-i\nu'_n\varphi) \tag{6.36}$$

of the field corresponds to the wave, propagating in the direction in which φ increases, and damping due to both the absorption in the cylinder and the radiation, inevitable with propagation over the convex surface.

Since the functions J_{ν_n} and N_{ν_n} are real at $\nu''_n = 0$, (6.34) has no real roots. It follows from (6.34) that

$$w = -\frac{2}{\pi ka}\frac{1}{[J'_{\nu_n}(ka)]^2 + [N'_{\nu_n}(ka)]^2}\,, \tag{6.37}$$

and at $\nu''_n = 0$ would be $w' < 0$. The undamped waves would exist only if the surface radiated the energy compensating the radiation losses. The impedance with $w' < 0$ is nonrealizable (see (6.7)).

The quantity $-\nu_n''$ increases quickly with the number n. Hence, the waves of higher numbers are quickly damped and the total field consists mainly of the wave with $n = 0$.

At $ka \gg 1$ the value $|\nu_0|$ has the order ka. For large argument and indices, the asymptotics of the cylindrical functions have a complicated form described by the Airy function, so that there are no simple formulas for ν_n even in the asymptotic domain.

Qualitatively, the radiation losses of the surface wave can by explained by the fact that its linear velocity increases when moving from the surface, and becomes larger than c at a certain distance. The wave becomes quick, but such a wave is always damped.

We consider the wave described by the function (6.33) for the field E, and find an expression for the energy in the two fluxes, one incoming into the cylinder and other outgoing to infinity. We observe the domain bounded by the segment of the cylinder boundary $r = a$ from $\varphi = 0$ to $\varphi = \Delta\varphi$, $\Delta\varphi \ll 1$, two half-planes $\varphi = 0$ and $\varphi = \Delta\varphi$, and the segment of the cylindrical surface $r = b$ between these half-planes. Further, we will suppose that $kb \to \infty$, more exactly, that $kb \gg \nu_n$, so that the quantity b will disappear in the final result.

The flux of the Poynting vector incoming into this volume through part of the half-plane $\varphi = 0$ from $r = a$ to $r = b$ is

$$S_\varphi\big|_{\varphi=0} = \mathrm{Re}\int_a^b E H_r^* \,\mathrm{d}r = \frac{\nu_n'}{k}\int_a^b \left|H_{\nu_n}^{(2)}(kr)\right|^2 \frac{\mathrm{d}r}{r} . \tag{6.38}$$

Here and further, the factor $c/(2\pi)$ is omitted. Such a flux multiplied by $\exp(2\nu_n''\Delta\varphi)$ goes out through the half-plane $\varphi = \Delta\varphi$. The difference of these fluxes is

$$-\frac{2\nu_n'\nu_n''}{k}\Delta\varphi\cdot\int_a^b \left|H_{\nu_n}^{(2)}(kr)\right|^2 \frac{\mathrm{d}r}{r} . \tag{6.39}$$

The integrand in (6.39) decreases quickly as r^{-2}, and the upper limit in the integral can be substituted by ∞.

The energy incoming into the cylinder is

$$S_{-r}\big|_{r=a} = \mathrm{Re}(e h_\varphi^*) a\cdot\Delta\varphi = a\cdot\Delta\varphi\frac{w'}{|w|^2}\left|H_{\nu_n}^{(2)}(ka)\right|^2 . \tag{6.40}$$

The outgoing radiated energy to radial infinity is

$$\lim_{b\to\infty} S_r\big|_{r=b} = \mathrm{Re}\left\{\frac{i}{k}H_{\nu_n}^{(2)}(kb)\frac{\mathrm{d}}{\mathrm{d}r}\left[H_{\nu_n}^{(2)}(kr)\right]\Big|_{r=b}\right\} b\cdot\Delta\varphi . \tag{6.41}$$

At $kb \gg |\nu_n|$, the Hankel asymptotics holds

$$H_{\nu_n}^{(2)}(kr) \simeq \sqrt{\frac{2i}{\pi kr}}\exp(-ikr)\exp(i\pi\nu_n/2) , \tag{6.42}$$

which gives

$$S_r\big|_{r\to\infty} = \frac{2}{\pi k}\Delta\varphi\cdot\exp(-\pi\nu_n'') . \tag{6.43}$$

The ratios of the fluxes (6.40) and (6.43) to the flux (6.38) give the relative losses (per unit length) in the cylinder and on the radiation, respectively.

It is known that the flux of the Poynting vector through the surface of a domain having no sources or absorbing media is zero (the energy conservation law), hence

$$-2\nu'_n \nu''_n \int_a^b \left| H^{(2)}_{\nu_n}(kr) \right|^2 \frac{\mathrm{d}r}{r} = ka \frac{w'}{|w|^2} \left| H^{(2)}_{\nu_n}(ka) \right|^2 + \frac{2}{\pi} \exp(-\pi \nu''_n) \,. \quad (6.44)$$

Of course, this relation can also be obtained without referring to the energy considerations, but immediately from the Bessel equation for the Hankel function, together with the radiation condition and boundary condition (6.34) at $r = a$. We outline respective derivations.

We temporarily denote $H = H^{(2)}_{\nu_n}(kr)$. This function satisfies the equation

$$\frac{1}{k^2} \frac{\mathrm{d}^2 H}{\mathrm{d}r^2} + \frac{1}{k^2 r} \frac{\mathrm{d}H}{\mathrm{d}r} + \left(1 - \frac{\nu_n^2}{k^2 r^2} \right) H = 0 \,. \quad (6.45)$$

The complex-conjugated function H^* satisfies the same equation when ν_n is replaced by ν_n^*. It follows from these two equations that

$$\left[\left(\nu_n^* \right)^2 - \nu_n^2 \right] \frac{H H^*}{r} = \frac{\mathrm{d}}{\mathrm{d}r} \left[r \left(H \frac{\mathrm{d}H^*}{\mathrm{d}r} - H^* \frac{\mathrm{d}H}{\mathrm{d}r} \right) \right] . \quad (6.46)$$

Integrating this equation from $r = a$ to $r = b$ and passing to the limit at $b \to \infty$ under the boundary condition at $r = a$ and asymptotics (6.42), we obtain the above equality (6.44).

The dependence of the field of the surface wave on r is defined by the function $H^{(2)}_{\nu_n}(kr)$. Near the surface the field can be described by several terms of the Taylor series

$$H^{(2)}_{\nu_n}(kr) = H^{(2)}_{\nu_n}(ka) + k\xi\, H^{(2)\prime}_{\nu_n}(ka) + \dots \,, \quad (6.47)$$

where $\xi = r - a$ is the distance from the surface, $k\xi \ll 1$. According to (6.34), the two-term approximation can be rewritten in the form $H^{(2)}_{\nu_n}(kr) = H^{(2)}_{\nu_n}(ka)(1 + ik\xi/w)$. Consequently,

$$\left| \frac{H^{(2)}_{\nu_n}(kr)}{H^{(2)}_{\nu_n}(ka)} \right| = \left(1 + k\xi \frac{w''}{|w|^2} \right) . \quad (6.48)$$

The waves arising at the incidence of the E polarized wave onto the cylinder are analogous to those arising at the incidence of the H polarized wave onto the impedance plane. They have a nonzero component H_φ in the direction of propagation. At $w'' < 0$ such a wave is a surface one, that is, its field decreases when moving away from the cylinder. According to (6.48), the decrease has the form $1 - \xi/d$, where $d = \lambda/(2\pi) \cdot |w|^2/(-w'')$; it is the same as for the waves of this type on the impedance plane. The parameter d does not depend on the wave number.

The next term of the Taylor series (6.47) is proportional to $(k\xi)^2$ and contains the second derivative of $H^{(2)}_{\nu_n}(kr)$. According to (6.45), this quantity contains ν_n, so that the coefficients at $(k\xi)^2$ are different for waves of different numbers. For the waves with higher numbers, the value $-\operatorname{Re}\nu_n^2$ is large, their fields decrease rapidly when moving from the surface at a distance where the quadratic term is already essential.

The functions $H^{(2)}_{\nu_n}(kr)$ corresponding to different ν_n, that is, to different roots of (6.34), have properties facilitating the use of series. They satisfy the following 'orthogonality' conditions

$$\int_a^\infty H^{(2)}_{\nu_n}(kr)H^{(2)}_{\nu_m}(kr)\frac{\mathrm{d}r}{r} = N_n^2\delta_{nm}\,. \tag{6.49}$$

This is not the usual Hermitian orthogonality, the 'norm' N_n :

$$N_n^2 = \int_a^\infty \left[H^{(2)}_{\nu_n}(kr)\right]^2\frac{\mathrm{d}r}{r} \tag{6.50}$$

is complex.

The property (6.49) is true for any value of w in (6.34). It is derived in a similar way to the diffraction problem on the metallic cylinder, when $w = 0$ (see (7.25) in [36]). From (6.45) for $H^{(2)}_{\nu_n}(kr)$ and a similar one for $H^{(2)}_{\nu_m}(kr)$ the identity

$$\begin{aligned}&\left(\nu_n^2-\nu_m^2\right)\frac{1}{r}H^{(2)}_{\nu_n}(kr)H^{(2)}_{\nu_m}(kr)\\ &\quad= -\frac{\mathrm{d}}{\mathrm{d}r}\left[r\left(H^{(2)}_{\nu_n}(kr)\frac{\mathrm{d}H^{(2)}_{\nu_m}(kr)}{\mathrm{d}r} - H^{(2)}_{\nu_m}(kr)\frac{\mathrm{d}H^{(2)}_{\nu_n}(kr)}{\mathrm{d}r}\right)\right],\end{aligned} \tag{6.51}$$

analogous to (6.46) follows. After integrating this identity from 0 to ∞ and using the conditions at $r = a$ and asymptotics for both $H^{(2)}_{\nu_n}(kr)$ and $H^{(2)}_{\nu_m}(kr)$, we obtain (6.49).

6.2.2.2 The Watson series. Amplitudes of surface waves

We find the field E satisfying (6.31) in the form of a series in terms of the functions

$$\Phi_n(\varphi)H^{(2)}_{\nu_n}(kr)\,. \tag{6.52}$$

The function $\Phi_n(\varphi)$ solves the equation

$$\frac{\mathrm{d}^2\Phi_n}{\mathrm{d}\varphi^2} + \nu_n\Phi_n = 0\,; \tag{6.53}$$

it is a linear combination of the functions $\exp(\pm i\nu_n\varphi)$, and is continuous for all φ. In order that the series of functions (6.52) satisfies condition (6.32), the derivatives $\mathrm{d}\Phi_n/\mathrm{d}\varphi$ must be discontinuous at $\varphi = 0$. Since (6.53) is homogeneous, the functions $\Phi_n(\varphi)$ are defined to within a constant factor. We can norm them by the condition

$$\left.\frac{\mathrm{d}\Phi_n}{\mathrm{d}\varphi}\right|_{\varphi=2\pi} - \left.\frac{\mathrm{d}\Phi_n}{\mathrm{d}\varphi}\right|_{\varphi=0} = 1\,. \tag{6.54}$$

In all expressions below, $0 \le \varphi \le 2\pi$.

The functions $\Phi_n(\varphi)$ satisfying (6.53), (6.54) and the continuity condition $\Phi(2\pi) = \Phi(0)$, are

$$\Phi_n(\varphi) = \frac{1}{1 - \exp(-2\pi i \nu_n)} \frac{1}{2 i \nu_n} \left[\exp(-i \nu_n \varphi) + \exp(i \nu_n (\varphi - 2\pi))\right] . \quad (6.55)$$

The first term in the square brackets describes the wave propagating from $\varphi = 0$ in the direction of increasing φ, the second term describes the wave propagating from $\varphi = 2\pi$ in the opposite direction. The amplitude of the first wave at $\varphi = 0$ and the second one at $\varphi = 2\pi$ are the same and equal to the value $(2 i \nu_n)^{-1}$ multiplied by the first factor from (6.55). This factor can be expressed in the form of the series

$$\frac{1}{1 - \exp(-2\pi i \nu_n)} = 1 + \exp(-2\pi i \nu_n) + \exp(-4\pi i \nu_n) + \dots . \quad (6.56)$$

Being multiplied by $(2 i \nu_n)^{-1}$, the terms of this series are the amplitudes of the wave which passed the angles 2π, 4π, etc., that is, the wave enveloping the cylinder (in the same or the opposite direction) once (the second term in (6.56)), twice (the third term) and so on.

The total field $E(r, \varphi)$ has the form

$$E(r, \varphi) = \sum_{n=0}^{\infty} A_n \left[2 i \nu_n \Phi_n(\varphi)\right] H_{\nu_n}^{(2)}(k r) . \quad (6.57)$$

According to (6.55), the function $2 i \nu_n \Phi_n(\varphi)$ is an aggregate of the waves of both directions, including an infinite number of those enveloping the cylinder. The fields of these waves are normed by (6.54). The coefficients A_n are found from the condition (6.32) in which $F(r)$ is determined from the demand that, near the cylinder, the incident wave is normed by (6.29).

Substituting (6.57) into (6.32) and using (6.51), we obtain

$$2 i \sum_{n=0}^{\infty} \nu_n A_n H_{\nu_n}^{(2)}(k r) = \mathcal{F}(r) . \quad (6.58)$$

Due to the orthogonality conditions (6.49), we have

$$A_n = \frac{1}{2 i \nu_n} \frac{1}{N_n^2} \int_a^{\infty} \mathcal{F}(r) H_{\nu_n}^{(2)}(k r) \frac{\mathrm{d} r}{r} . \quad (6.59)$$

It remains to find the function $\mathcal{F}(r)$. Similar to the field $E(r, \varphi)$, the field $E^{\text{in}}(r, \varphi)$ is created by the same 'discontinuity' (namely, axial current) proportional to $\mathcal{F}(r)$. In order to calculate the field created by this current in the absence of the cylinder, that is, in the free space, we introduce the Green function

$$G(\vec{r}, \vec{r}\,') = \frac{i}{4} H_0^{(2)}(k |\vec{r} - \vec{r}\,'|) . \quad (6.60)$$

Here $\vec{r}\,'$ is the point with coordinates $\hat{r}$, $\hat{\varphi}$ where the field E^{in} is to be found. The function G satisfies the equation

$$\Delta G + k^2 G = \delta(|\vec{r} - \vec{r}\,'|) , \quad (6.61)$$

where Δ is the two-dimensional Laplacian with respect to variables r, φ, whereas the coordinates $\hat{r}$, $\hat{\varphi}$ are the parameters. The function $E^{\text{in}}(r,\varphi)$ satisfies the same equation with zero on the right-hand side, which should be held in the entire volume except for the half-plane $\varphi = 0$. Both these equations lead to the equality

$$E^{\text{in}} \cdot \delta(|\vec{r} - \vec{r}\,'|) = E^{\text{in}} \cdot \Delta G - G \cdot \Delta E^{\text{in}} . \tag{6.62}$$

This equality should be integrated over the domain in the plane $z = const$, bounded by the loop enclosing the ray $\varphi = 0$, and by the infinitely remote circle. The functions E^{in} and G are analytical in this domain, and therefore the second Green theorem can be applied. Since the integral over the remote circle is zero, the contour integral in the right-hand side is reduced to the integral over the loop. The integral of the first term $E^{\text{in}} \cdot \partial G/\partial N$ taken over the loop equals zero as well, because the function E^{in} is continuous at the ray and G is continuous, together with its derivatives.

The second term under the integral has the form $G \cdot \partial E^{\text{in}}/\partial N$. After integrating and taking into account that $\partial/\partial N = \pm r^{-1}\partial/\partial\varphi$, we obtain

$$E^{\text{in}}(\hat{r},\hat{\varphi}) = \frac{i}{4}\int_a^{\infty} H_0^{(2)}(k|\vec{r} - \vec{r}\,'|)\frac{\mathcal{F}(r)}{r}\mathrm{d}r , \tag{6.63}$$

where the point $\vec{r}$ lies on the ray.

We assume that $\mathcal{F}(r) = C\delta(r - r_0)r_0$. Then the jump of the component H_r equals $i/k \cdot C\delta(r - r_0)$. This means that the fields $\vec{E}$, $\vec{E}^{\text{in}}$ are created by the current thread, lying in the plane $\varphi = 0$ at a distance r_0 from the cylinder axis. The amplitude of the current is proportional to the coefficient C. The argument in the Hankel function is $k[r_0^2 + \hat{r}_0^2 - 2r\hat{r}\cos\hat{\varphi}]^{1/2}$. We pass to the limit $r_0 \to \infty$, so that the field E^{in} near the cylinder becomes the plane wave. At $kr_0 \gg 1$ and $kr_0 \gg (ka)^2$, the argument of the Hankel function near the cylinder is $kr_0 - k\hat{r}\cos\hat{\varphi}$. Substituting this function with its asymptotics (6.42), we have

$$E^{\text{in}}(\hat{r},\hat{\varphi}) = C \cdot \frac{i^{3/4}}{4}\sqrt{\frac{2}{\pi r_0}}\exp(-ikr_0)\exp(ik\hat{r}\cos\hat{\varphi}) . \tag{6.64}$$

The coefficient C is found from the requirement that the product of all factors by the function $\exp(ik\hat{r}\cos\hat{\varphi})$ equal unity. At $r_0 \to \infty$, the amplitude of the current should increase as $r_0^{1/2}$ in order that the amplitude of the field near the cylinder remains equal to unity.

Calculate the amplitudes of two opposite surface waves, excited on the impedance cylinder with $w'' < 0$ when the incident wave is the plane E polarized wave having the form (6.29) near the cylinder. Substituting the same function $\mathcal{F}(r)$ into (6.59) and assuming $kr_0 \gg |\nu_n|$, so that the asymptotics (6.42) are applicable for the function $H_{\nu_n}^{(2)}(kr_0)$, we obtain

$$A_n = -\frac{2}{\nu_n N_n^2}\exp\left(\frac{i\pi\nu_n}{2}\right) . \tag{6.65}$$

The squared norm N_n^2 can be expressed by the values of the function $H_\nu^{(2)}(ka)$ and its derivative with respect to ν at $\nu = \nu_n$. The derivations given in [36], formula (7.30), for the case $w = 0$, are easily extended to the case $w \neq 0$.

The solution to the problem of diffraction of an H polarized wave on the impedance cylinder in the form of the Watson series can be easily obtained from that for the E polarized wave by substituting w by w^{-1} in all the expressions. Indeed, if only the component H_z^{in} of the magnetic field differs from zero in the incident field, whereas $E_z^{\text{in}} = 0$, then $e_\varphi = w h_z$ on the cylinder boundary. Since $E_\varphi = i/k \cdot \partial H_z/\partial z$, the boundary condition for H_z has the form

$$\left(H_z + \frac{i}{wk} \frac{\partial H_z}{\partial z} \right)\bigg|_{r=a} = 0 \,. \tag{6.66}$$

In the surface wave, H_z has the same form (6.33). The equation for ν_n differs from (6.34) for the substitution $w \to w^{-1}$:

$$H_{\nu_n}^{(2)}(ka) + \frac{i}{w} H_{\nu_n}^{(2)\prime}(ka) = 0 \,. \tag{6.67}$$

The wave arising at this polarization are of E type; they exist at $w'' > 0$. The field near the boundary, at $k\xi \ll 1$, $\xi = r - a$, is proportional to $1 + iwk\xi$, and its modulus is proportional to $1 - w''k\xi$. When moving away from the boundary, at $w'' > 0$ the field decreases at the same rate as the field of this type on the impedance plane; the effective thickness of the wave is $d = \lambda/(2\pi w'')$.

6.2.2.3 Complex impedance providing radar invisibility of circular cylinder

When the plane wave falls onto a body of bounded size, a scattered field arises, which, at large distances from the body, has the form of a spherical wave (cylindrical in two-dimensional problems). The amplitude of the wave scattered in the direction opposite to that of the incident wave is called the amplitude of the reflected wave. The protection against one-positional radar consists in choosing the impedance that provides zero values of this amplitude or makes it as small as possible.

In the model problem concerning the normal incidence of the plane wave on the impedance plane, a similar quantity, namely, the amplitude of the reflected plane wave equals zero if the impedance is equal to unity. The absence of the reflection is explained by the fact that the wave completely penetrates into the surface layer and is absorbed there. The impedance should have a zero imaginary part. If $w'' \neq 0$, then, according to (6.9), the reflection coefficient differs from zero.

When the plane wave falls onto a body of finite size, that is, onto a nonflat surface, it is impossible, in general, to produce an absence of reflection by the unit impedance (but see Section 6.2.2.4). The structure of the incident wave near the surface does not coincide with that of the normally falling plane wave, and the field does not penetrate completely into the surface layer with $w = 1$.

However, the incident plane wave can excite the surface waves on a nonflat surface. These waves take away part of the energy of the incident wave and, while propagating, this energy partially penetrates into the surface, and partially radiates (see (6.44)). The radiation does not increase the amplitude of the reflected wave, it goes out in other directions.

The surface waves can exist only on surfaces with nonzero imaginary part of the impedance. The absence of a reflected wave is provided by two mechanisms, namely, the absorption of the energy in the surface layer and its leaving in the form of a surface wave. Such joint action is possible only on a surface with a complex impedance, $w' \neq 0$, $w'' \neq 0$.

In this subsubsection we find such an impedance for the circular cylinder. Since the amplitude of the back wave is simply expressed for any w, this inverse problem can be solved in a logically natural way, by equating its expression to zero and finding w as a root of the obtained equation. As follows from the above consideration, this root is complex.

It would be possible to obtain the equation for w by setting $\varphi = 0$ in (6.57) and equating the series to zero. However, this way of solving is inconvenient for the inverse problem, because the required value w is involved in the series by means of the auxiliary values ν_n that must themselves be found from the transcendental equations (6.34) or (6.67) containing w as a parameter. The solution in the form of the Watson series describes the physical meaning of the diffraction process on the impedance cylinder and the role of the surface waves. However, the inverse problem can be more conveniently solved with the aid of another expression of the field, namely, in the form of the *Rayleigh series*.

Each term of this series for the diffracted field contains the product $H_n^{(2)}(kr)$ $\cos(n\varphi)$, $n = 0, 1, 2, \ldots$, that is, outside the cylinder it is an analytical function and satisfies the wave equation and the radiation condition. The boundary condition is satisfied only by the sum of two such series for diffracted and incident fields. The value w is involved in the coefficients of the series for a diffracted field. In contrast to the Watson series, the Rayleigh series converges slowly at $ka \gg 1$, but this fact is not too significant for modern computers. For the impedance cylinder, the Rayleigh series is constructed by the mentioned method of separation of the variables (see, i.e. (7.1.2) in [36]).

We start with the E polarization case. The incident field (6.29) is developed in the Fourier series

$$E^{\text{in}}(r,\varphi) = \sum_{n=0}^{\infty} B_n^{\text{in}} J_n(kr)\cos(n\varphi)\,, \tag{6.68}$$

where

$$B_n^{\text{in}} = \frac{2i^n}{1+\delta_{0n}}\,. \tag{6.69}$$

In contrast to the Watson series, the technique which uses the excitation current at a finite distance, is not needed when deriving the Rayleigh series, since the incident field does not satisfy the radiation condition: it is the plane wave coming from infinity.

The diffracted field is found in the form

$$E^{\mathrm{dif}}(r,\varphi)=\sum_{n=0}^{\infty}B_n^{\mathrm{dif}}H_n^{(2)}(kr)\cos(n\varphi)\,. \tag{6.70}$$

Substituting these series and their derivatives with respect to r at $r=a$ in the boundary condition (6.30), we obtain a system of functional equations valid for all φ. Due to the linear independence of the functions $\cos(n\varphi)$ and their completeness in the class of even functions (all functions considered here are of this kind), the coefficients at corresponding functions on both sides can be equated, which gives

$$B_n^{\mathrm{dif}}=\frac{(-1)^n}{1+\delta_{0n}}\frac{J_n(ka)+iwJ_n'(ka)}{H_n^{(2)}(ka)+iwH_n^{(2)\prime}(ka)}\,, \tag{6.71}$$

where the prime denotes the function derivatives with respect to the total argument.

The terms of series (6.70) begin to decrease quickly at n larger than $2ka$. At a large distance from the cylinder, $r\gg a$, the Hankel functions and their derivatives can be replaced by their asymptotics (6.42), according to which these functions are proportional to $\exp(-ikr)/(kr)^{-1/2}$. At these distances the scattered field has the form of a divergent cylindrical wave $F(\varphi)\exp(-ikr)/(kr)^{-1/2}$, where $F(\varphi)$ is the scattering pattern. Putting $\varphi=0$, (i.e. $\cos(n\varphi)=1$ for all n), we obtain the expression for the amplitude of the reflected wave

$$F^w=\sum_{n=0}^{\infty}B_n^{\mathrm{dif}}\exp(i\pi n/2)\,, \tag{6.72}$$

where an nonessential factor is omitted. For convenience, we use the value of the impedance as the upper index in this quantity. The required value of w is a root of the equation

$$F^w=0\,. \tag{6.73}$$

Before analyzing the solution to this equation, we establish a connection between two forms of the solution to the same diffraction problem, in the form of series (6.57) and (6.70), respectively. This connection is based on the analogy between the denominators of the coefficients B_n^{diff} (6.71) in (6.70) and the left-hand side of (6.34) for the numbers ν_n in (6.57). The series (6.57) can be obtained from (6.70) and (6.68) by using the technique of asymptotic summation. We outline backgrounds of this technique without describing it in detail.

Introduce a function of the complex variable ν, defined and single-valued on the full complex plane, which at $|\nu|\to\infty$ either decreases, or increases more slowly than $\nu^{-1}\exp(|\nu|)$, whereas at the points $\nu=n$, $n=0,1,\dots$ equals the nth term of the Rayleigh series. We assume (without proof) that such a function exists. The integral of this function divided by $(-1)^n\sin(\pi\nu)$, taken over the remote circle of the complex plane is zero, because the integrand tends to zero quickly enough at

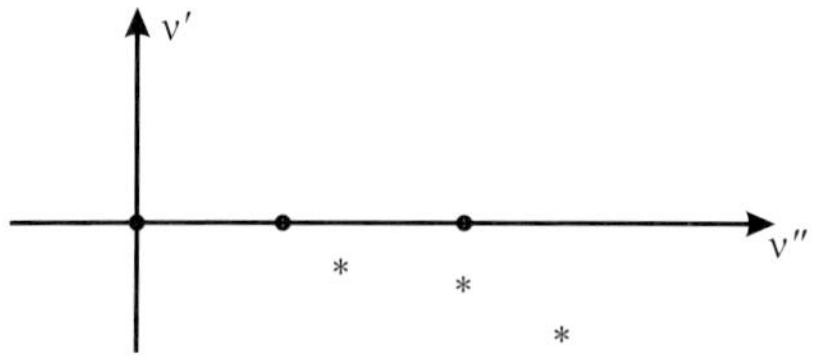

Figure 6.4 The Rayleigh (•) and Watson (∗) poles in the plane of the complex variable ν.

$|\nu| \rightarrow \infty$. By the Cauchy theorem, the sum of the residuals in the roots of the integrand equals zero. The residuals at $\nu = n$ (where $\sin(\pi\nu)$ in the denominator is zero) are equal to the corresponding terms of the Rayleigh series, and the sum of these residuals equals this series. Thus, this series equals the sum of the rest of the residuals with the opposite sign. The nth pole is located at the point of the plane ν, where the denominator of B_n^{diff} is zero. These poles coincide with the roots of (6.39). The residuals at these poles are the corresponding terms of the Watson series, and its sum is this series itself. In this way the Rayleigh series is transformed to the Watson series (see Figure 6.4).

Equation (6.73) has the solutions w with $w' > 0$ and $w' < 0$. The latter is physically unrealizable, and we do not consider it here. The real and imaginary part of the roots versus the frequency are shown in Figure 6.5.

The larger the value $-w''$, the more significant is the role of the surface waves (and the larger part of the energy is radiated in other directions) in the mechanism providing the absence of reflection. The closer w' is to unity, the larger the role of the absorption in the material of the cylinder surface. The first process is significant if the curvature radius a of the surface is small, the second one dominates if a is large, that is, if the surface is close to a flat one. At $ka < 1$, that is, at $a < \lambda/2\pi$, the imaginary part of the impedance is not small ($-w''$ is of the order of 0.3) and w' is not close to unity ($1 - w'$ is of the order of 0.5). In this range ('low frequencies') the role of the surface waves is especially significant. At higher frequencies ($ka > 1$), their role is less visible. However, as can be seen from Figure 6.5, the process of producing the surface waves remains significant also at large frequencies, that is, at smaller curvature. Only at $ka > 6$, that is, at $a > \lambda$, the optimal impedance w becomes approximately equal to unity, that is, the same as that for the flat surface.

In order to characterize the measure of protection of the impedance surface at frequencies different from that for which the impedance provides full radar invisibility, we introduce the quantity $|F^w/F^0|$ (the *visibility factor*), where F^0 is the value of the scattering pattern in the back direction for the metallic cylinder ($w = 0$) of the same shape. The value $F^0(ka)$ is calculated by (6.71) and (6.72) after substituting $w = 0$. This factor characterizes the protection effectivity for the metallic body owing to the substitution of its surface with the optimal impedance one.

In Figure 6.6, the value $|F^w/F^0|$ is represented for three values of w as functions of the frequency. The solid line relates to the case $w = 1$ (optimum for the flat surface). The effective protection in a wide region is reached only for large frequences, approximately for $ka > 6$. The other two curves describe the frequen-

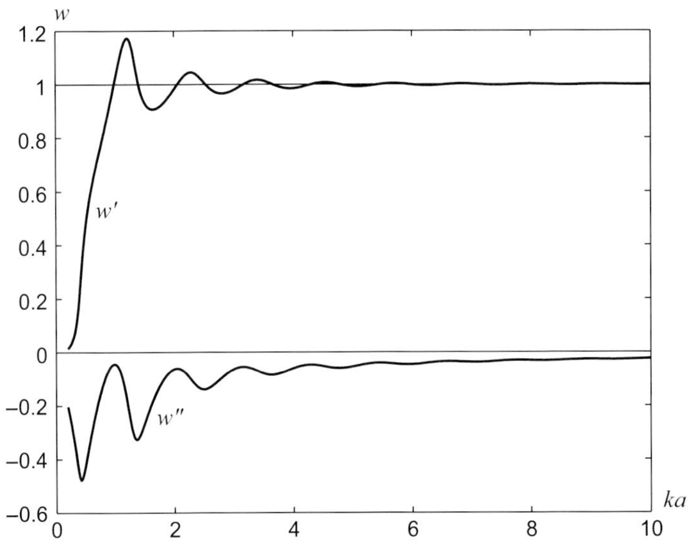

Figure 6.5 Optimal complex impedance of a circular cylinder; *TM(E)* polarization case.

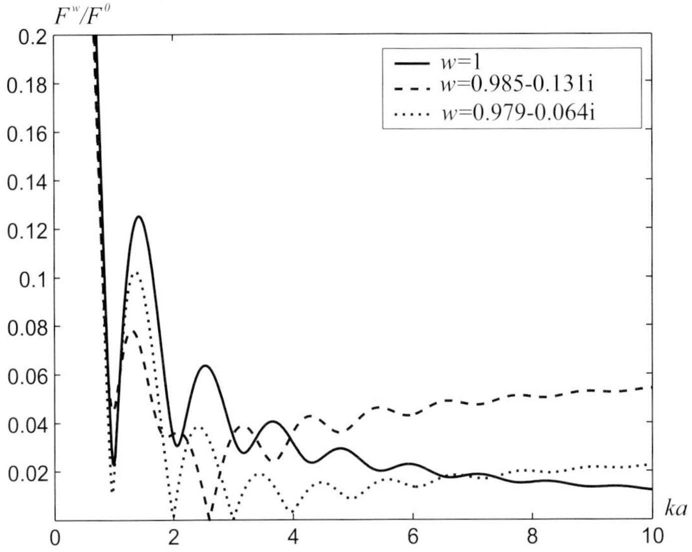

Figure 6.6 Frequency dependence of the visibility factor for a circular cylinder; *TM(E)* polarization case.

cy dependences of the visibility factor for two complex impedances. The first one, $w = 0.985 - 0.131i$, is the solution to (6.73) for $ka = 2.6$, that is, it provides zero visibility at this frequency. The second one, $w = 0.979 - 0.064i$, corresponds to $ka = 3.0$. The visibility for these impedances is lower than for $w = 1$ up to frequencies of the order $ka = 5$. The dependence of the visibility factor on the fre-

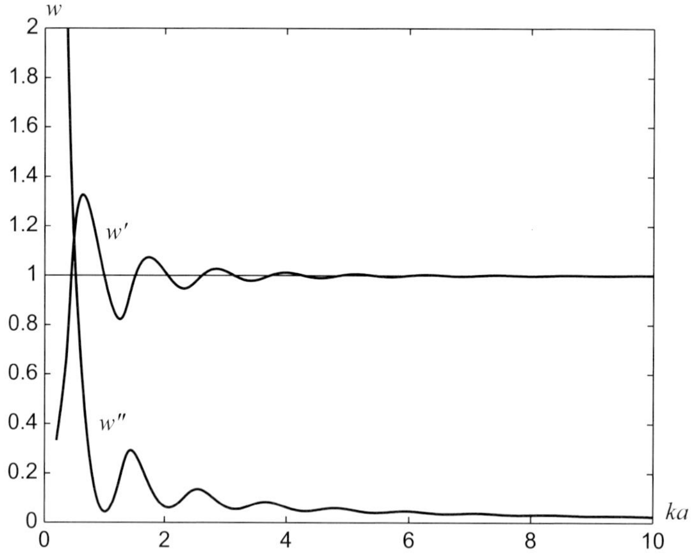

Figure 6.7 Optimal complex impedance of a circular cylinder; *TE(H)* polarization case.

quency is not critical: the process providing invisibility at a certain frequency and little visibility at close frequencies has no resonant character.

The above results are obtained under the assumption that the impedance does not depend on the frequency. In the general case, this dependence exists and, for instance, for the goffer it is not small. The dependence was considered in Section 6.2.2.3, and a system was described, allowing manipulation of it. Using the described approach as well as other methods, the frequency range, in which the visibility remains small (being zero at a certain frequency), can be extended.

All the results obtained for the E polarization can be rewritten for the H polarization by substituting w with w^{-1} in all expressions. The impedance providing invisibility for the H polarization, that is, satisfying (6.73) after this substitution in (6.71), has a positive imaginary part, $w'' > 0$. Recall that it is the condition at which the surface waves exist for this polarization.

The values w', w'' providing invisibility for this polarization at other frequencies are given in Figure 6.7. Figure 6.8 shows the visibility factor versus the frequency for $w = 0.997 - 0.131i$, being optimal at $ka = 2.6$; and $w = 0.993 - 0.055i$, being optimal at $ka = 3.2$. Besides the sign of w'', the behavior of the curves in these figures is similar to that for E polarization.

For E polarization, the cylinder scatters significantly more energy, and, in particular, creates the reflected field significantly larger than for H polarization. Therefore, if the polarization of the incident wave is not fixed, the cylinder with the impedance which provides a protection for the E polarization (e.g. in the form of the azimuthal goffer with $w'' < 0$), is more protected.

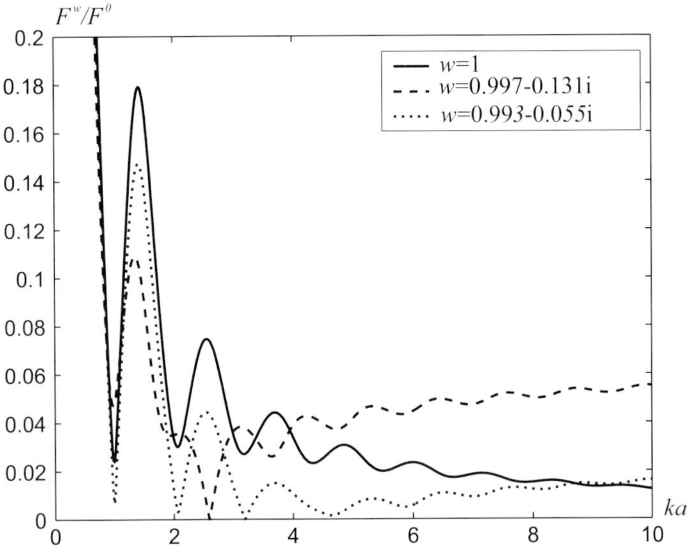

Figure 6.8 Frequency dependence of the visibility factor for a circular cylinder; *TE(H)* polarization case.

6.2.2.4 Impedance sphere. Mutual cancellation of reflected fields at $w = 1$

The plane wave falling onto the sphere with unit impedance does not excite the wave reflected in the backward direction. This is a partial case of the known theorem (see, e.g. [98]), which claims that, if a body has a symmetry axis of the fourth order and the impedance $w = 1$, and the plane wave falls onto the body in the direction of this axis, then the amplitude of the reflected wave is zero. The waves generated by the currents induced by the incident wave on the surface are mutually canceled. Here we illustrate the compensation process using the example of a sphere.

We introduce the spherical coordinate system $\{R, \theta, \varphi\}$ with the z-axis in the opposite direction to the incident wave (Figure 6.9), so that the value $\theta = 0$ corresponds to the direction to the source. The angle φ is counted from the direction of the vector $\vec{E}$ in the incident wave. The equation of the boundary is $R = a$.

The spherical components of the fields are expressed by the two scalar functions (Debye potentials) $U(R, \theta, \varphi)$, $V(R, \theta, \varphi)$. The angular components involve the derivatives of both functions. Here we give the expressions for two components (two others are expressed in similar way):

$$E_\theta = \frac{1}{R}\frac{\partial^2(RU)}{\partial R \partial\theta} - \frac{ik}{R\sin\theta}\frac{\partial(RV)}{\partial\varphi}, \tag{6.74a}$$

$$H_\varphi = -\frac{ik}{R}\frac{\partial(RU)}{\partial\theta} + \frac{1}{R\sin\theta}\frac{\partial^2(RV)}{\partial R\partial\varphi}. \tag{6.74b}$$

The radial component E_R is expressed by the potential U and its derivative with respect to R. Similarly, H_R is expressed by the functions U and V and their derivatives with respect to R.

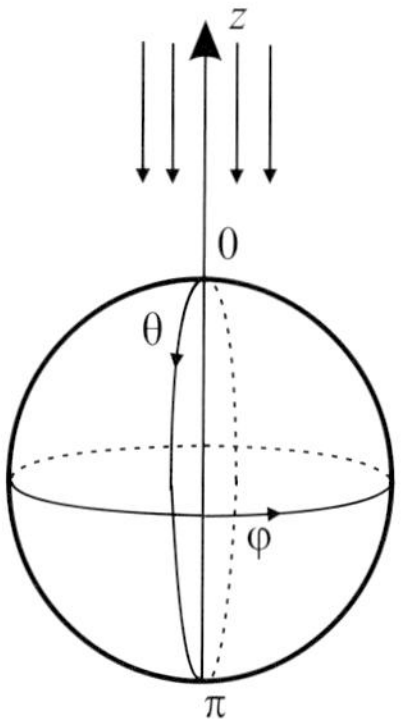

Figure 6.9 Impedance sphere.

The component E_R^{in} of the incident wave is proportional to $\cos\varphi$ and therefore the potential U^{in} is proportional to $\cos\varphi$. Similarly, the potential V^{in} is proportional to $\sin\varphi$.

The functions U and V satisfy the wave equation. The potentials U^{in} and V^{in} can be written as a series by the partial solutions of this equation (cf. (6.68)):

$$U^{\text{in}} = \cos\varphi \sum_{n=1}^{\infty} A_n^{\text{in}} \frac{1}{\sqrt{R}} J_{n+1/2}(kR) P_n^{(1)}(\cos\theta)\,, \tag{6.75a}$$

$$V^{\text{in}} = \sin\varphi \sum_{n=1}^{\infty} B_n^{\text{in}} \frac{1}{\sqrt{R}} J_{n+1/2}(kR) P_n^{(1)}(\cos\theta)\,; \tag{6.75b}$$

here $P_n^{(1)}$ are the associated Legendre functions. Let us check that all $A_n^{\text{in}} = B_n^{\text{in}}$ for the example of the component E_θ^{in}. Only under this equality, E_θ^{in} is proportional to $\exp(ikz)$ on the ray $\theta = 0$ at $kR \gg 1$, and the wave in the opposite direction is absent from the incident wave. To prove this, it is sufficient to substitute (6.75) into (6.74a) and put $\theta = 0$. This gives

$$E_\theta^{\text{in}} = \frac{\cos\varphi}{R} \sum_{n=1}^{\infty} \left.\frac{\partial P_n^{(1)}}{\partial\theta}\right|_{\theta=0} \left\{ A_n^{\text{in}} \frac{\mathrm{d}\left[\sqrt{R} J_{n+1/2}(kR)\right]}{\mathrm{d}R} + ikB_n^{\text{in}}\left[\sqrt{R} J_{n+1/2}(kR)\right] \right\}. \tag{6.76}$$

Here it is taken into account that $P_n^{(1)}(\cos\theta)\cdot\sin^{-1}\theta = -\partial P_n^{(1)}(\cos\theta)/\partial\theta$ at $\theta = 0$ for all n. At $kR \gg 1$, the value $\sqrt{R} J_{n+1/2}(kR)$ is proportional to $\sin(kR + \alpha)$, where α is a certain constant, and the expression in the braces equals (to within the factor k) $A_n^{\text{in}} \cos(kR + \alpha) + iB_n^{\text{in}} \sin(kR + \alpha)$. At $A_n^{\text{in}} = B_n^{\text{in}}$ it is proportional to $\exp(ikR + i\alpha)$ and describes the wave propagating at $\theta = 0$ in the direction $-z$, that is, the incident wave; if $A_n^{\text{in}} \neq B_n^{\text{in}}$ the field would involve the addend $\exp(-ikR - i\alpha)$, that is, the back wave. At $\theta = 0$, the coordinate R coincides with z which completes the proof.

The components of the diffracted field depend on φ in the same way as those of the incident one, so that the potentials U^{diff} and V^{diff} can be expanded into a series similar to (6.75). The Bessel functions in these series should be replaced by the Hankel ones, as follows

$$U^{\text{diff}} = \cos\varphi \sum_{n=1}^{\infty} A_n^{\text{diff}} \frac{1}{\sqrt{R}} H_{n+1/2}^{(2)}(kR) P_n^{(1)}(\cos\theta) , \tag{6.77a}$$

$$V^{\text{diff}} = \sin\varphi \sum_{n=1}^{\infty} B_n^{\text{diff}} \frac{1}{\sqrt{R}} H_{n+1/2}^{(2)}(kR) P_n^{(1)}(\cos\theta) . \tag{6.77b}$$

The entire fields are expressed by the potentials $U^{\text{in}} + U^{\text{diff}}$ and $V^{\text{in}} + V^{\text{diff}}$. On the surface, at $R = a$, these components are related as

$$E_\theta = -w H_\varphi , \quad E_\varphi = w H_\theta . \tag{6.78}$$

Substituting the expression for the components of the potentials into the first of these relations, we obtain a functional equation valid for all θ. Equating independently the coefficients at $\partial P_n^{(1)}/\partial\theta$ and at $P_n^{(1)}/\sin\theta$ on both sides of this equation, we obtain the expressions for the coefficients A_n^{diff} and B_n^{diff}. The same coefficients are obtained when using the second relation of (6.78). The Rayleigh series for the diffraction problem on the impedance sphere is constructed using this method.

The expressions for A_n^{diff} and B_n^{diff} are interchanged when replacing $w \to w^{-1}$. This can be easily verified by comparison of (6.78) with the expressions for the fields through the potentials. At $w = 1$, we have $A_n^{\text{diff}} = B_n^{\text{diff}}$ for all n.

It follows from this equation that the back reflection is absent, because at the ray $\theta = 0$ the diffracted field decreases faster than R^{-1}, that is, the scattering pattern in this direction is zero. Verify this assertion for the component E_θ^{diff}. After substituting $A_n^{\text{diff}} = B_n^{\text{diff}}$ into the expression for this component at $\theta = 0$ we obtain, similar to (6.76),

$$E_\theta^{\text{diff}} = \frac{\cos\varphi}{R} \sum_{n=1}^{\infty} \left.\frac{\partial P_n^{(1)}}{\partial\theta}\right|_{\theta=0} \left\{ A_n^{\text{in}} \frac{\mathrm{d}[\sqrt{R} H_{n+1/2}^{(2)}(kR)]}{\mathrm{d}R} + ikB_n^{\text{in}}[\sqrt{R} H_{n+1/2}^{(2)}(kR)] \right\} . \tag{6.79}$$

At $kR \gg 1$, the expression $\sqrt{R} H_{n+1/2}^{(2)}(kR)$ is asymptotically proportional to $\exp(-ikR)$, and the values in the braces are zero for all n.

It turns out that, for the sphere, the terms in the back-scattered field created by the surface currents expressed by the potentials U and V, respectively, are mutually canceled.

In the conclusion of this section, we note that the mechanism of mutual cancellation providing the radar invisibility of the body, does not work for any deformation distorting the body symmetry with respect to the direction of the incident wave. Two other mechanisms, namely, the absorption in the material and the arising surface waves are less critical to deformations of the surface geometry and violation of the body orientation. They can be applied to bodies of practically any shape. This will be shown in the next section for the example of a cylinder with elliptical cross-section.

6.2.3
Impedance providing invisibility of the cylinder with arbitrary cross-section

6.2.3.1 Variable impedance

The sphere and the circular cylinder are the only the bodies for which the scattering pattern does not depend on their orientation with respect to the incident wave. Due to this fact the radar invisibility is provided by creating a constant (the same on the entire surface) impedance. As will be shown later, almost any body can be made invisible by creating an appropriate impedance on it. However, the impedance is variable, as it is different at different points of the boundary.

The fields induced on the entire surface of the body participate in creating the scattered field. Only at a higher frequency do the fields in a small neighborhood of the 'flash point' mainly contribute to this field. The location of this area depends on the body orientation. The impedance providing a small or zero reflection for any orientation of the body should be matched with its curvature in this area. Of course, it should be variable.

Let the body, for instance, be a 'half-cylinder', that is, the cylinder with a half-disk shaped cross-section. If the body is oriented in such a way that the wave falls onto the flat part of the surface, then for the reflection to be small (for a high frequency), the impedance of this part should be close to unity. If the wave falls onto the convex part of the surface, then the reflection is small if the impedance is close to that providing invisibility of the cylinder with the same curvature.

At middle and small frequencies the impedance distribution on the surface is defined not only by the curvature of a certain part of it. The scattering is an integral process, and the mathematical technique should correspond to its nature.

6.2.3.2 System of integral equations

The field induced on the surface of the body can be found if the impedance distribution is known. After this, the amplitude of the reflected wave can be calculated. The first problem is reduced to an integral equation for the field. The requirement (6.73) that the back wave is absent, leads to the second integral equation. Then all integrals are taken over the body surface. The mathematical problem for determining the impedance, which provides the invisibility, consists of these two equations with respect to the two unknown functions describing distributions of the field and the impedance on the surface. In fact, the field distribution is an auxiliary function of this problem. It depends on the body orientation.

We derive the respective formulas for the two-dimensional problem and restrict ourselves to the E polarization case. Let a wave with electric field having only the z-component, fall onto a cylinder of arbitrary cross-section with the directrix along the z-axis. The tangential components of the fields on the surface are connected by relation (6.28), in which the indices are omitted:

$$e = wh\,. \tag{6.80}$$

Here h is the only tangential component H_s of the magnetic field, $H_s = -i/k \cdot \partial E/\partial N$ (see (6.28)), s is the coordinate along the cross-section contour S; the normal $\vec{N}$ is assumed to be directed outside S. The orientation of the cylinder, with respect to the direction of the source of the incident wave, is characterized by the angle Ω between this direction and a fixed direction connected with the contour (e.g. the reference origin of the s-coordinate). The functions to be found are $h(s, \Omega)$ and $w(s)$.

We temporarily assume that the source of the incident field E^{in} (and, consequently, of the total diffraction field E) is located at the point $\vec{r}_0$ at a finite distance from the body, and this source is described by the δ-function. This assumption simplifies the derivation of the equation and, as will be shown, does not restrict its generality. Everywhere outside S the field E satisfies the equation

$$\Delta E + k^2 E = \delta(|\vec{r} - \vec{r}_0|)\,, \tag{6.81}$$

the radiation condition, and the boundary condition (6.80) on S. The incident field E^{in} satisfies the same equation, the radiation condition, and it is analytical in the entire plane $z = const.$

We introduce the Green function (6.60) with the source at the point $\vec{r}\,'$, located outside S. It satisfies equation (6.61). From (6.81) and (6.61) it follows that

$$G \cdot \Delta E - E \cdot \Delta G = G \cdot \delta(|\vec{r} - \vec{r}_0|) - E \cdot \delta(|\vec{r} - \vec{r}\,'|)\,. \tag{6.82}$$

The quantities $E(\vec{r}, \vec{r}_0)$, $E^{\text{in}}(\vec{r}, \vec{r}_0)$, and $G(\vec{r}, \vec{r}\,')$ are functions of the first argument; the second is a parameter, it describes the location of the 'source' of these functions. We integrate (6.82) over the entire domain outside S. Since the both functions E and G satisfy the radiation condition, the integral over the remote circle is zero, and only the integral over S remains,

$$\int \left(E \frac{\partial G}{\partial N} - G \frac{\partial E}{\partial N} \right) \mathrm{d}s = G(\vec{r}_0, \vec{r}\,') - E(\vec{r}\,', \vec{r}_0)\,. \tag{6.83}$$

Here it is taken into account that the outward normal to the integration domain is $-\vec{N}$. In the case when the body is absent, the left-hand side of (6.83) is zero. Since the Green function is invariant with respect to interchanging of its arguments, and $E(\vec{r}\,', \vec{r}_0) = E^{\text{in}}(\vec{r}\,', \vec{r}_0)$ in this case, we have

$$E^{\text{in}}(\vec{r}\,', \vec{r}_0) = G(\vec{r}\,', \vec{r}_0)\,. \tag{6.84}$$

Besides, this equality follows immediately from the definitions of the both functions.

Equality (6.83) can be rewritten in the form

$$E(\vec{r}\,', \vec{r}_0) = E^{\text{in}}(\vec{r}\,', \vec{r}_0) + \int \left[-e(s, \vec{r}_0) \frac{\partial G(s, \vec{r}\,')}{\partial N} + ikh(s, \vec{r}_0) G(s, \vec{r}\,') \right] \mathrm{d}s\,. \tag{6.85}$$

With accuracy to a nonessential factor, the surface value of the magnetic field h equals the density of the surface electric field. The surface value of the field e can be called the '*magnetic field*'. In these terms, relation (6.85) means that the field arising at diffraction on an impedance surface is the sum of the incident field and the fields of the two currents induced on the surface: the magnetic field (the first summand under the integral) and the electric field (the second summand under the integral). Both currents radiate as in a vacuum, because the function G involved in (6.85) is the Green function of free space.

Note that a Green function which is more complicated than (6.60) can be introduced by subordinating it to the condition $G = 0$ or $\partial G/\partial N = 0$ on the contour S. Then only one surface-induced current, the magnetic or electric, will participate in creating the field $E(\vec{r}\,', \vec{r}_0)$, in the first and second case, respectively. However, these currents will not radiate in free space, but in the presence of the body, on which their radiation is diffracted. The integrand will contain only one summand. Since the equality (6.84) is not valid for such a Green function, the first summand on the right-hand side of (6.85) should be kept in the form of the function G.

The considerations given after (6.85) have only a methodical sense, we now return to our problem.

In order to obtain the first of the mentioned equations for the functions $w(s)$ and $h(s, \Omega)$, we apply relation (6.85) to the case when the point $\vec{r}\,'$ reaches the point s' on the contour C. In this case the first summand in the integral obtains a strong singularity at $s = s'$. As the double layer potential, the integral obtains the additional summand $-e(s')/2$, and the rest of the integral should be considered in the principal value sense. As a result, we have

$$\frac{1}{2} w(s') h(s', \Omega) = e^{\text{in}}(s', \Omega) + \int h(s, \Omega) \left[ik G(s, s') - w(s) \frac{\partial G(s, s')}{\partial N} \right] ds \,. \quad (6.86)$$

Here relation (6.80) is taken into account, and the argument $\vec{r}_0$ is replaced by the orientation angle Ω, $0 \le \Omega < 2\pi$.

It is easy to check that the limiting values of the function $\partial G(s, \vec{r}\,')/\partial N$ in the integrand are finite and the same on both sides of point s'. This fact is especially important for numerical calculations.

The expression for the amplitude of the field scattered in the direction of the source is obtained from (6.85) after substituting $\vec{r}\,'$ with $\vec{r}_0$. The difference $E(\vec{r}\,', \vec{r}_0) - E^{\text{in}}(\vec{r}\,', \vec{r}_0)$ is the diffracted field at the point $\vec{r}\,'$, created by the source located at $\vec{r}_0$. If these points coincide, then this difference is the field scattered in the opposite direction to the incident wave. If the point $\vec{r}_0$ is located in the far-field zone ($r_0 \gg ka^2$), then the incident field is a plane wave, and the scattered field is a cylindrical wave. The latter has the form $F^w \exp(-ikr) \cdot (kr)^{-1/2}$, where F^w is the amplitude of the scattered wave. In this definition, the amplitude is normed in a different way from that in Section 6.2.2.3. Instead of the assumption that the incident field near the cylinder is the plane wave (6.29) of the unit amplitude, here we assume that this field is created by the source of the unit amplitude (see (6.81)) and the value of F^w depends on $\vec{r}_0$. Also, we consider only (6.73) or the ratio F^w/F^0, where F^0 is the amplitude of the wave scattered by the metallic body of the same geometry, and hence the normalization of F^w is nonessential.

The radar invisibility condition (6.73) means that the integral addend in (6.85) is zero. In this addend, the Green function G can be replaced by E^{in} due to (6.84) and $\partial G/\partial N$ by ikH^{in} due to (6.28). Then the required condition takes the form

$$\int \left[w(s)h(s,\Omega)h^{\text{in}}(s,\Omega) - h(s,\Omega)e^{\text{in}}(s,\Omega)\right] ds = 0 \tag{6.87}$$

for all Ω, $0 \le \Omega < 2\pi$. Equalities (6.86) and (6.87) make up the equation system for the variable impedance $w(s)$ providing the radar invisibility of the body for any orientation.

After the function $h(s)$ has also been determined from this system, the visibility factor $|F^w/F^0|$ introduced in Section 6.2.2.3 can be calculated. In order to obtain the expression for this factor, we consider the diffracted fields E^0 and E, satisfying the nonhomogeneous wave equation (6.81) and the difference fields $E^0 - E^{\text{in}}$ and $E - E^0$, satisfying the corresponding homogeneous equation. Combining the equations for E^0 and $E^0 - E^{\text{in}}$ (similar to when deriving equation (6.82)), and integrating over the domain exterior to S, we obtain

$$E^0(\vec{r}_0,\vec{r}_0) - E^{\text{in}}(\vec{r}_0,\vec{r}_0) = ik\int e^{\text{in}}h^0 ds\,; \tag{6.88}$$

here $e^0 = 0$, $\partial E^0/\partial N = ikh^0$ on S is assumed. The left-hand side of (6.88) is the field scattered from the metallic body. It is proportional to F^0. In the same way, from the equations for E and $E - E^0$ we obtain

$$E(\vec{r}_0,\vec{r}_0) - E^0(\vec{r}_0,\vec{r}_0) = -ik\int eh^0 ds\,. \tag{6.89}$$

The left-hand side of (6.89) is proportional to $F^w - F^0$. After dividing (6.88) by (6.89), we obtain the required expression for the visibility factor for a cylinder of arbitrary cross-section

$$\left|\frac{F^w}{F^0}\right| = 1 - \frac{\int eh^0 ds}{\int e^{\text{in}}h^0 ds}\,, \tag{6.90}$$

where $e = wh$.

Relation (6.90) is valid for any frequency. Of course, $|F^w/F^0|$ equals zero for the frequency, at which the functions $w(s)$, $h(s,\Omega)$, and $h^0(s,\Omega)$ are determined from the above integral equations. For other frequencies this coefficient, calculated at the same function $w(s)$, differs from zero and depends on the body orientation.

The results of this subsection can easily be extended onto the three-dimensional vector problem. In this problem the fields do not satisfy the wave equations but the system of Maxwell equations. The Green function is the solution to this system for the field created by an elementary dipole, either electric or magnetic. The Green theorem used for transformation of the integral over the area into that over the contour, should be replaced by the Lorentz lemma in integral form. The system obtained will consist of two equations of the type (6.86) and (6.87), with integration over the body surface. The derivations arriving at this system are similar to those leading to (6.86) and (6.87), but they are more cumbersome.

6.2.3.3 Numerical solution of the integral equation system

Integral equations (6.86) and (6.87) are nonlinear, because they involve the product of the functions $w(s)h(s, \Omega)$. They are also nontypical: the integration in them is made over the one-dimensional closed contour S, whereas one of the unknown functions, namely, $h(s, \Omega)$ depends on two variables. This type of equation does not often occur in the literature, and they are likely not investigated theoretically. Even the question about the existence of solutions to this type of equation is apparently still open.

If the system of equations (6.86) and (6.87) has no solution, then the complex impedance providing zero back-scattering does not exist. In this case the problem should be replaced by an optimization one consisting, for instance, in minimization of the functional

$$L_1(w, h) = \int_{\Omega} \left[\int_S w(s)h(s, \Omega)H^{\text{in}}(s, \Omega)\text{d}s - \int_S h(s, \Omega)E^{\text{in}}(s, \Omega)\text{d}s \right]^2 \text{d}\Omega \tag{6.91}$$

with respect to the functions $w(s)$, $h(s, \Omega)$ satisfying the nonlinear equation (6.86).

Even if the equation system (6.86) and (6.87) is solvable, it is obviously ill-posed because (6.86) being considered as a linear equation with respect to $w(s)$ (when $h(s, \Omega)$ is known) has a smooth bounded kernel. It is known that solutions to this type of integral equation are unstable (see, e.g. [41]). Additional computational difficulties arise at resonant frequencies of the interior domain bounded by S, and in their neighborhoods. All these problems require developing special methods for solving the equation system (6.86) and (6.87).

For the circular cylinder, the system can be solved by an iterative method, at each step of which first the integral equation (6.86) is solved with respect to $h(s, \Omega)$ at w taken from the previous iteration, and then the next approximation to w is calculated explicitly from (6.87). As an initial approximation for w we can take $w = 1$. The process converges at any frequency. Of course, the results coincide with those obtained in Section 6.2.2.3 by the separation of variables.

Unfortunately, this simple iterative method does not converge when applying to problems with complicated body shapes. For the case of an elliptical cylinder with the half-axes a and b, its second substep was modified. Instead of solving (6.87) at a given $h(s, \Omega)$, we calculate the next approximation to $w(s)$ by minimizing the sum of the mean-square discrepancies of (6.86) and (6.87). Namely, the functional to be minimized in this substep has the form

$$L(w) = L_1(w) + L_2(w) \tag{6.92}$$

where $L_1(w)$ coincides with $L_1(w, h)$ given in (6.91) (with $h(s, \Omega)$ calculated in the first substep) and

$$L_2(w) = \int_S \int_\Omega \left\{ \frac{1}{2} w(s) h(s, \Omega) - E^{\mathrm{in}}(s, \Omega) - \int_S h(s', \Omega) \left[ik G(s, s') - w(s') \frac{\partial G(s, s')}{\partial n_{s'}} \right] \mathrm{d}s' \right\}^2 \mathrm{d}\Omega \, \mathrm{d}s \,. \quad (6.93)$$

The minimization problem for the functional $L(w)$ is reduced to a linear integral equation with respect to $w(s)$. The derivations are simple but the result is awkward and is not given here. The initial approximation for $w(s)$ can be chosen as the known solution of system (6.86) and (6.87) for a body with similar geometry. For instance, in the case of an elliptical cylinder, such a body can be a cylinder with the previous ratio of the axes, starting from $b/a = 1$.

This algorithm does not converge unconditionally either. This is because its first substep (solving (6.87) with respect to $h(s, \Omega)$ at given $w(s)$) does not necessarily decrease the value of the functional $L(w)$. For the example considered here the convergence was observed at frequencies that differ from the resonant ones by more than 15% of the distance between them, and at the axes ratio b/a decreasing from 1 to 0.2. Developing an effective algorithm for nonlinear equation systems of the type (6.86) and (6.87) is an important computing problem.

6.2.3.4 Cylinder of elliptical of cross-section

In this subsection we present the numerical results for the infinite cylinder with elliptic cross-section obtained by solving the integral equation system (6.86), (6.87). We consider the cylinder with an axes ratio b/a varying from 1.0 (circular cylinder) to 0.2 (almost the band-shaped one) in the frequency range from $ka = 1.5$ to $ka = 3.9$ (a is the large half-axis). This range is located between the first and second resonant frequencies of all ellipses considered. This fact provides the convergence and stability of the iterative procedure described above.

Figure 6.10 shows the distribution of the real (a) and imaginary (b) parts of $w(\varphi)$ providing zero back reflection for elliptic cylinders with different axes ratio at fixed frequency $ka = 2.3$. The direction $\varphi = 0$ coincides with the half-axis a; due to the symmetry, the results are given for the first quadrant only.

At $b/a = 1$ the values of w', w'' are constant, independent of φ. They coincide with those from Figure 6.5 corresponding to this frequency; in particular, $w'' < 0$. However, already at $b/a = 0.8$ (which corresponds to the eccentricity $\varepsilon = 0.6$), w'' becomes positive in the part of the contour adjoint to the ends of the large axis. It remains negative only for $\varphi > 0.3\pi$ (all numbers are approximate). The real part of the impedance is smaller than that for the circular cylinder at $|\varphi| < 0.2\pi$. At larger angles (closer to the ends of the small axis), w' is essentially larger. This effect is apparent for cylinders with larger eccentricities of cross-section. For $b/a = 0.4$ ($\varepsilon = 0.92$), w'' is negative only at $\varphi > 0.38\pi$, that is, near the ends of the small axis where w' equals several units.

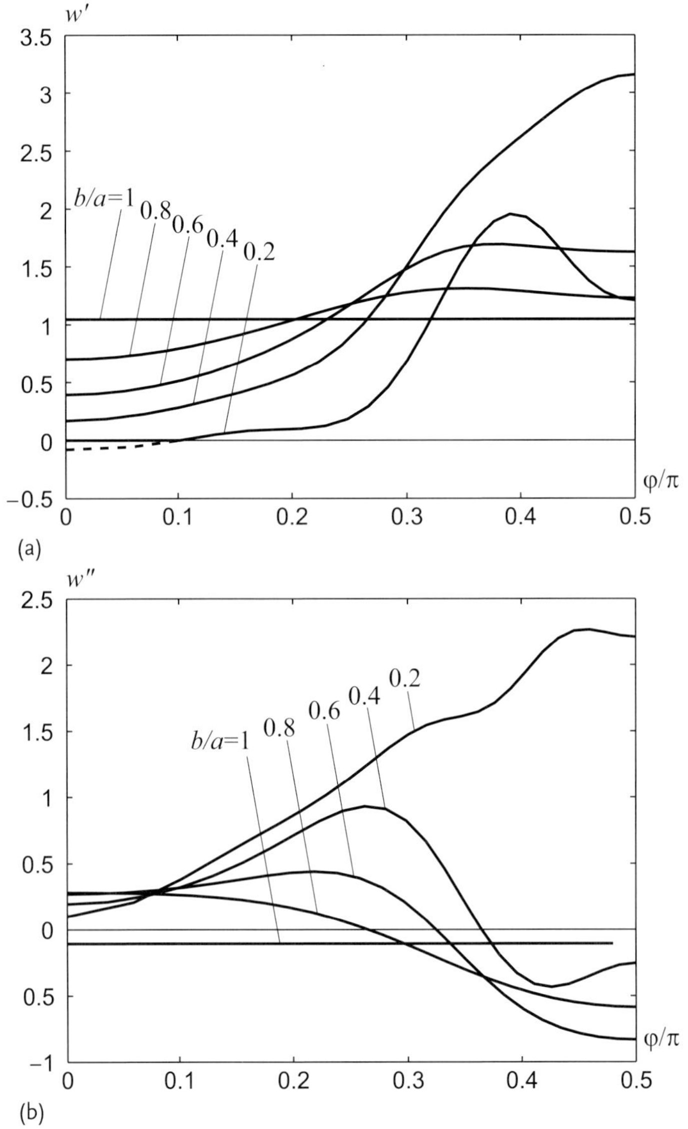

Figure 6.10 Optimal impedance distribution for elliptical cylinder with different axes ratio; $ka = 2.3$; E polarization case.

The processes providing the absence of back reflection for the elliptic cylinder are more complicated than those for the circular one. In particular, they depend on the cylinder orientation and consist in the surface waves arising on segments with $w'' < 0$ and their fast radiation at segments with $w'' > 0$.

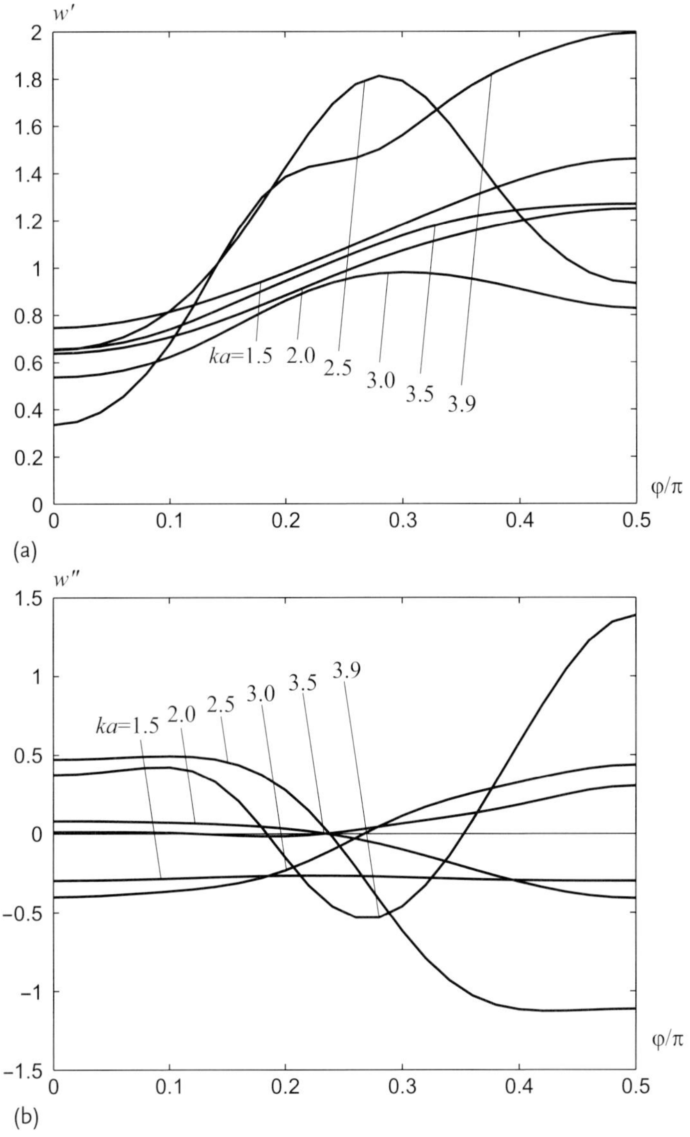

Figure 6.11 Optimal impedance distribution for elliptical cylinder of different frequencies at $b/a = 0.8$; E polarization case.

The very oblate cylinder (with an eccentricity of about 0.98) cannot be made invisible for all orientations. In this case the solution to the above equation system has $w'(s) < 0$ for the fragments of the contour adjoint to the ends of the large axis (see the dotted segment at $\varphi < \pi/10$ of the corresponding line in Figure 6.10a). As was shown above, such an impedance cannot be realized. The reflection from the 'band' can be made small, but it cannot be fully annihilated for all orientations.

The simplest (but not the best) way to deal with this situation is to set $w' = 0$ on these parts of the surface. In fact, we should not solve (6.86) and (6.87) exactly in this case, but find $w = w' + iw''$ as a function minimizing the discrepancy of (6.87) with the two constraints: equation (6.86) and the condition $w' \geq 0$.

If the impedance is realized by the goffer (azimuthal for this polarization), then the function w'' can be easily realized by making its depth variable in accordance with (6.22). The function w' is determined by the losses in material filling the goffer, and it can also be varied by choosing its values in different parts of the surface and its electrodynamic parameters.

The numerical results obtained for different frequencies at fixed axes ratio $b/a = 0.8$ are given in Figure 6.11. The two first resonant frequencies of the interior domain with this boundary are $k_1 a = 2.72$, $k_2 a = 4.09$. At frequencies close to them ($ka = 2.5$ and $ka = 3.9$) the impedance providing invisibility is distributed very nonuniformly on the surface. However, the impedance is noticeably varying on the surface at other frequencies as well.

The proposed technique can be extended to the case when the protection should be provided against multi-directional detection. In this case, instead of solving (6.87), one should minimize the sum (or integral) of the squared modulus of the scattered wave amplitude over the selected directions. Of course, this amplitude cannot be made equal to zero in all these directions, including the backward direction.

6.2.4
Matched impedance. Invisible screen

If the body orientation is fixed, then the scattering amplitude can be made small by creating the impedance matched with the field of the incident wave on a part of the surface. The matching consists in equating the ratio of the fields $\vec{e}$ and $\vec{h}$ defined by the impedance with that of the fields $\vec{e}^{\,\text{in}}$ and $\vec{h}^{\,\text{in}}$ in the incident wave. If such an impedance could be created on the entire surface of the body, then at the incidence of this wave, the scattered field would not arise at all.

In this sense, the impedance $w = 1$ on the plane is matched with the field of the normally falling plane wave because the ratio of the fields E^{in} and H^{in} equals unity for this wave. The reflection does not arise at the incidence of such a wave onto the plane. The matching considered in this subsection is a generalization of this elementary case on arbitrary bodies and arbitrary incident waves.

If the impedance is matched with the field of the incident wave, then the scattered field is equal to zero. The fields $\vec{e}$, $\vec{h}$ are the sums of the fields $\vec{e}^{\,\text{in}}$ and $\vec{h}^{\,\text{in}}$ (i.e. the fields which would exist on the surface S in the absence of the body) and the fields $\vec{e}^{\,\text{diff}}$ and $\vec{h}^{\,\text{diff}}$ of the scattered wave. The two last fields satisfy the homogeneous Maxwell equations and the radiation condition. On the surface, $\vec{e}$ and $\vec{h}$ are connected by the condition

$$\vec{e} - \left[\vec{N} \cdot w\vec{h}\right] = 0 \tag{6.94}$$

equivalent to (6.3). Here $\vec{N}$ is the unit vector of the outward normal to S, the impedance w can be a tensor and, in general, it is variable and dependent on the point on the surface, $w = w(s)$.

This condition can be rewritten in the form

$$\vec{e}^{\,\mathrm{diff}} - \left[\vec{N} \cdot w\vec{h}^{\,\mathrm{diff}}\right] = -\left\{\vec{e}^{\,\mathrm{in}} - \left[\vec{N} \cdot w\vec{h}^{\,\mathrm{in}}\right]\right\} . \qquad (6.95)$$

By the above definition, the impedance w is matched with the incident field if it satisfies the equation

$$\vec{e}^{\,\mathrm{in}} - \left[\vec{N} \cdot w\vec{h}^{\,\mathrm{in}}\right] = 0 . \qquad (6.96)$$

If the impedance is matched on the entire surface, then, according to (6.95), the scattered field satisfies the boundary condition

$$\vec{e}^{\,\mathrm{diff}} - \left[\vec{N} \cdot w\vec{h}^{\,\mathrm{diff}}\right] = 0 . \qquad (6.97)$$

The homogeneous equation system for E^{diff} and H^{diff} with the homogeneous boundary condition, the radiation condition and the additional demand (6.7) (or analogous one for the tensor impedance) has only zero solution.

The reference to requirement (6.7) is necessary. There is a countable set of the functions $w_n(s)$, $n = 0, 1, \ldots$ for which the field in the infinite domain outside a closed surface on which condition (6.97) holds, does not equal zero. The corresponding fields E_n^{diff}, H_n^{diff} are the eigenfunctions of the homogeneous problem, describing the eigenoscillations of the infinite domain; they make up a complete system of the vector fields [33]. We call such fields the *generalized eigenoscillations*. All the functions $w_n(s)$ have their negative real part on at least a part of the surface. In the eigenoscillations, the energy outgoes from the body to infinity. The set of such functions is convenient when solving diffraction problems. However, these functions are not realizable, if the surface S is not a *laser* one, that is, if its impedance satisfies condition (6.7).

This condition also makes impossible the fulfilment of equalities (6.96), that is, the matching of the impedance on the entire surface simultaneously.

If the impedance $w(s)$ is matched with the incident field, that is, condition (6.96) holds, then the normal component of the Poynting vector in the absence of the body is

$$S_N = -\frac{c}{8\pi} \operatorname{Re}\left(w\vec{h}^{\,\mathrm{in}} \cdot \vec{h}^{\,\mathrm{in}*}\right) . \qquad (6.98)$$

It has the same sign as the value $-w'$ and an analogous one for the tensor impedance. There are no bodies or sources inside S, hence, the integral of S_N over the entire surface (the flux of $\vec{S}$ through the surface) equals zero. On a part of the surface (where $S_N < 0$, i.e. 'energy flows into the volume') we have $w' > 0$, and on the other part (where 'energy flows out of the volume') $w' < 0$. The impedance of the body can be matched with the incident field only on the part of the surface where $S_N < 0$. Keeping in mind the geometro-optical limit, this part can be conveniently called the illuminated part. The other part where $S_N > 0$ can be called the shaded one. Matching of the impedance is possible only on the illuminated part of the body. This noncomplete matching does not provide an absence of scattering; however, it does lead to its essential decrease.

We illustrate this assertion on the model example of diffraction on an infinite cylinder. In this case we can use the explicit expressions for back-scattering. According to (6.85), the left-hand side of (6.86) is proportional to the amplitude of the reflected wave, that is

$$F^w \sim \int_S h(s)\left[w(s)h^{\text{in}}(s) - e^{\text{in}}(s)\right] \mathrm{d}s\,. \tag{6.99}$$

Here the dependence on the orientation is omitted, since the impedance is found with the assumption that the orientation is fixed. If $w(s)$ is found from (6.96) having the form $e^{\text{in}} - wh^{\text{in}} = 0$ in the scalar case, then the kernel of this equation equals zero on the illuminated part of the surface, and the integral is taken only over the shaded part. If $w = 0$ on this part (metal), then

$$F^w \sim \int_{S_{\text{shad}}} h(s)e^{\text{in}}(s)\mathrm{d}s\,. \tag{6.100}$$

According to (6.99), radiating in free space, the surface 'currents' $h^{\text{in}}(s)$ and $e^{\text{in}}(s)$ create the scattered field (see the text after (6.85)), that is, the waves radiated by them do not undergo diffraction on the body. The absence of the integral over the illuminated part in F^w means that the fields created by both 'currents' cancel each other in the direction of the incident field source. This relates also to all other scattering directions. The back-scattering is created only by the electric 'current' on the shaded side. The assertion that it is small in comparison with the reflection from the metallic body, that is, with

$$F^0 \sim \int_S h(s)e^{\text{in}}(s)\mathrm{d}s\,, \tag{6.101}$$

is based on the fact that the current induced on the shaded part of the body boundary is small in comparison with that induced on its illuminated part.

The scattering remains small also for small variations of the orientation with respect to the orientation for which the matched impedance was calculated.

The absence of scattering from the body with matched impedance can be explained without reference to the formulas. We may apply the logical scheme similar to that used when explaining the Archimedean expulsive force that acts on the body submerged in a fluid. The pressure of the fluid onto the body boundary does not depend on the material filling the volume. The fluid presses onto the boundary in exactly the same way as it pressed onto the fluid filling this volume before the body was submerged. The pressure equals the weight of the fluid and compensates it.

The fields $\vec{e}$ and $\vec{h}$ in the incident wave on the virtual boundary of the body that will be 'submerged' in this field do not depend on what is located on the 'other side' of this boundary (surface). Before submerging the body, the tangential fields on 'that side' were the same, and the scattered field was absent. The fields on the boundary of the submerged body are the same. This assertion is valid, since, due to the definition of the quantity 'matched impedance', the ratio $\vec{e}/\vec{h}$ (tensor) on 'that side' is the same as before. In other words, the hypothesis that $\vec{e}$ is continuous, leads to the fact that $\vec{h}$ is continuous as well (a consequence of the matching), hence both fields are continuous and scattering is absent.

Consider equation (6.96) for w. It should hold on the 'illuminated' part of the surface, and its solution satisfies condition (6.7) there. According to (6.96), $\vec{e}^{\,\mathrm{in}}$ is perpendicular to $w\vec{h}^{\,\mathrm{in}}$. If the vectors $\vec{e}^{\,\mathrm{in}}$, $\vec{h}^{\,\mathrm{in}}$ are not perpendicular, then w should be a tensor, that is, the impedance should be anisotropic. The degree of anisotropy depends on the angle $\alpha \leq \pi/2$ between these vectors.

We next find an expression for the minimal anisotropy. Let the axes x, y of the local coordinate system be directed along the principal axes of the tensor w. We denote its principal values by w_1 and w_2, $w_1/w_2 \leq 1$. According to (6.18), the components of the fields $\vec{e}^{\,\mathrm{in}}$, $\vec{h}^{\,\mathrm{in}}$ are connected by the relations

$$e_x^{\mathrm{in}} = -w_1 h_y^{\mathrm{in}}\,, \quad e_y^{\mathrm{in}} = w_2 h_x^{\mathrm{in}}\,. \tag{6.102}$$

We define the orientation of the axes x, y with respect to the given vectors $\vec{e}^{\,\mathrm{in}}$, $\vec{h}^{\,\mathrm{in}}$ by the angle β between the x-axis and the vector $\vec{e}^{\,\mathrm{in}}$. Then the angle between the y-axis and the vector $\vec{h}^{\,\mathrm{in}}$ is $\beta - \alpha$. Relations (6.102) lead to the following equation system:

$$e^{\mathrm{in}} \cos\beta = -w_1 h^{\mathrm{in}} \sin(\beta - \alpha)\,, \tag{6.103a}$$

$$e^{\mathrm{in}} \sin\beta = w_2 h^{\mathrm{in}} \cos(\beta - \alpha)\,. \tag{6.103b}$$

At $\alpha = \pi/2$ the obvious result follows from this system: $w_1 = w_2 = e^{\mathrm{in}}/h^{\mathrm{in}}$, and the angle β can be arbitrary. The ratio w_1/w_2 is real at any α.

Therefore the three quantities w_1, w_2, and β are connected by two equations (6.103). The angle β can be found if the ratio w_1/w_2 is given. This ratio should be smaller than a certain value dependent on α. If this requirement is violated, then the angle β calculated from (6.103) becomes complex. We find this limiting value of w_1/w_2.

From (6.103) it follows that β satisfies the equation

$$\frac{w_1}{w_2} \tan\beta \cdot \tan(\beta - \alpha) = -1\,. \tag{6.104}$$

It is the quadratic equation with respect to $\tan\beta$. Its roots are real if the discriminant $D = \tan^2\alpha \cdot (w_2/w_1 - 1)^2 - 4w_2/w_1$ is positive. It is positive if w_2/w_1 is smaller than the smallest root of the equation $D = 0$, considered as the equation with respect to w_2/w_1. Its root is $\tan^2(\alpha/2)$. Therefore, if $\alpha \neq \pi/2$, then the ratio of the principal values of the matched tensor impedance should be smaller than $\tan^2(\alpha/2)$.

It is convenient to characterize the anisotropy of the impedance by the ratio of the difference of its principal values to their sum. If the impedance solves (6.96),

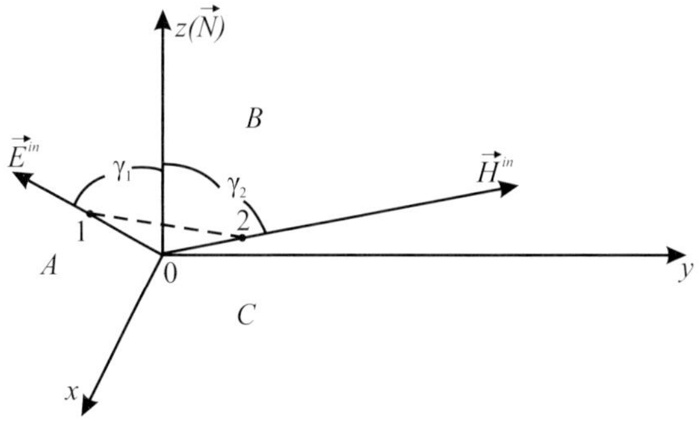

Figure 6.12 Checking relation (6.106).

then the anisotropy fulfills the condition

$$\frac{w_2 - w_1}{w_2 + w_1} \geq \cos\alpha \,. \tag{6.105}$$

The minimal anisotropy equals $\cos\alpha$. It is realizable if the principal axes of the tensor are directed along the bisectors of the angles between $\vec{e}^{\,\text{in}}$ and $\vec{h}^{\,\text{in}}$, that is, if $\beta = \alpha/2$ and $w_1/w_2 = \tan^2(\alpha/2)$.

The inverse problem may consist not in finding the properties of the surface of the body on which the diffraction occurs, but in finding its shape. The screen protecting a certain body against radar, which has a fixed location, should have the impedance matched with the field of the wave falling onto it. The plane surface with $w = 1$ is such a screen, but is not unique. In fact, the screen can have any shape. It does not create the scattered field if its impedance is matched. The shape of the screen can be chosen in such a way that the impedance is isotropic. For this purpose, according to (6.105), it is necessary for the projections of the vectors $\vec{E}^{\text{in}}$, $\vec{H}^{\text{in}}$ onto the tangent plane (i.e. the vectors $\vec{e}^{\,\text{in}}$, $\vec{h}^{\,\text{in}}$) to be perpendicular at each point of the surface. Such a surface can be constructed (and not uniquely) for any two three-dimensional vectors. It is sufficient to take any plane containing the vector $\vec{E}^{\text{in}}$ (the plane *A* in Figure 6.12) and then project the vector $\vec{H}^{\text{in}}$ onto it and construct the second plane (plane *B*) containing the vector $\vec{H}^{\text{in}}$ and its projection. The planes *A* and *B* are orthogonal. The plane perpendicular to the line of intersection of the the planes *A* and *B* (the plane *C*) is the required one, since the projections of both vectors onto it are orthogonal.

We denote by γ_1 and γ_2 the required angles between the normal to the plane *C* and the vectors $\vec{E}^{\text{in}}$, $\vec{H}^{\text{in}}$, respectively, and by δ, the angle between these two vectors. These three angles are connected by the relation

$$\cos\gamma_1 \cos\gamma_2 = \cos\delta \,. \tag{6.106}$$

It follows from this equality that $\gamma_1 \leq \delta$, $\gamma_2 \leq \delta$, and $\gamma_1 + \gamma_2 > \delta$. Two angles are connected by only one condition. This means that there exists a one-parameter set of planes tangent to the required surface at the given point. The question about the existence of such a surface in the general case has probably not yet been investigated.

In order to check relation (6.106), we introduce the Cartesian coordinate system originating at point 0 with the axes x, y, z directed along the lines of intersection of the planes A and C, B and C, and A and B, respectively. Then $z = 0$ on the required plane, the vector $\vec{E}^{\rm in}$ lies in the plane $y = 0$, and $\vec{H}^{\rm in}$ lies in the plane $x = 0$. We mark off the unit segments in the planes A and B, starting at the point O and parallel to these vectors. The coordinates of their ends (points 1 and 2 in Figure 6.12) are: $x_1 = \sin\gamma_1$, $y_1 = 0$, $z_1 = \cos\gamma_1$; $x_2 = 0$, $y_2 = \sin\gamma_2$, $z_2 = \cos\gamma_2$. The squared length of the segment between these points is $\sin^2\gamma_1 + \sin^2\gamma_2 + (\cos\gamma_1 - \cos\gamma_2)^2$. However, this segment is the base of the isosceles triangle with the angle δ at the top, and its squared length is $2(1 - \cos\delta)$. Equating these two values gives (6.106).

Of course, the examples given in this subsection do not exhaust all the possibilities of 'handling' (using impedance surfaces) the properties of the fields arising by diffraction on the bodies located in free space. Similar problems concerning diffraction on the impedance surfaces bounding open volumes will be considered in the next subsection.

6.2.5 Waveguides and resonators with impedance walls

6.2.5.1 Regular waveguides

The process of wave diffraction is 'momentary'. Having no exact definition, this term describes an intuitive meaning about the 'collision' of the electromagnetic wave with an obstacle. In a similar nonmathematical sense, the propagation of waves in the waveguide is a 'continuous' process. When propagating, the waves collide with the walls. As a results of their multiple interaction, the notion of the *eigenwave* (*waveguide mode*) as a key object of waveguide theory, arises. Such waves exist in any regular (i.e. uniform along its entire length) waveguide. The relation between the field components in the mode is constant, the *complex amplitude* is common for all components and depends on the z-coordinate along the waveguide as $\exp(-ihz)$, $h = h' + ih''$.

In this section, certain results of regular waveguide theory (see, e.g. [36]), necessary for formulation of the inverse problems considered here, are summarized.

All components of the mode are represented by the two scalar functions $\Phi(x, y)$, $\Psi(x, y)$ of the transverse coordinates. Both functions satisfy the same equation

$$\Delta\Phi + \alpha^2\Phi = 0\,, \tag{6.107a}$$

$$\Delta\Psi + \alpha^2\Psi = 0\,, \tag{6.107b}$$

where Δ is the two-dimensional Laplace operator. On the contour S of the cross-section they satisfy the boundary conditions

$$\alpha^2\Phi = -w_1\left(-ih\frac{\partial\Psi}{\partial s} + ik\frac{\partial\Phi}{\partial N}\right)\,, \tag{6.108a}$$

$$-ih\frac{\partial \Phi}{\partial s} - lk\frac{\partial \Psi}{\mathrm{d}N} = w_2\alpha^2\Psi\,. \qquad (6.108b)$$

Here s is the coordinate along the contour S, N is the interior normal to S. If S is the circle (circular waveguide), then $\vec{N} = -\vec{r}$, $\vec{s} = a\vec{\varphi}$, where a is the circle radius. Conditions (6.108) are the boundary conditions $e_z = -wh_\varphi$, $e_\varphi = w_2h_r$ written by the functions Φ, Ψ and their derivatives. The relation between the tangential components of the fields is the same (6.18) as in the problem about the plane, only the Cartesian system (x, y, z) transforms into (z, s, N).

The numbers α^2 are the eigenvalues of the homogeneous problem (6.107), (6.108a). The *wave numbers* h are connected with the 'frequency' k and the numbers α^2 by the relation $h^2 = k^2 - \alpha^2$. The quantities α^2 make up a countable set of complex values α_j^2, $j = 0, 1, 2, \ldots$, to each of them one or several pairs of the 'eigenfunctions' $\Phi^j(x, y)$, $\Psi^j(x, y)$ and 'eigenwave numbers' h_j correspond. Usually, the dependence of these functions on each of the transverse coordinates are characterized by one number, n or q, so that j is an aggregate of the two numbers $j \equiv \{n, q\}$.

In the waveguide with metallic walls, that is, at $w_1 = w_2 = 0$, the boundary conditions (6.108) take the form

$$\Phi = 0\,, \qquad (6.109a)$$

$$\frac{\partial \Psi}{\partial N} = 0\,, \qquad (6.109b)$$

in which the both functions are not mutually connected. This means that the problems (6.107a), (6.109a) and (6.107b), (6.109b) are independent. The set of numbers α_j^2 splits into two subsets, one of which contains the eigenvalues of the first problem, while the second one contains the eigenvalues of the second problem. For the modes of the first problem, $\Psi^j \equiv 0$ which leads to $H_z = 0$. Similarly, for the modes of the second problem $\Phi^j = 0$ which leads to $E_z = 0$. The modes of the first subset are called E modes (or TH modes), the modes of the second subset are called H modes (or TE modes). All the eigenvalues α_j^2 for the metallic waveguides are real and positive. Several of them are smaller than k^2 and for these $h_j^2 > 0$; they are called the *traveling modes*. The rest eigenvalues are larger than k^2, so that $h_j^2 < 0$; such modes are called *damping modes*.

6.2.5.2 Circular waveguides with an azimuthal goffer

Here we consider two problems related to 'circular' waveguides, that is, waveguides with a circular cross-section. These problems are solved by substituting smooth metallic walls by walls with an azimuthal goffer. Such a goffer connects only the components e_z and h_φ; the boundary conditions (6.108) with $w_2 = 0$ are fulfilled on it.

The geometrical parameters of the goffer determine the impedance w_1. The field in the goffer grooves on the cylindrical surface is described not by the trigonometric functions as in (6.20), but by the cylindrical Bessel and Neumann functions in which the coordinate $-r$ occurs instead of z. The ratio of the tangential components averaged over the period are given by an expression similar to (6.22) with the ratio of the cylindrical functions instead of the trigonometric ones. If ε and μ are real, then this ratio is imaginary.

Strictly speaking, this ratio is not the impedance, since the field in the grooves depends on the angular coordinate φ, exactly as the field outside the goffer. If this field is proportional either to $\cos(n\varphi)$, $n = 0, 1, 2, \dots$ or to $\sin(n\varphi)$, $n = 1, 2, \dots$, then the index n is in the functions J_n, N_n, and the ratio of the fields depends on n. However, below we keep the notion of the 'impedance' and notation w_1 for this quantity, although this impedance is different, for instance, for the modes analogous to E_{01} and E_{11} in the metallic waveguide.

In the first of the two problems considered below we deal with the field of the symmetric E mode, that is, with the mode E_{01}. The electric field of this mode has nonzero longitudinal component E_z on the axis (i.e. at $r = 0$). In several electronic tubes a beam of electrons is moving along their axis. The action of these tubes is based on the energy interchange between this beam and the field of the propagating wave.

For this interchange to be effective, the electrons and the electromagnetic wave should move with close velocities. The velocity of the electrons is smaller than the velocity of light, c. The problem is to slow the wave sufficiently so that its phase velocity $c \cdot k/h$ also becomes smaller than c. To this end, its propagation factor h must be larger than k. Since $h_j = k^2 - \alpha_j^2$, the conditions must be created under which $\alpha_j^2 < 0$.

In a waveguide with smooth metallic walls, all the modes are fast, $\alpha_j^2 > 0$. On the other hand, a slow surface wave can propagate along the impedance plane (see Section 6.2.1.2). For the considered polarization, this is possible for $w_1'' > 0$. If w_1'' grows, then the field of the wave concentrates more strongly near the plane, and the wave velocity decreases. It is expected that, if the metallic walls of the waveguide are substituted with the impedance ones with $w_1'' > 0$, then the field of the mode is concentrated near the walls and the waveguide nature of the field reveals itself to a smaller degree. Creating the impedance with sufficiently large w_1'', we can slow the mode of the waveguide by these walls.

This intuitive consideration is easily checked by simple calculations. For the symmetric mode (with $\partial/\partial s \equiv 0$), the boundary conditions (6.108) with $w_2 = 0$ give $\Psi \equiv 0$, and

$$\alpha^2 \Phi = i k w_1 \frac{\partial \Phi}{\partial r} \tag{6.110}$$

on S. For the considered mode, $\Phi(r) = J_0(\alpha r)$, and the equation for α has the form

$$\frac{\alpha}{k} \frac{J_0(\alpha a)}{J_1(\alpha a)} = w_1'' ; \tag{6.111}$$

here, for simplicity, $w_1' = 0$ is assumed, that is, there are no losses in the walls. We are interested in the first root of this equation. At $w_1'' = 0$ this root satisfies the equation $J_0(\alpha a) = 0$, the wave is the E_{01} mode of the waveguide with smooth walls. As w_1'' increases, the root of (6.111) decreases; for sufficiently large positive w_1'' we have the root $\alpha_j^2 < 0$. It corresponds to a wave analogous to E_{01}, but a slow one.

The second problem is connected with the need to weaken the transformation of the mode types in an irregular waveguide. It can also be solved by creating an azimuthal goffer. When colliding with an irregularity, the incident wave (with $j = 0$) generates other eigenwaves (with $j \neq 0$) and passes to them a part of its energy. If the incident wave is the only traveling wave, that is, $h_j^2 < 0$ for $j \neq 0$, then all other waves damp at a distance of the order $|h_j|^{-1}$ from the irregularity, and the field in the waveguide is not noticeably distorted. However, if at least one arisen wave is traveling (e.g. $h_1^2 > 0$), then it can perturb the field significantly. This effect is especially strong if the waves of this type arising on the irregularities located at different places (see dashed lines in Figure 6.13) are added in-phase or almost in-phase. The phase differences of the waves arising at different places are caused by the difference in the phase velocities of the waves with $j = 0$ and $j = 1$. At the segment between the two irregularities, the part of the parasitic mode, which arises on the left irregularity, obtains the phase $h_1 l$, where l is the distance between the irregularities. The phase of the second part arising on the right irregularity starts from the value $h_0 l$. The phase difference is $|h_0 - h_1| \cdot l$. The first multiplier is proportional to the difference $\alpha_1^2 - \alpha_0^2$. The smaller this difference, the larger the total amplitude of the wave $j = 1$ and the field distortion in the waveguide caused by the transformation of the wave $j = 0$ into the wave $j = 1$.

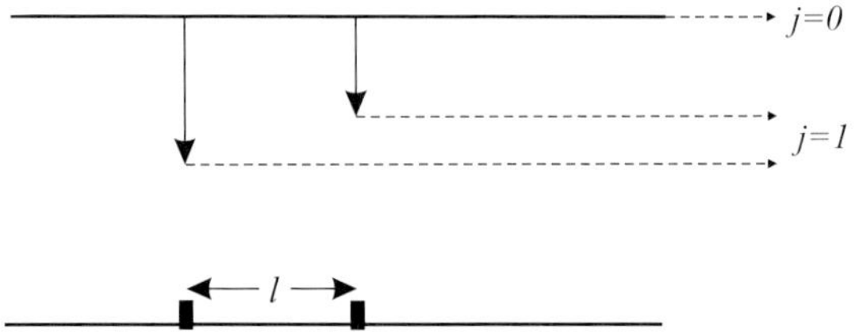

Figure 6.13 Adding waves arising at two irregularities.

In certain waveguides with metallic walls the eigenvalues of two different modes coincide, that is, $\alpha_1^2 = \alpha_0^2$. This eigenvalue is degenerated and several waves of different types correspond to it. In this case the transformation of the wave $j = 0$ into the wave $j = 1$ is especially strong, since $h_1 - h_0 = 0$. If the field distortion caused by this process is undesirable, then degeneration should be avoided. The problem is to find the impedance providing the largest difference in the eigenvalues α_1^2 and α_0^2.

In circular waveguides with smooth metallic walls, the waves H_{01} and E_{11} are degenerated. The eigenvalues α_1^2 and α_0^2 satisfy the same equation $J_1(\alpha a) = 0$. At any irregularity disturbing the axial symmetry of the waveguide, the wave H_{01} ($j = 0$) generates the wave E_{11} ($j = 1$). This means that a transformation of the wave H_{01} into E_{11} occurs.

The azimuthal goffer does not affect the structure or the phase velocity of the wave H_{01}. The field of this wave is described by the function $\Psi = J_0(\alpha_0 r)$ (at $\Phi \equiv 0$), where α_0 satisfies the equation $J_0'(\alpha a) = 0$, equivalent to $J_1(\alpha a) = 0$.

In the impedance waveguide, the wave analogous to E_{11} differs from that in the waveguide with metallic walls. In the metallic waveguide, the wave E_{11} is described by the function $\Phi = J_1(\alpha_1 r)\cos\varphi$ (at $\Psi \equiv 0$), where α_1 satisfies the equation $J_1(\alpha a) = 0$. There is another wave, H_{11}, having a field with the same angular dependence as in E_{11}. This field is described by the function $\Psi = J_1(\alpha_1 r)\sin\varphi$ (at $\Phi \equiv 0$), its eigenvalue is a root of the equation $J_1'(\alpha a) = 0$.

In the impedance waveguide, there are also two eigenwaves having this angular dependence on the field components. They are equivalent to the modes E_{11} and H_{11} of the metallic waveguide and are described by the pair of functions

$$\Phi = A \cdot J_1(\alpha r)\cos\varphi\,, \quad \Psi = B \cdot J_1(\alpha r)\sin\varphi\,. \tag{6.112}$$

The eigenvalues α and the ratio of the coefficients A/B are different for these two pairs. Both values depend on the impedance. For a small impedance, the value α of the first pair is close to the eigenvalue of the wave E_{11}, and $B \ll A$. In the second pair, α is close to the eigenvalue of the wave H_{11}, and $A \ll B$.

Substituting (6.112) into the boundary condition (6.108) leads to a homogeneous linear equation with respect to A and B. Equating its determinant to zero gives the following transcendental equation for the eigenvalues α:

$$w_1'' J_1^2 - \frac{\alpha}{k} P J_1 J_1' - P w_1'' (J_1')^2 = 0\,, \tag{6.113}$$

where $P = k^2 a^2 \alpha^2 / h^2$; the common argument αa of the Bessel functions is omitted for simplicity. In (6.113) it is assumed that there are no losses in the walls, that is, $w_1' = 0$, so that $w_1 = i w_1''$. The ratio of the coefficients obtained from the above linear equation system is

$$\frac{B}{A} = -\frac{h}{\alpha} \cdot \frac{1}{ka} \cdot \frac{J_1}{J_1'}\,. \tag{6.114}$$

At $w_1'' = 0$ (metallic waveguide), (6.113) splits into two equations. One of them is $J_1(\alpha a) = 0$ and its root is α_1. This is the equation for the E_{11} wave. For $w_1'' \ll 1$, (6.113) can be approximately written as

$$J_1(\alpha a) = -\frac{k}{\alpha_1} w_1'' J_1'(\alpha_1 a)\,. \tag{6.115}$$

Its root is

$$\alpha = \alpha_1 - \frac{k}{\alpha_1 a} w_1''\,. \tag{6.116}$$

The larger the impedance, the more efficiently the degeneracy between the waves H_{01} and 'E_{11}' (i.e. the wave analogous to E_{11} in the metallic waveguide) is removed.

In the general case (for any w_1''), (6.113) also splits into two equations, for the waves H_{01} and 'E_{11}', respectively. The equation for 'E_{11}' has the form

$$J_1 - \frac{1}{w_1''}\frac{P\alpha}{2k}J_1'\left[1-\sqrt{1+\frac{4k^2}{P\alpha^2}(w_1'')^2}\right] = 0\,. \qquad (6.117)$$

In particular, $J_1 = 0$ for $w_1'' = 0$. Equation (6.117) allows observation of the variation in the eigenvalue of this wave and, in this way, the process of removing the degeneracy between the waves E_{01} and E_{11} by varying w_1'' from zero to infinity. In the limiting case, $w_1'' = \infty$, this equation takes the form $J_1(a\alpha) = -ka\alpha/h \cdot J_1'(a\alpha)$ which follows from (6.114).

6.2.5.3 **Impedance transformer of the field structure**

Here we deal with a problem which, in a certain sense, is opposite to that considered above. In this problem, it is necessary to find the parameters of an irregular segment of the waveguide, with given field structure at the segment input and desired field at the output. Such a problem arises in different technical applications, in which the generator sends the energy in the form of a field of a certain structure, but the receiving device can work only with a field of another structure. In Section 2.1.2 this problem was considered as a phase optimization one. Here we attempt to solve it in another way.

In the irregular waveguide there are no eigenwaves (i.e. the waves of the same structure in the entire waveguide). However, a regular waveguide (the so-called *reference waveguide*) can be associated with any cross-section of the irregular one, which has the same parameters as the irregular waveguide in this cross-section. Then the field in any cross-section, $z = const$, of the irregular waveguide can be expressed as the sum of the eigenwaves of the reference one. The amplitude A_j of the jth wave in this sum depends on z, the waves transform into each other and energy interchange occurs between them. The energy interchange between the waves of numbers j and m is described by the coupling coefficient S_{jm}. These coefficients are determined by the condition that the variation of A_j on the segment Δz, caused by the existence of the mth wave, is $\Delta A_j = S_{jm}A_m\Delta z$. These quantities are coefficients in the infinite system of the ordinary differential equations of first order for the amplitudes $A_j(z)$. The end conditions for these amplitudes are found from the requirement that the fields on the waveguide input and output have given structures.

It is assumed that only the impedance of the walls is different in different cross-sections, whereas their shape (i.e. the contour $S(z)$) remains the same. The impedance of the walls is either isotropic, or anisotropic with one of the principal axes of the tensor directed along the contour and the other one parallel to the z-axis. The two-dimensional vectors $\vec{e}$ and $\vec{h}$ (i.e. the projections of the vectors $\vec{E}$ and $\vec{H}$ onto the plane tangent to the surface) are connected by the relations $e_z = -w_1 h_s$, $e_s = w_2 h_z$, where w_1 and w_2 depend on z and s.

The coefficients S_{jm} are proportional to the following integrals taken over the cross-section contour:

$$S_{jm} \sim \int_S \left(h_s^j h_s^m \frac{\partial w_1}{\partial z} - h_z^j h_z^m \frac{\partial w_2}{\partial z} \right) \mathrm{d}s \tag{6.118}$$

(see (1.38) in [88]).

If w_1 and w_2 do not depend on φ in the waveguide of the circular cross-section (S is the circle), then the irregularity (dependences of w_1 and w_2 on z) couples only the waves with the same angular dependences. For instance, the waves H_{0q} are coupled only with the waves H_{0p}, the fields of which have other dependences on r. If it is necessary to provide the coupling between the waves with different angular dependences, then a variable impedance on the cross-section contour must be created. If the waves with fields proportional to $\cos(n_1\varphi)$ and $\cos(n_2\varphi)$ have to be coupled, then the impedance should depend on φ as $\cos((n_1 - n_2)\varphi)$ or $\cos((n_1 + n_2)\varphi)$.

The wavenumbers $h_j(z)$ also depend on z. They are calculated as $h(z) = [k^2 - \alpha_j^2(z)]^{1/2}$, where $\alpha_j(z)$ are the eigenvalues of the homogeneous problem (6.108) for the potential functions, in which all the values depend on z.

If a wave, arising at the cross-section $z = z_1$, would not interchange its energy with other waves, then its amplitude would vary so that the energy flux transferred by it is kept, and its phase is varied by the law

$$\exp\left[-i \int_{z_1}^{z} h_j(z) \mathrm{d}z \right] \tag{6.119}$$

generalizing the law of varying phase $\exp(-ih_j(z - z_1))$ in the regular waveguide. If energy interchange between the waves is taken into account, then variations of their amplitudes and phase are connected by more complicated laws described by the ordinary differential equation system.

In applications to the field transformer in the form of an irregular waveguide with variable impedance, the inverse diffraction problem formulated here consists in determining the functions $w_1(s, z)$ and $w_2(s, z)$ when the input and output fields are given. The coefficients $S_{jm}(z)$ and $h_j(z)$ of the differential equation system for the functions $A_j(z)$ depend on $w_1(s, z)$ and $w_2(s, z)$. At the input and output of the waveguide, the fields are expressed as the sums of the fields of the eigenwaves. The coefficients in these sums are the values of A_j at the beginning and end of the waveguide segment.

Even finding a partial solution to this problem (see Chapter 3 in [36]) should involve physical considerations based on the relations (6.118) and (6.119). Above we have mentioned only the formulation of the general problem.

6.2.5.4 **Resonators with impedance walls**

The closed resonator is an empty volume bounded by walls with certain physical properties (metallic, impedance, semi-transparent, etc.). The resonator is characterized by a set of complex eigenfrequencies $k_j = k_j' + ik_j''$, $j = 1, 2, \ldots$. They are defined by the size and shape of the resonator as well as (if the walls have impedance) by the value of the impedance and its distribution over the walls. In fact, the real part k' of the eigenfrequency coincides with the resonant frequency, that is, with the frequency of the excitation device, at which the amplitude of the oscillation is maximal. The imaginary part k''_j defines the width of the 'resonant curve', it is inversely proportional to the oscillation quality (Q-factor), $Q = k'_j/(2k''_j)$. The energy of the oscillation system decreases by $\exp(-2\pi/Q_j)$ times per period.

Each eigenfrequency is associated with a certain eigenoscillation and, in particular, with a certain distribution of the field $\vec{h}(s)$ on the walls. The larger the field of the jth oscillation on a segment of the walls, the more strongly the impedance of this segment affects the value of the eigenfrequency of the resonator. This fact will allow the handling (in a certain range) of the eigenvalues of different oscillations by choosing the impedance distribution on the walls. Usually, the problem consists either in increasing the difference $k'_j - k'_m$ between the resonant frequencies of different oscillations in order to extend the frequency band, in which the resonator is equivalent to the one-frequency oscillating contour, or in increasing the losses (i.e. k''_j) of a certain oscillation in order to decrease its influence on the other oscillations.

The numbers k_j are the eigenvalues of the homogeneous problem consisting of the Maxwell equation system

$$\operatorname{rot} \vec{E}^j + i k_j \vec{H}^j = 0\,, \tag{6.120a}$$

$$\operatorname{rot} \vec{H}^j - i k_j \vec{E}^j = 0 \tag{6.120b}$$

in the resonator and the boundary conditions $e_t = -w_1 h_\tau$, $e_\tau = w_2 h_t$ on its closed boundary S (cf. (6.18)); here $\{\vec{t}, \vec{\tau}, \vec{N}\}$ is the right triple of the Cartesian unit vectors, N is the interior normal to the surface. The impedances w_1, w_2 can be variable on the resonator walls.

If the impedance is small, approximate expressions for k'_j and k''_j can be obtained for any resonator, independently of the resonator shape and the impedance distribution on the walls. From the boundary conditions we have

$$\int_S \left[\vec{E}\vec{H}^*\right]_{-N} ds = \int_S \left(w_1|h_\tau|^2 + w_2|h_t|^2\right) ds\,. \tag{6.121}$$

Combining the equations (6.120), we obtain the equality

$$i k \operatorname{div}\left[\vec{E}\vec{H}^*\right] = k^2\left|\vec{H}\right|^2 - \left|\operatorname{rot} \vec{H}\right|^2\,; \tag{6.122}$$

here and below we omit the index j in the eigenvalues and the eigenoscillation fields. Applying the Stokes theorem to (6.121) and integrating (6.122) over the resonator volume V gives

$$k^2 = \frac{\int_V \left|\operatorname{rot} \vec{H}\right|^2 dV}{\int_V \left|\vec{H}\right|^2 dV} + \frac{ik}{\int_V \left|\vec{H}\right|^2 dV} \int_S \left(w_1|h_\tau|^2 + w_2|h_t|^2\right) ds\,. \tag{6.123}$$

Let $|w_1| \ll 1$, $|w_2| \ll 1$. Then, it is easy to check that $k'' \ll k'$. Upto an accuracy of terms of the first order, it follows from (6.123) that

$$k'' = \frac{\int_S \left(w_1'|h_\tau^0|^2 + w_2'|h_t^0|^2\right) \mathrm{d}s}{2\int_V \left|\vec{H}^0\right|^2 \mathrm{d}V}\,, \tag{6.124}$$

where $\vec{h}^0$, $\vec{H}^0$ are the magnetic fields (on the surface and in the volume, respectively) of the similar eigenoscillation in the resonator with smooth metallic walls ($w_1 = w_2 = 0$). Damping of the eigenoscillation is caused by energy penetration into the resonator walls. It grows as the real part of the impedance increases, especially on the wall segments where the current is maximal.

With the same accuracy, (6.123) gives the expression for the real part of the eigenfrequency

$$k' = k^0 - \frac{\int_S \left(w_1''|h_\tau^0|^2 + w_2''|h_t^0|^2\right) \mathrm{d}s}{2\int_V \left|\vec{H}^0\right|^2 \mathrm{d}V}\,. \tag{6.125}$$

Here it is taken into account that the first addend on the right-hand side of (6.123) is stationary with respect to the small variations of $\vec{H}$. This means that, if $\vec{H}$ differs from $\vec{H}^0$ in the term of the first order of $|w|$, then this addend differs from k^0 in terms of the second order of $|w|$ dropped in (6.125).

According to (6.125), the replacement of metallic walls by impedance ones with $w'' > 0$ reduces the eigenfrequency. Such a replacement as if moves aside the walls; it is equivalent to an enlargement of the resonator volume. The impedance with $w'' < 0$ increases the eigenfrequency.

If the eigenfrequencies k_j^0 and k_m^0 of two different oscillations in a resonator with metallic walls are close, then they can be moved apart, for instance, by making the walls partially corrugated. The higher frequency can be increased if we create the impedance with $w'' < 0$ on the wall segments where the current of this oscillation is large. If the current of the second oscillation is small, then its eigenfrequency varies a little. It can be also decreased by creating an impedance with $w'' > 0$ on the corresponding part of the wall.

6.3
Metallic short-periodical array

In this subsection another class of inverse problems is investigated, in which the polarization properties of the incident and diffracted fields are essential. The field transformation is provided not by the impedance surfaces, but by the semi-transparent ones constructed from short-periodical arrays. Several requirements on the diffracted field, in particular, its polarization structure, can be met by choosing the location of arrays and their geometrical parameters.

6.3.1 Equivalent boundary conditions

In this subsection we consider a periodical array of metallic wires with period p small in comparison with the wavelength λ. Such an array is a semi-transparent screen with a large anisotropy. Since $kp \ll 1$, the diffraction on this array does not generate *side spectra*, that is, the plane waves directed differently from the incident and reflected waves. In a layer with a thickness of the order p, the fields vary quickly, at distances of the same order. At larger distances from the array the fields are smoothed; the scale of distances where they vary noticeably, is λ, $\lambda \gg p$. In the array neighborhood, the wave equation, to which the Cartesian components of the fields are subordinated, can be substituted by the Laplace equation. When moving away from the array at distances larger than p, the fields and their derivatives tend to their limiting values. These values are connected by linear relations independent of the field value and structure far from the array. This fact allows us to substitute the array with an equivalent thin film. The values of the tangential components of the fields on different sides of the equivalent film are connected by the same relation as the above limiting values on the sides of the array. In the other words, the tangential components of the fields on both sides of the film are the limiting values for the fields existing in the domain into which the array is immersed.

The theory of the short-periodical array is constructed in a similar way to that of the short-periodical goffer. The fields at 'static infinity' are the limiting values for the fields in the entire exterior domain, and they are the fields on the surface of the equivalent film. The difference between the equivalent boundary conditions for the goffer and for the array, consists in the fact that in the first case the field exists only on one side of the goffer, whereas in the second case it exists on both sides of the array. The boundary conditions for the array connect the values of the tangential components of the fields on both sides of the equivalent film.

We introduce the Cartesian coordinates (s, t, N), where the axes s and t lie in the film plane, parallel and perpendicular to the wires, respectively, whereas the N-axis is directed normally to the film, so that the orths $\vec{s}, \vec{t}, \vec{N}$ make up the right triple. Different sides of the array (and the film) are marked by '+' and '−' in such a way such that $\vec{N}$ is directed from '+' and '−' (Figure 6.14).

The two independent polarizations are supported by the two pairs of boundary conditions on the film. In one of the polarizations, only the tangential components $e_s^{\pm}$, $h_t^{\pm}$ exist on the film and they are connected by these conditions; in the other polarization, the conditions connect the nonzero components $e_t^{\pm}$, $h_s^{\pm}$ on the film. The first polarization ('E polarization') is excited when $E_t = 0$, $H_s = 0$ in the incident wave, and the second one ('H polarization') is excited when $E_s = 0$, $H_t = 0$ in the incident wave. The boundary conditions do not couple these polarizations. The anisotropy of the array reveals itself by the fact that the boundary conditions are different for different polarizations.

The assertion that the array does not couple the polarizations and that the equivalent boundary conditions do not depend on the exterior field (i.e. they are just *boundary* conditions) is valid only in the case when the exterior field does not de-

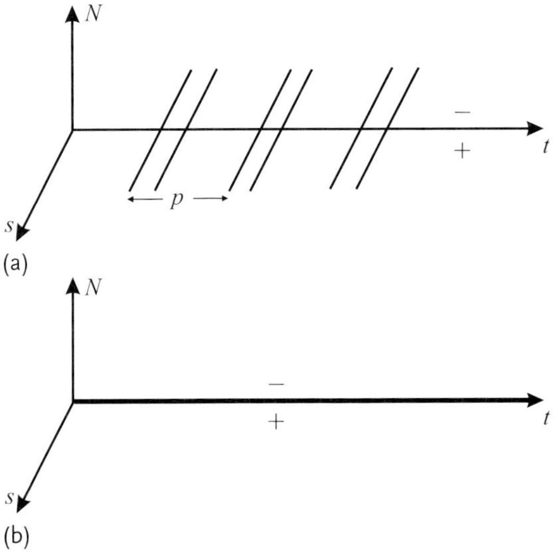

Figure 6.14 Geometry of the array (a) and equivalent film (b).

pend on the *s*-coordinate. If this condition is not fulfilled, then the mathematical technique and the main formulas (see 6.127, (6.129) and (6.134) below) are more complicated, but the physical image ('ripples' in the layer of the thickness p and the smoothed field outside this layer) remains. Below we assume, for simplicity, that $\partial/\partial s \equiv 0$.

6.3.2 Electric polarization

Electric polarization is when the electric vector is parallel to the wires and the slots between them. The simplest theory of this polarization is based on the assumption that the wave falling onto the array is completely reflected, that is, the array behaves in the same way as a solid metallic plane. The tangential component e_s of the field $\vec{E}$ equals zero on this plane. In this approximation the boundary conditions have the form

$$e_s^+ = 0\,, \quad e_s^- = 0\,. \tag{6.126}$$

These conditions do not take into account the fact that the electrical field in the slots, although not large, is not equal to zero. The theory based on this assumption cannot describe the situation which occurs when the wires are very thin, that is, when the metal occupies only a small part of the period. This theory does not describe the small (but finite) transparency of the array for this polarization. The theory based on the boundary condition (6.126) and the analogous condition (6.133) for magnetic polarization correctly describes the property of anisotropy, and it is sufficient to describe devices which use just this property. The theory describing the array transparency for this polarization should take into account the shape and density of the wires.

The Cartesian component $E_s(t, N)$ satisfies the Laplace equation near the wires, and equals zero on their boundary; it is a periodical function of t with the period p. If the wire cross-section is symmetrical (which is assumed below), then the solution of such a problem is real (more precisely, in-phase, which in our case is the same) and symmetrical with respect to the plane $N = 0$, that is, $E_s(t, -N) = E_s(t, N)$. Consequently, the limiting values $\lim_{|N| \gg p} E_s^{\pm}$ of the electric field are the same on both sides of the array, i.e.

$$e_s^{+} = e_s^{-} . \tag{6.127}$$

This 'boundary' condition, valid far from the array, is weaker than (6.126) and it is not sufficient to describe the properties of the equivalent film. This condition should be complemented by the condition for the limiting values of the component

$$H_t = \frac{i}{k} \frac{\partial E_s}{\partial N} \tag{6.128}$$

of the magnetic field. This component is asymmetrical with respect to the plane $N = 0$, $H_t(-N, t) = -H_t(N, t)$. This property is inherent also for its limiting values at $|N| \gg p$, that is, $h_t^{+} - h_t^{-} \neq 0$. Near the wires the component E_s varies rapidly, on the scale of p, so that $\partial E_s / \partial N$ is of the order $p^{-1} E_s$, and the ratio $\partial E_s / \partial N$ to E_s is real. Consequently, the difference $h_t^{+} - h_t^{-}$ is much larger than $e_s^{\pm}$ and differs from it in phase by the factor i, that is, the magnetic field is shifted in phase by $\pi/2$ with respect to the electric one. The large factor in the ratio between $\partial E_s / \partial N$ and E_s is of order $(kp)^{-1}$, that is, the ratio of the wavelength to the period of the array.

The phase shift between the magnetic and electric fields has key significance for polarization transformers, considered below. In fact, it is caused by the static nature of the field E near the array, which, in turn, is connected with the smallness (in comparison with λ) of the parameter p defining the scale of the inhomogeneity of the body on which the field is diffracted.

The second (in addition to (6.127)) equivalent boundary condition for the field of the E polarization on the short-periodical ($kp \ll 1$) array has the form

$$e_s = -i P_s (h_t^{+} - h_t^{-}) , \tag{6.129}$$

where P_s is a real coefficient containing the small factor kp. This coefficient depends on the shape and relative size of the wires.

If $kp \to 0$, then P_s tends to zero as well. As follows from (6.129), in the limit $e_s = 0$, and the system of conditions (6.127) and (6.129) transforms into condition (6.126). However, if the wires are very narrow and occupy a small part of the period, then P_s may be not very small. If the ratio of the width of the wires to the period tends to zero, so that the array 'disappears', then P_s tends to infinity. It follows from (6.129), that $h_t^{+} = h_t^{-}$, which, together with (6.127) means that the field does not 'notice' the array.

In order to calculate the value P_s, it is necessary to solve the mentioned static problem and find the ratio of the limiting values (at $|N|/p \to \infty$) of the solution and its derivative with respect to N. This problem can be solved, for instance, by the method of conform transformations. A simple formula is obtained for the array consisting of the flat bands. Then

$$P_s = -\frac{p}{\lambda} \ln\left(\sin\frac{\pi q}{2}\right) , \tag{6.130}$$

where q is the ratio of the band width to the period [89]. If this ratio tends to zero, then the logarithm is very large, so that P_s is also large, and (6.126) gives $h_t^+ = h_t^-$. This means that the array influences the field weakly. If q is not close to zero, then $P_s \ll 1$, and the array is opaque for this polarization.

The boundary conditions (6.127) and (6.129) allow one to find the coefficients of transmission and reflection of the plane wave falling normally onto the array with the wires parallel to the vector $\vec{E}$. If a wave of unit amplitude falls onto the array from the side '+', then

$$e_s^+ = 1 + R_s\,, \quad e_s^- = T_s\,, \tag{6.131a}$$

$$h_t^+ = 1 - R_s\,, \quad h_t^- = T_s\,, \tag{6.131b}$$

where R_s and T_s are the transmission and reflection coefficients, respectively. Substituting these expressions into (6.127) and (6.129), we have

$$R_s = -\frac{1}{1 - 2i P_s}\,, \tag{6.132a}$$

$$T_s = -\frac{2i P_s}{1 - 2i P_s}\,. \tag{6.132b}$$

If the wires in not too thin, then $P_s \ll 1$ and the transmitted wave has a small amplitude; it is shifted in phase with respect to the incident wave by an angle close to $\pi/2$, and $T_s \approx -2i P_s$. These two properties of the short-periodical array are used in the polarization transformers described below.

6.3.3
Magnetic polarization

If the electric field is perpendicular to the wires and slots of the array, then the array perturbs the field weakly. The currents flowing in it are small; in the slots they transform into the capacitance currents (displacement currents). In zero approximation such currents can be neglected, both components E_t and H_s have no discontinues on the equivalent film, so that the equivalent boundary conditions for this polarization have the form

$$e_t^+ - e_t^- = 0\,, \quad h_s^+ - h_s^- = 0\,. \tag{6.133}$$

The array as if is absent.

These boundary conditions are not valid if the slots between the wires are very small, so that the capacity current perpendicular to the slots is not small. In this case the transparency of the array is smaller than unity, and if the slots disappear, then the transparency of the array is zero. The more precise boundary conditions

$$e_t^+ - e_t^- = 0\,, \tag{6.134a}$$

$$e_t^\pm = i P_t(h_s^+ - h_s^-) \tag{6.134b}$$

are valid in this limiting case as well. The real coefficient P_t depends on the shape and relative thickness of the wires and contains the large multiplier $(kp)^{-1}$, that is, the ratio of the wavelength to the period. The coefficient P_t is small only if the slots are very small. Then condition (6.134b) transforms to $e_t^\pm = 0$ and the array is opaque for this polarization.

If the wires have the form of bands, then the explicit expression for P_t can be obtained, as follows

$$P_t = \frac{\lambda}{4p} \frac{1}{\ln\left(\cos \frac{\pi q}{2}\right)}\,. \tag{6.135}$$

Formulas (6.134a) and (6.135) are obtained by the same method as (6.129) and (6.130). Near the array, the component H_s satisfies the Laplace equation, the Neumann condition on the boundary of the wires and the periodicity condition with respect to t. This component varies fast, within the scale of the order p, and the electric field $E_t = -ik^{-1}\partial H_s/\partial N$ is $(kp)^{-1}$ times larger than the magnetic one. Formula (6.134b) describes the relation between the limits to which the field H_s and its derivative tend when $|N|/p \to \infty$.

Let the plane wave fall normally to the array with the wires perpendicular to its field $\vec{E}$. Then the coefficients of transmission R_t and T_t are found from an equation system similar to (6.131). According to (6.134), these coefficients are

$$R_t = \frac{i(2P_t)^{-1}}{1 - i(2P_t)^{-1}}\,, \tag{6.136a}$$

$$T_t = \frac{1}{1 - i(2P_t)^{-1}}\,. \tag{6.136b}$$

Under the usual conditions, $P_t \gg 1$, $T_t \approx 1$, that is, the array is almost fully transparent for this polarization.

6.3.4
Short-periodical grid

An example of the isotropic semi-transparent screen is a grid consisting of two short-periodical identical arrays of metallic wires. The arrays are located perpendicular to each other. The theory of the grid is more complicated than that of the array, because it needs to solve the three-dimensional static problem about the field

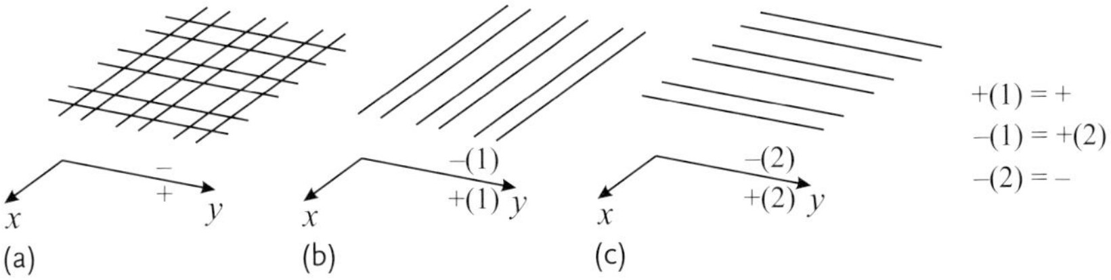

Figure 6.15 The grid (a) as a covering of two arrays (b) and (c).

near the grid, whereas for the array such a problem was two-dimensional. The grid with not very thin wires is almost opaque for both polarizations. The currents flowing in the wires of each array are proportional to the projection of the field $\vec{E}$ of the incident wave onto the respective wires.

We construct the approximate theory of the grid, which does not take into account the influence of the wires of one direction onto the currents flowing in the wires of the other direction. This approximation is equivalent to the assumption that the grids lie in two different parallel planes, the distance between them being large in comparison with the period p. The transmission of the wave through each array is described by formulas (6.127), (6.129) and (6.134) applied to the isolated arrays. Namely, the field transmitted through the first array is the incident wave for the second one. The equivalent boundary conditions on a certain flat film couple the components of the fields on the input side '+(1)' of the first array with the components of the fields on the output side '−(2)' of the second array (see Figure 6.15).

In the planes of the arrays, we introduce the Cartesian coordinates (x, y) directed along the wires. The x-axis coincides with the wires of the first array, and the y-axis coincides with the wires of the second array.

We analyze the transmission of the plane wave with components E_x, H_y through the grid, using in turn the results for each array. For the first array, the incident wave is E polarized, and

$$e_x^{+(1)} = e_x^{-(1)}\,; \quad e_x^{\pm(1)} = -i P_s\left(h_y^{+(1)} - h_y^{-(1)}\right). \tag{6.137}$$

For the second array, the incident wave is H polarized, and

$$e_x^{+(2)} = e_x^{-(2)}\,; \quad e_x^{\pm(2)} = i P_t\left(h_y^{+(2)} - h_y^{-(2)}\right). \tag{6.138}$$

In the chosen approximation, the fields in domains '−(1)' and '+(2)' coincide, $h_y^{+(2)} = h_y^{-(1)}$, $e_x^{+(2)} = e_x^{-(1)}$. Eliminating them from the above equations, we obtain the relations between the fields on both sides of the grid

$$e_x^{+} = e_x^{-}\,; \quad e_x^{\pm} = -\frac{i}{1/P_s - 1/P_t}\left(h_y^{+} - h_y^{-}\right). \tag{6.139}$$

If neither the wires nor the slots between them are very narrow, then $P_s \ll 1$, $P_t \gg 1$, and (6.139) coincides with (6.129). As it was expected, the grid reflects the E polarized wave approximately in the same manner as the array reflects the wave of this polarization. It is easy to see that this fact is true if the field $\vec{E}$ in the incident wave is directed along the y-axis.

6.4
Creation of a field of circular polarization

In this section we give the background of the theory of devices creating a wave of circular polarization. Their action is based on two properties of the short-periodical array. First, the coefficient of transmission of the normally falling wave of H polarization is close to unity if the slots between the wires are not exponentially narrow, that is, with $P_t \gg 1$. Second, the wave of E polarization is reflected almost fully and the transmitting wave of small amplitude obtains the additional phase $\pi/2$ on the array with not exponentially narrow wires, that is, with $P_s \ll 1$.

6.4.1
Transformation of linear polarization into circular

If the above two properties hold, then the coefficient P_s in expression (6.132) is small (proportional to p/λ), $P_s \ll 1$, whereas the coefficient P_t in the expression (6.136) is large (proportional to λ/p), $P_t \gg 1$. Hence, the transmission coefficients (6.132b) and (6.136b) for the E and H polarized waves, respectively, can be calculated by the approximate formulas

$$T_s = -2i P_s , \tag{6.140a}$$

$$T_t = 1 . \tag{6.140b}$$

Let the linear polarized wave with electric vector $\vec{E}$ making up the angle α with the direction of the wire (i.e. with the s-axis of the chosen Cartesian system) normally fall onto the array. Consider the polarization of the wave transmitted through the array. The amplitude of the wave equals unity. The incident wave can be expressed as a sum of the two waves linearly polarized in the directions s and t. The amplitude of these waves are $\cos\alpha$ and $\sin\alpha$, respectively. According to (6.140), after transmission through the array, the amplitudes of the waves become $-2i P_s \cos\alpha$ and $\sin\alpha$. They make an elliptical polarized wave. If $\sin\alpha = 2P_s \cos\alpha$, that is,

$$\tan\alpha = 2P_s \tag{6.141}$$

(which is equivalent to $\alpha = 2P_s$ at $P_s \ll 1$), then the transmitted wave has circular polarization. Indeed, in this case, the decisive fact is that the phase shift $\pi/2$ takes place between the two components E_s and E_t of the electrical field in the transmitted wave. Therefore, the amplitude of E_s obtains the factor i, whereas the amplitude of E_t remains real (see (6.140)). The moduli of these amplitudes are the same.

The appearance of the phase shift in the transmitted wave can be explained by the fact that the wave passes through the domain (vicinity of the array) in which the fields vary much faster than in the plane wave. In the plane wave, the ratio of the gradient of the field to the field itself has the order λ^{-1}. Near a narrow slot this ratio is much larger, it has the order of the inverse period of the array or inverse width of the wires. In this domain, the term $k^2 E_s(t, N)$ is small in the wave equation, and E_s satisfies the Laplace equation. Both the function E_s and its gradient are real (in-phase). The magnetic field H_t is *in-quadrature* with the electric field E_s (see (6.128)). The magnetic field on the metal and in the slot has the same order as this field in the incident wave, and it is in-phase with the electric one in this wave. Therefore, the electric field in the slot is shifted in the phase by $\pi/2$ with respect to the electric field of the incident wave. For a single hole in the metallic screen, this fact is a known property of solutions of the appropriate integral equation. For the array, it leads to the mentioned phase shift of the transmitted plane wave of E polarization, and, hence, to transformation of the polarization. Probably, in the same way, some other known cases of phase 'jumps' by $\pi/2$ can be explained, which occur, for instance, when passing through the focus, reflecting from the critical cross-section in the waveguide or from the surface with $\varepsilon = 0$ in a layer with variable ε, etc.

6.4.2
Resonant antenna and transmission resonator of circular shape

The transmitted wave of the circular polarization, arising at the incidence of the linear polarized plane wave onto the array, has a small amplitude and carries a small part of the energy (equal approximately to $2\alpha^2$). The greater part of the energy is reflected.

In the real devices of resonant type, the reflected wave must again fall on the array, in just the same phase, and this process should be multiply repeated. In other words, the array should be a part of the resonator wall. The eigenoscillation of the resonator is a linear polarized standing wave, and the wires of the array make up a certain angle with the vector $\vec{E}$ of this oscillation. The oscillation of the desired structure can be formed either by using an additional array (as in Figure 6.16) or by placing the metallic walls in an appropriate way.

The device shown schematically in Figure 6.16 is one of the possible types of resonant antenna or transmission resonant transformer of the polarization. The polarization is transformed on array 3, playing the role of the antenna aperture. The distance between arrays 1 and 3 equals approximately $\lambda/2$ and array 2 is located approximately in the middle. The middle array is needed to form the eigenoscillation with the desired polarization (in Figure 6.16 the vector $\vec{E}$ is directed vertically).

If array 1 is replaced by the solid screen (with possible exciting elements), then the device acts as a resonant antenna of circular polarization. If array 1 has a finite (not small) transparency for the waves with the field $\vec{E}$ parallel to the wires, then the system is a transmission resonator. This means that it almost fully reflects the waves of frequencies not close to the eigenfrequency of the resonator formed by the arrays 1 and 3. In a narrow frequency range around the eigenfrequency, the transmission resonator is almost fully transparent, and, owing to the properties of the array 3, it transforms linear into circular polarization.

Different types of device are possible, which use the properties of the cavity resonators and the arrays with wires directed at an angle to the vector $\vec{E}$ in its eigenoscillation. In particular, in the antenna version of the device, the input metal-

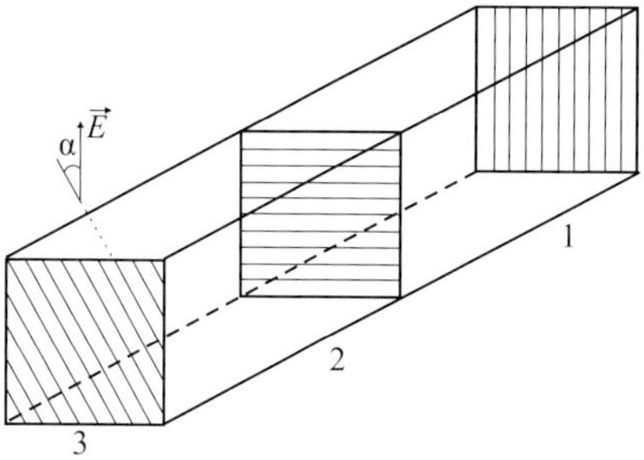

Figure 6.16 Transmission resonator as a polarization transformer.

lic surface can be goffered. Then an eigenoscillation with desired polarization and field structure can be created without using the middle array. The efficiency of the system is larger if the wires of array 3 (and the grooves of the goffer) are not straight, but curved, matched with the spatial structure of the field of eigenoscillation.

The angle α may be not small. At any α, the transmission resonator is fully matched at its resonant frequency, and the radiated field has circular (not elliptic) polarization. The larger α, the faster the ellipticity of the radiation grows when moving away from the resonant frequency; however, then the reflection increases slowly, the matching of the resonator remains for a larger frequency shift (see [91] and [96]).

Transformation of the linear polarization into elliptic one is performed also by reflection from the goffer (see Figure 6.1). The wave polarized along the grooves reflects from their crest and does not penetrate almost inside them, whereas the wave polarized across the grooves reaches their bottom. A phase difference δ arises between the two components of the reflected wave. The polarization ellipticity depends on δ and the angle α between the field $\vec{E}$ of the incident wave and the groove direction. According to relation (2.15) from [36], the squared eccentricity is

$$e^2 = \frac{2V}{1+V}\,, \quad V = \sqrt{1 - \sin^2(2\alpha)\sin^2\delta}\,. \tag{6.142}$$

At $\alpha = \pi/4$, we have $e^2 = 2\cos\delta/(1 + \cos\delta)$. The circular polarization occurs when $\alpha = \pi/4$; $\delta = \pi/2, 3\pi/2$, which corresponds to $w = \pm i$. According to (6.22), at $q = 1$ (infinitely thin partitions between the grooves) and $\varepsilon = \mu = 1$, w gets this value if the (minimal) groove depth is $\lambda/4$. Then the difference between the distances overcame by the reflected waves of two polarizations is $\lambda/4$. These conditions ($\alpha = \pi/4$, the groove depth is $\lambda/4$) also follow from the elementary physical consideration.

If the assembly of elements used when constructing passive radioengineering devices is more diverse, then the possibility for transformation of the electromagnetic fields into the field of the desired structure is greater. In this chapter only two types of these elements were considered: the impedance and semi-transparent (anisotropic) surfaces. There are several other elements that can be used when constructing and optimizing various devices.

Epilogue: Ethical Aspects of Scientific Work

There are two ways to point to the road.
First, to go ahead and lead the people from in front.
Second, to stand at a place and point the direction with the hand.
Below, the second of them is applied.

Introduction

The basic idea of these notes can be formulated as follows: 'The probability that the scientist will obtain an interesting result is larger if he is unselfish, generous and friendly in his contacts with the colleagues'. Somewhat cynically, this thought can be expressed very briefly as: 'virtue is favorable'.

As a rule, any clearly formulated thought is either paradoxical, or trivial. The one mentioned above is evidently trivial. It has been repeated many times in the literature (usually irrespective of the scientists) and in philosophical concepts. All religions claim – in various forms – that it is better to be good than bad.

Below, an attempt is made to show that the specific character of the scientific profession makes this assertion especially important and real for scientists and that following the basic moral criteria is one of the premises of successful scientific work. The author tries to specify these criteria for people of this profession. Certainly, this specification is very subjective and reflects only his observation and personal experience.

The above premise is neither sufficient nor necessary. There are numerous exceptions to the formulated rule. There are scientists, immoral in their attitude to other scientists, who nevertheless have obtained interesting results. Of course, as any generalization, this statement will have exceptions. Many scientists work independently, informing the scientific community only about the final results. Great scientists often obtain very important results alone, without contacting anyone. These notes do not relate to them.

In order not to block up the assertion with numerous clauses and specifications, we do not mention such exceptions below. Although a long good road does not always lead to the target most quickly, this does not mean that it is better to move

Phase Optimization Problems. O. O. Bulatsyk, B. Z. Katsenelenbaum, Y. P. Topolyuk, and N. N. Voitovich

ISBN: 978-3-527-40799-6

more directly, especially if this requires passing a smelly bog. Going directly, one can either reach the target more quickly, or get stuck in the bog and arrive nowhere at all.

Any team of people will work better if they are friendly, trustworthy and ready to help each other. This concerns scientific workers as well. However, scientific work has some peculiarities and, using these, the exclusive meaning of the type of interaction between members of a scientific team will be explained. These peculiarities follow from the fact that the most successful scientific work is collective work, in the sense that it requires frequent and long discussions between scientists. Participation in such discussions is an important part of the scientist's activity.

Before passing on to the description of these peculiarities and to the ethical problems they create, we make some general remarks concerning the text below. The author uses such words as 'it is necessary', 'it follows', etc., and expounds his considerations as the absolute truth. The clauses like 'it seems', 'probably', etc., are assumed, but omitted for brevity. The author understands the subjectivity of his position. He only asserts that the peculiarities of scientific work which he describes, occur often, and that following the well-known ethical principles for this work is both possible and useful.

The word 'science' is understood as fundamental science. A greater number of scientists deal with applied sciences or work in large teams carrying out design projects. In fundamental science the purpose of investigation consists in studying certain phenomena, and the work efficiency depends on the success of small groups, and therefore, on the individual characteristics of their members. The institutes with thousands of employees engaged in solving specific problems, are functioning according to the laws of complex mechanisms. Their efficiency depends on the coordination of all their parts. The rules which are essential for fundamental science, relate only to this science itself.

Scientific contacts

The collective nature of scientific work, i.e., the discussion of a certain problem by several people, results, first, in the fact that the authorship of an idea or result is often not very clear. This question can be resolved in a conflict-free manner only when appealing to some moral criteria. The decision about authorship cannot be based on the formal status of the participants in the discussion. As a rule, the real hierarchy in the group is always known to its members, but it cannot be formalized. Very often it does not coincide with the hierarchy of degrees, ranks and positions. This is one of the essential differences of a scientific group from a usual work collective in which, as in the army, the opinion of a 'senior' is more important than the opinion of a 'junior'.

The second peculiarity of the collective nature of scientific work consists in the fact that the contact between the scientists, whose area of interest may not be the same or even related, is effective, often extremely so. Such contact demands skill in accepting other points of view, methods of research, systems of images and other principles of idealization. The question of scientific tolerance and the admission of one's own nonexclusiveness is a moral one.

The last, third, peculiarity consists in the fact that active, i.e., creative participation in the discussion of another person's problems does not formally affect the status of the participant. This participation is 'giving' without immediate compensation.

Below, the classification of various types of scientific contact as essential elements of scientific work, is attempted. The problem formulation, the methods for its solution, and the results obtained are discussed within this context. New interesting publications are also considered. The efficiency of these contacts does not increase (and sometimes even decreases) as the number of participants increases.

1) **Two–three persons participate in the discussion.** The meeting is not fixed in any way and is not regulated. One of the participants reports on his work – the successes and, first of all, the difficulties. By asking questions, others force him to formulate the essence of his successes or failures more precisely. Sometimes advice given by one of the participants, or the reference to a little-known publication is of paramount importance. Very often such a discussion seems to end with nothing. However, even in this case, it can be useful not only for the speaker, but for all the participants. Usually, the result of such an 'unproductive' discussion reveals itself after a while when one of the participants appears with a new idea. In the author's opinion, this type of informal discussion is the most effective form of collective scientific work.
2) **A microseminar of the whole group.** A reasonable size for this group is ten to twelve people. The active interest of all the members of the group in the work of all the other members is possible only in such groups, and it is this interest which creates the scientific team. Certainly, if a complex experiment is carried out then the group should be larger, it should also contain engineers and technicians. The work of such a seminar will already be formalized and held on a regular basis, every one or two weeks. As a rule, finished projects or information on new ones will be discussed there.
   Such a seminar is run by the head of the group, whose role in the whole scientific process is very important. The leader must discuss the current state of the work of all the participants and be well informed about their successes and difficulties. Whenever possible, he should resolve all conflicts or, at least, settle the disputes. At the same time, the leader will have their own problems, successes and, perhaps failures.
3) **A big seminar.** This may involve a city or a whole institute (of it is large). The most interesting completed projects or sometimes planned ones, are reported in these seminars. Scientists who attended large conferences share their impressions. As a rule such a seminar is organized on a monthly basis. The reports do not cause long discussions, and consist mainly of information staring. It is especially helpful for small groups and young scientists, who need a moral encouragement.

4) **Conferences and symposia.** These may be international. They are very formalized, crowded events. Proceedings of the conferences are published in advance, the presentations often have a formal character. The aim of conferences is to facilitate contact between scientists from different cities and countries. These conferences could have a very important role, involving scientists from different fields of interest in the discussion of a concrete problem. It is already well known, that the points of growth in science are the touch points of different sciences. Scientific contact is effective and useful for the scientist (not only for a beginner) when other scientists give their time, their vision of the problem, their knowledge, their ideas. The key point in this is the verb 'to give'. However, plenty of ordinary, though qualitative reports result in the conferences being divided into sections so that scientists from different specialities may have no contact with each other. Now large conferences are verbal analogues of scientific journals, with a wide spectrum of topics.

The head = the coauthor?

We begin with a description (without comment) of two episodes from the recent history of science. The first of these was told more than half a century ago by the author's supervisor Academician M. Leontovich; the second is a well-known one.

About a hundred years ago a young scientist, L. Mandelstam, created the theory of light dispersion from the surface of a fluid. The theory consists of the thermodynamic calculation of the surface shape (it is nonplanar owing to thermal fluctuations) and the calculation of the scattering pattern of the plane light wave falling onto the surface. In the optical part of the paper, the author confined himself to the calculation of the pattern in the plane of incidence. About ten years later, a paper was published in which the pattern was calculated for all other directions. Its author claimed that the method applied in the optical part of Mandelstam's paper, was inapplicable for this purpose, and used another, more complicated, method. In 1925, Mandelstam (already an academician) charged his postgraduate students M. Leontovich and A. Andronov with the task of making a full optical calculation using his old method. They made it and published the result. Twenty years later, the former students, while sorting the papers of their former teacher, who had since passed away found this calculation there. It was made by him some months before he gave the task to them.

The director of a research institute has attributed his name to the list of coauthors of all the papers published by employees of 'his' institute. The list of his scientific publications has increased by more than one hundred items yearly.

The question is: can the supervisor of PhD students or the head of a group of scientists consider himself the coauthor of the work (report, paper, monograph), made by them? We compare the arguments for both positive and negative answers.

A. Yes, he can. He formulated the problem; this requires erudition and experience, as well as an approximate vision of what the solution would be. He prompted the method to be applied. He participated in frequent discussions and gave advice, often quite effectively. He could do this work himself. He gave time and attention to the work, and facilitated it by providing the labware, access to powerful computers, etc. He 'personally' got the grant for payment of the work, and, in order to receive grants, one must have many of one's own publications.

B. No, he cannot. He only carried out his duty, confirmed his right to supervise scientists, i.e., his scientific and social status. He helped in the work but it was done by others. He should set an example, creating in the collective an atmosphere of mutual, impartial support. A young scientist supervised by him should have a good paper in his biography as this will help him in the future not to be discouraged by failures which are inevitable in the work of any scientist.

The author is sure that the arguments of group B are more convincing than those of group A. As in the majority of ethical problems, the choice between A and B is the choice between a direct, immediate benefit (one more paper, participation in a prestigious conference, etc.) and an uncertain, difficult to formalize benefit in the future (the respect of a few, but not all, colleagues, perhaps the foundation of a scientific school, etc.). It is impossible to prove that B is better than A. One can either agree with this or not. Any dispute leads to extended repetition of the arguments of the type given above.

In the author's opinion, the term 'head' is unsuitable, since it is impossible to teach an adult person anything, it is only possible to help them to study something, so it is impossible to supervise the intellect of an other person, only to help them with their thinking.

We conclude with the story, which took (or could take) place. A young scientist carried out some very interesting research. After his report at the laboratory seminar, the head of the laboratory gave him some reasonable advice about writing a paper and then attributed his name as one of the coauthors, providing in this way its submission to a prestigious international conference. The director of the institute also agreed to be a coauthor, which guaranteed the inclusion of this work in the conference program. For various reasons the young scientist did not attend the conference, and the presentation was read by the head of the laboratory. When preparing the proceedings, the conference organizers decided to reduce the number of coauthors of all the papers to only two people: the speaker and one more person. It was assumed that, in this way, the 'coauthors' would be removed. The head of the laboratory could not exclude the name of the director. As a result, in many references to this work, the name of the real author – the young scientist – is obviously not mentioned.

Colleagues, collective, scientific school

Scientist N has obtained an interesting result or, on the contrary, cannot find the solution to a problem in any way, or cannot understand a paradox which resulted, etc. He needs to consult with somebody or at least to talk about his problems. He asks scientist M to help him. Scientist M can either refuse, guided by reasons A, or agree, guided by reasons B.

A. I will waste time, receiving nothing in return. I will interrupt the work, which I am doing all the time for the solution of my own problem. The authorship of the advice which I, probably, can give, will not belong to me. I must share the knowledge which N does not have, and therefore I will cease to be a monopolist. If N successfully solves the problem, he will have more chances to attend the conference than I have.

B. I will learn a lot of new things. The method of consideration used by N, may suggest to me a new idea for solving my own problem. I will know about a new article in the magazine, which I have not yet read. Now I have come to deadlock in the work, stepping aside a little might help me. When I need to discuss the work with somebody, N will not refuse me.

The choice between A and B is a question of ethics. In the author's opinion, the choice of the arguments of group B is more moral and, at the same time, this choice is expedient. The last statement cannot be proved. As for any moral choice, it cannot be justified logically.

If M agrees to help the colleague, then not only will the probability of obtaining a good result by N increase, but also the successes of M will probably increase. Also, these two scientists will become scientific colleagues.

If such an attitude is established between the majority of the participants in an administrative division (10–12 people), this a division will form a scientific collective. There is a good criterion that the collective exists: activity of its participants in the seminars. The reasons 'I am not engaged in it, therefore it is not interesting to me' or 'I am not engaged in it, therefore it is interesting to me' characterize the absence or presence of the communication between the scientist and the collective. In forming the scientific collective, the role of the head is very important. This role might really belong to one of the participants, having no administrative rights.

A collective can also be formed by the scientists who do not belong to an administrative division. The so-called Bourbaki group was a grotesque but real example of the scientific collective. Several young French mathematicians have published their works under this common pseudonym. The name belonged to the person, whose monument had been erected in the city where the scientists lived. Under this name, they have even published a multi-volume textbook. They were connected by a nontrivial approach to the exposition of the mathematics, by the stronger than usual formalization of the material. The author does not know of any other example of the collective pseudonym in science. In Russian literature there was a collective name Koz'ma Prutkov, but its creators used this pseudonym only for a joke. One of the participants of this group, writer Aleksey Tolstoy, published his trilogy and other literary works under his own name.

Usually, the process of self-purification is inherent in the scientific collective. The climate ingrained in it, is uncomfortable to the persons preferring the reasons of type A, and they gradually leave the collective.

The scientific collective successfully working for many years can become the scientific school. This is a group of scientists united by common subjects, common ethical principles and, as a rule, common 'scientific roots'. The seniors (by age) were involved in a scientific group in their youth and the following generation were taught by them or joined them.

Belonging to a scientific school does not necessarily begin in youth. It is not fixed in any way and, in fact, depends on the scientist. It means, in particular, that the members of this school are scientific colleagues, i.e., they can discuss their scientific problems with any other members of the school. This also involves scientists of different age and different scientific and social status. Of course, snobbery (in Russian – zaznaistwo), a cancerous growth in any collective, particularly scientific, is completely absent in these contacts. You may even be three times an academician (there are some in existence), but you should admit your mistake if a post-graduate student shows you that you are wrong.

In the author's opinion, the term 'scientific school' cannot be precisely defined. The efficiency of the scientific school largely benefits from the absence of formal attributes. This removes many problems which complicate the activity of the more formalized scientific communities.

In conclusion of these notes we mention a question, which lies somewhat beyond the topics stated in the title, but which contains an essential moral component. Should one deal with problems having a quite easily achievable solution, but which are not interesting enough (choice A), or with difficult problems which, with a high probability, cannot be solved at all, but the solution of which would be of great interest (choice B)?

In Strugatskikh's story 'Monday begins from Saturday' an employee of the Research Institute of Magic tried to solve only the problems which were proved to be unsolvable. He chose a limiting form of the variant B. Of course, such a choice can be presumed only by the scientist working in magic conditions. However, sometimes the possibility of a choice also occurs for scientists working in other, absolutely non-magic, institutes. How should they respond, by choice A or choice B?

But this is another story.

Boris Z. Katsenelenbaum

P.S. Over more than sixty years the author has published almost two thousand pages containing formulas. Here he publishes seven pages without them. However, these pages have a direct relation to what is written in this book and in the previous two thousand pages.

References

References

1 Ferwerda, H.A. (1978) Problem of the wave front phase reconstruction according to amplitude distribution and coherent functions, in *Inverse Scattering Problems in Optics* (ed. H.P. Baltes), Springer, Berlin.

2 Kuznetsova, T.I. (1988) On the phase problem in optics. *Sov. Phys. Usp.*, **31** (14), 364–371.

3 Klibanov, M.V. (1985) On uniqueness of definition of a finite function by modulus of its Fourier transform. *Doklady Akademii Nauk SSSR*, **285**, 278–280, in Russian.

4 Hauptman, H.A. (1990) The phase problem of X-ray Crystallography, in *Signal Processing, Part II: Control Theory and Applications* (eds F.A. Grunbaum, J.W. Helton and P. Khargonecar), Springer, New York, pp. 257–273.

5 Tetelbaum, S.I. (1948) Optimal form of antenna for wireless transfer of power. *Sb. nauch.-tekh. st. In-ta elektrotekhniky AN USSR*, 120–126, in Russian.

6 Tartakovsky, L.B. and Tikhonova, V.K. (1959) Synthesis of linear radiator by given distribution of amplitude. *Raditekhnika and Elektronika*, **4** (12), 2016–2019, in Russian.

7 Ramm, A.G. (1968) Optimal solution of the antenna synthesis problem. *Doklady AN SSSR*, **180**, 1071–1074, in Russian.

8 Choni, Yu.I. (1968) On antenna system synthesis according to given amplitude radiation pattern. *Radioelektronika*, **11** (12), 1325–1327, in Russian.

9 Bakhrakh, L.D. and Kremenecky, S.D. (1974) *Synthesis of Radiating Systems. Theory and Calculation Methods*, Sovetskoe Radio, Moscow, in Russian.

10 Dunber, A.S. (1958) On the theory antenna beam scanning. *J. of Applied Physics*, **13** (5), 31–39.

11 Precup, R. (2002) *Methods in Nonlinear Integral Equations*, Kluwer, Dordrecht.

12 Masujima, M. (2005) *Applied Mathematical Methods in Theoretical Physics*, Wiley-VCH Verlag GmbH, Weinheim.

13 Bauschke, H.H., Combettes, P.L. and Luke, D.R. (2002) Phase retrieval, error reduction algorithm, and Fienup variants: A view from convex optimization. *Journal of the Optical Society of America*, **19** (7), 1334–1345.

14 Fienup, J.R. (1982) Phase retrieval algorithms: a comparison. *Appl. Opt*, **21**, 2758–2769.

15 Vainberg, M.M. and Trenogin, V.A. (1974) *Theory of Branching of Solutions of Non-linear Equations*, Noordhoff International Publishing, Leyden.

16 Deuflhard, P. (2004) Newton Methods for Nonlinear Problems. Affine Invariance and Adaptive Algorithms, in *Series Computational Mathematics*, vol 35, Springer, Berlin.

17 Voitovich, N.N. (1985) Synthesis of closed plane antenna with limitations

Phase Optimization Problems. O. O. Bulatsyk, B. Z. Katsenelenbaum, Y. P. Topolyuk, and N. N. Voitovich

ISBN: 978-3-527-40799-6

at field in near zone. *Radio Eng. and Electron. Phys.*, **30** (6), 11–15.
18 Vasil'kiv, Ya.V., Koval'chuk, A.M. and Savenko, P.A. (1984) Synthesis of nulls in the beam pattern of a linear array. *Radiophysics and Quantum Electronics*, **27** (1), 73–78.
19 Savenko, P.A. and Anokhin, V.J. (1997) Synthesis of amplitude-phase distribution and shape of a planar antenna aperture for a given power pattern. *IEEE Trans. Antennas and Propag*, **45** (4), 744–747.
20 Voitovich, N.N., Zamorska, O.F. and Germanyuk, R.I. (1998) Optimization of plane antennas with semitransparent aperture. *Electromagnetics*, **18** (5), 481–494.
21 Voitovich, N.N. and Topolyuk, Yu.P. (1989) Synthesis of axisymmetrical resonator with spherical outside surface starting from a given amplitude directivity pattern. *J. Comm. Tech. Electron.*, **34** (15), 25–30.
22 Andriychuk, M.I. and Topolyuk, Yu.P. (1990) Synthesis of conformal resonator antennas in terms of a given amplitude directivity pattern. *J. Comm. Tech. Electron.*, **35** (14), 116–121.
23 Bondarenko, N.G. and Talanov, V.I. (1964) Some equations of theory of quasioptical systems. *Izv. Vuzov, Radiofiz.*, **7** (2), 313–327, in Russian.
24 Voitovich, N.N., Katsenelenbaum, B.Z., Korshunova, E.N., Pangonis, L.I., Pereyaslavets, M.Yu., Sivov, A.N. and Shatrov, A.D. (1989) *Electrodynamics of Antennas with Semitransparent Surfaces: Methods of Constructive Synthesis*, Nauka, Moscow, in Russian.
25 Andriychuk, M.I., Voitovich, N.N., Savenko, P.A. and Tkachuk, V.P. (1993) *Synthesis of Antennas according to Amplitude Directivity Pattern: Numerical Methods and Algorithms*, Naukova Dumka, Kiev, in Russian.
26 Savenko, P.A. (2002) *Nonlinear Problems of Radiating Systems Synthesis (Theory and Methods of Solution)*, IAPMM NASU, Lviv, in Ukrainian.
27 Voitovich, N.N. (1972) Antenna design for given pattern of amplitude of the radiation (Semenov's method). *Radio Eng. Electron. Phys.*, **17** (12), 2000–2005.
28 Voitovich, N.N., Gis, O.M. and Topolyuk, Yu.P. (1999) Mean square approximation of compactly supported functions with free phase by functions with bounded spectrum. *Dopovidi NAN Ukrainy*, **3**, 7–10, in Ukrainian.
29 Voitovich, N.N., Topolyuk, Yu.P. and Reshnyak, O.O. (2000) Approximation of compactly supported functions with free phase by functions with bounded spectrum. *Fields Institute Communications*, **25**, 531–541.
30 Katsenelenbaum, B.Z. and Semenov, V.V. (1967) Synthesis of phase correctors shaping a specified field. *Radio Eng. and Electron. Phys.*, **12**, 223–231.
31 Voitovich, N.N., Kazantsev, Yu.N. and Tkachuk, V.P. (1984) Formation of a predetermined radiation pattern with the help of a quasi-optical line. *Radio Eng. and Electron. Phys.*, **29** (6), 13–18.
32 Voitovich, N.N., Tkachuk, V.P. and Kazantsev, Yu.N. (1989) Synthesis of quasi-optical radiating system. *Proc. of Int. Symp. on Antennas and Propagation (ISAP-89), Tokyo*, **4**, 893–896.
33 Agranovich, M.S., Katsenelenbaum, B.Z., Sivov, A.N. and Voitovich, N.N. (1999) *Generalized Method of Eigenoscillation in Diffraction Theory*, Wiley-VCH Verlag GmbH, Berlin.
34 Denisov, G.G., Samsonov, S.V. and Sobolev, D.I. (2006) Two-dimensional realization of a method for synthesis of waveguide converters. *Radiophysics and Quantum Electronics*, **49** (12), 961–967.
35 Thumm, M. and Kasparek, W. (2002) Passive high-power microwave components. *IEEE Trans. Plasma Science*, **30** (3), 755–786.
36 Katsenelenbaum, B.Z. (2006) *High-Frequency Electrodynamics*, Wiley-VCH Verlag GmbH, Weinheim.
37 Andriychuk, M.I. (2007) Transformation of field in a regular waveguide via phase correctors. *Proc. XII-th Sem./Workshop Direct and Inverse Probl. Electromagn. & Acoust. Wave Theory (DIPED-2007), Lviv*, pp. 63–66.

38 Vlasov, S.N. and Orlova, I.M. (1974) Quasioptical transformer which transforms the waves in a waveguide having a circular cross-section into a highly directional wave beam. *Journal Radiophysics and Quantum Electronics*, **17** (1), 115–119.
39 Fox, A.G. and Li, T. (1961) Resonant Modes in a Maser Interferometer. *Bell. Syst. Tech. J.*, **40** (2), 453–488.
40 Gelfand, I.M. and Fomin, S.V. (1963) *Calculus of Variations*, Prentice-Hall, Englewood Cliff.
41 Tikhonov, A.N. and Arsenin, V.Y. (1977) *Solutions of Ill-Posed Problems*, Winston and Sons, Washington.
42 Savenko, P.A. (1979) Synthesis of linear antenna arrays with a prescribed amplitude pattern. *Radiophysics and Quantum Electronics*, **22** (12), 1045–1049.
43 Vainberg, M.M. (1972) *A Variational Method and the Monotone Operator Method in the Theory of Nonlinear Equations*, Nauka, Moscow, in Russian.
44 Morozov, V.A. (1966) On the solution of functional equations by the method of regularization. *Soviet Math. Dokl.*, **7**, 414–417.
45 Topolyuk, Yu.P. (2008) Stability of the minimizing sequences in variational problem with free phase. *Information Extraction and Processing*, **28** (104), 11–16, in Ukrainian.
46 Andriychuk, M.I. and Voitovich, N.N. (1985) Synthesis of a closed planar antenna with given amplitude pattern. *Radio Eng. and Electron. Phys.*, **30** (5), 35–40.
47 Voitovich, N.N. and Savenko, P.A. (1975) Branching of solutions of the antenna synthesis problem based on a specified amplitude radiation pattern. *Radio Eng. and Electron. Phys.*, **20**, 1–8.
48 Katsenelenbaum, B.Z. and Semenov, V.V. (1967) Synthesis of Phase Correctors Shaping a Specified Field. *Radio Eng. and Electron. Phys.*, **12**, 223–231.
49 Voitovich, N.N. and Semenov, V.V. (1968) Forming a field of prescribed structure. *Radiotekhnika i Elektronika*, **13** (7), 1213–1221, in Russian.
50 Voitovich, N.N. and Rovenchak, A.I. (1982) Modification of an iterative method for homogeneous problems. *Zhurnal Vychislitel'noy Matematiki i Matematicheskoy Fiziki*, **2**, 348–357, in Russian.
51 Yaroshko, S.M. (2002) Proof of modified method of successive approximations calculation of multiple characteristic numbers of a completely continuous operator. *Visnuk Lvivskogo univers. Ser. prukl. mat. inform.*, **5**, 51–60, in Ukrainian.
52 Semenov, V.V. (1969) Calculation method of multilens system forming of given field. *Radiotehnika i Electronika*, **14** (7), 1321–1323, in Russian.
53 Semenov, V.V. (1972) Two problems in antenna synthesis theory. *Radio Eng. Electron. Phys.*, **17** (1), 18–24.
54 Bulatsyk, O.O. and Voitovich, N.N. (2003) Analytic solutions to a class of nonlinear integral equation connected with modified phase problem. *Information Extraction and Processing*, **19** (95), 33–39, in Ukrainian.
55 Bulatsyk, O.O., Topolyuk, Yu.P. and Voitovich, N.N. (2009) Generalized nonlinear integral equation of Hammerstein type. *Direct and Inverse Problems of Electromagnetic and Acoustic Wave Theory (DIPED-2009). Proc. of Int. Seminar/Workshop, Lviv*, pp. 181–185.
56 Voitovich, N.N. (2003) Antenna synthesis by amplitude radiation pattern and modified phase problem, in *Electromagnetic Fields: Restrictions and Approximations* (ed. B.Z. Katsenelenbaum), Wiley-VCH Verlag GmbH, Berlin, pp. 195–233.
57 Lander, F.J. (1974) The Bezoitiant and inversion of Hankel and Toeplitz matrices. *Matematicheskie Issledovaniya*, Kishinev, **2** (32), 69–87, in Russian.
58 Fiedler, M. (1986) *Special Matrices and their Applications in Numerical Mathematics*, Martinus Nijhoff Publishers, Dordrecht, Boston, Lancaster.
59 Bulatsyk, O.O. (2003) Investigation of branching of solutions to nonlinear equation of the modified phase problem in the case of discrete Fourier transform. *Visnyk Lvivskogo univers. Ser. Prykl. Mat. Inform.*, **7**, 20–32, in Ukrainian.

60 Ilyin, B.A., Sadovnichiy, V.A. and Sendov, B.X. (1979) *Mathematical Analysis*, Nauka, Moscow, in Russian.

61 Levin, B.Ya. (1964) *Distribution of Zeros of Entire Functions*, Amer. Math. Soc., Providence.

62 Rudin, W. (1966) *Real and Complex Analysis*, McGraw-Hill Book Company, New York.

63 Voytovich, N.N. and Savenko, P.A. (1979) Synthesis of antennas from a specified amplitude pattern and related problems in quasioptics (review). *Radio Eng. and Electron. Phys.*, **24** (1), 1–12.

64 Gis, O.M. and Topolyuk, Yu.P. (2001) Mean square approximation of non-negative finite functions by the modulus of function with the finite spectrum (the case of vanishing solutions of approximation on support on the finite function). *Matematychni Metody i Fiziko-Mekhanichni Pola*, **44** (2), 48–54.

65 Prudnikov, A.P., Brychkov, Yu.A. and Marichev, O.I. (1992) *Integrals and Series*, vol 4, Gordon and Breach Science Publishers, New York.

66 Tikhonov, A.N., Leonov, A.S. and Yagola, A.G. (1995) *Nonlinear Ill-posed Problems*, Nauka, Moscow, in Russian.

67 Alber, Ya.I., Iusem, A.N. and Solodov, M.V. (1998) On the projected method for nonsmooth convex optimization in a Hilbert space. *Mathematical Programming*, **81** (12), 23–35.

68 Combettes, P.L. (2001) Quasi-Fejerian analysis of some optimization algorithms, in *Inherently Parallel Algorithms in Feasibility and Optimization and Their Applications* (eds D. Butnariu, Y. Censor, and S. Reich), Elsevier, New York, pp. 115–152.

69 Voitovich, N.N. and Topolyuk, Yu.P. (2000) Convergence of iterative method for problem with free phase in the case of isometric operator. *Direct and Inverse Problems of Electromagnetic and Acoustic Wave Theory (DIPED-2000), Proc. of V-th Int. Seminar/Workshop, Lviv-Tbilisi*, pp. 52–56.

70 Savenko, P.A. (2000) Numerical solution of one class of nonlinear problems of the theory of radiating systems synthesis. *Journal Vych. Math. And Math. Phys*, **40** (6), 929–939, in Russian.

71 Bulatsyk, O.O., Gis, O.M. and Voitovich, N.N. (2002) Branching of solutions of the nonlinear equations arisen in the modified phase problem. *Matematychni Metody i Fiziko-Mekhanichni Pola*, **45** (2), 64–74, in Ukrainian.

72 Voitovich, N.N. and Topolyuk, Yu.P. (2004) Convergence rate of iterative method for problem with free phase in case of isometric operator. *Direct and Inverse Problems of Electromagnetic and Acoustic Wave Theory (DIPED-2000), Proc. of IX-th Int. Seminar/Workshop, Lviv-Tbilisi*, pp. 14–17.

73 Vasilyev, F.P. (1981) *Solution Methods of Extremal Problems*, Nauka, Moscow, in Russian.

74 Colton, D. and Kress, R. (1998) *Inverse Acoustic and Electromagnetic Scattering Theory*, Springer, Heidelberg.

75 Voitovich, N.N. and Semenov, V.V. (1970) Quasioptical lines as feeders. *Radiotekhnika i Elektronika*, **15** (4), 697–704, in Russian.

76 Kusyi, O.V., Voitovich, N.N. and Zamorska, O.F. (2006) Recognition of irregular waveguide geometry using opposite directions technique. *Direct and Inverse Problems of Electromagnetic and Acoustic Wave Theory (DIPED-2006), Proc. of XI-th Int. Seminar/Workshop, Lviv-Tbilisi*, pp. 17–20.

77 Tupychak, I.V. and Voitovich, N.N. (2006) Real vanishing solutions to nonlinear equation related to modified phase problem. *Direct and Inverse Problems of Electromagnetic and Acoustic Wave Theory (DIPED-2006), Proc. of XI-th Int. Seminar/Workshop, Lviv-Tbilisi*, pp. 169–172.

78 Slepian, D. (1964) Prolate sheroidal wave functions. Fourier analysis and uncertainty, IV. *Bell. Syst. Tech. J.*, **43** (6), 3009–3057.

79 Voitovich, N.N. (1988) Synthesis of two-dimensional antenna arrays by generalized variable separation method. *Radiotekhnika i Elektronika*, **33** (12), 2637–2639, in Russian.

80 Balyash, Yu.G., Voitovich, N.N. and Yaroshko, S.A. (1989) Generalized sep-

aration of variables in problems of diffraction and antenna synthesis. *The 1989 URSI International Symposium on Electromagnetic Theory, Stockholm*, pp. 650–652.

81 Voitovich, M.M. and Yaroshko, S.A. (1999) A variational-iterative method for the generalized separation of variables in the solution of multidimensional integral equations. *J. Math. Sci. (New York)*, **96** (2), 3042–3046.

82 Biletskyy, V. (2010) An iterative method of generalized separation of variables for solving linear operator equations. *J. Numer. Appl. Math.*, **100** (1), 2–9.

83 Voitovich, N.N., Dogadkin, A.B. and Katsenelenbaum, B.Z. (1964) Lens-waveguide line. *Radio Eng. and Electron. Phys.*, **9** (9), 1412–1413.

84 Katsenelenbaum, B.Z. (2008) Radar protection of bodies with a complex surface. *J. Comm. Tech. and Electronics*, **53** (6), 639–641.

85 Katsenelenbaum, B.Z. and Voitovich, N.N. (2009) Reducing the backscattering via complex impedance coating. *IEEE Trans. Antennas Propagat.*, **57** (7), 2123–2129.

86 Katsenelenbaum, B.Z. (2009) Excitation of surface waves by diffraction on impedavce circular cylinder. *J. Comm. Tech. and Electronics*, **54** (2), 292–297.

87 Katsenelenbaum, B.Z. (2008) Matching the complex impedance of the body with the field of the incident wave. *Proc.XIIIth Sem./Workshop Direct and Inverse Probl. Electromagn. & Acoust. Wave Theory (DIPED-2008), Tbilisi*, pp. 11–14.

88 Katsenelenbaum, B., Marcader del Rio, L., Pereyaslavets, M., Sorolla Auza, M. and Thumm, M. (1998) *Theory of Nonuniform Waveguides*, IEE Series, London.

89 Nefyodov, Ye.I. and Sivov, A.N. (1977) *Electrodynamics of Periodical Structures*, Nauka, Moscow, in Russian.

90 Katsenelenbaum, B.Z. (1992) Resonant rectangular antenna of the circular polarization. *J. Comm. Tech. and Electronics*, **37** (9), 1586–1591.

91 Katsenelenbaum, B.Z., Sivov, A.N., Chuprin, A.D. and Shatrov, A.D. (1995) A new type of converter from linear to circular polarization based on a three-grating transmission resonator. *J. Comm. Tech. and Electronics*, **40** (1), 1–5.

92 Sivov, A.N. and Chuprin, A.D. (1994) Diffraction of plane wave on short-periodical array of hole cylinders with anisotropic conductivity of surface along helix lines. *J. Comm. Tech. and Electronics*, **39** (2), 111–119.

93 Alekseev, I., Fedyanovich, V. and Shestakov Yu. (1989) Transparency of rapid array of parallel overlapping metallic bands separated by dielestric layer. *J. Comm. Tech. and Electronics*, **39** (12), 2313–2322.

94 Korshunova, E.N., Sivov, A.N. and Shatrov, A.D. (1995) Frequency characteristics of resonant antennas synthesed by given excitation and radiation pattern. *J. Comm. Tech. and Electronics*, **40** (12), 1721–1730.

95 Sivov, A.N., Chuprin, A.D. and Shatrov, A.D. (1996) A method for the solution of inverse scattering problems in electromagnetics. *J. Comm. Tech. and Electronics*, **41** (1), 29–32.

96 Shatrow, A.D., Chuprin, A.D. and Sivov, A.N. (1995) Constructing the phase converters consisting of arbitrary number of translucent surfaces. *IEEE Transactions on Antennas and Propagation*, **43** (1), 109–113.

97 Pelosi, G. and Ufimtsev, P.Ya. (1996) The impedance-boundary condition. *IEEE Antennas and Propagation Magazine*, **38** (1), 31–35.

98 Weston, V.H. (1963) Theory of absorbers in scattering. *IEEE Trans on Antennas and Prop*, **11** (9), 578–584.

99 Bulatsyk, O.O., Topolyuk, Yu.P., and Voitovich, N.N. (2010) Finite-parametric solutions of a class of Hammerstein nonlinear integral equations related to phase problem. *J. Numer. Appl. Math.*, **100** (1), 10–28.

Index

Phase Optimization Problems. O. O. Bulatsyk, B. Z. Katsenelenbaum, Y. P. Topolyuk, and N. N. Voitovich

ISBN: 978-3-527-40799-6